Artificial Life Models in Hardware

Andrew Adamatzky · Maciej Komosinski
Editors

Artificial Life Models in Hardware

Editors
Andrew Adamatzky
University of the West of England
Bristol
UK

Maciej Komosinski
Poznan University of Technology
Poznan
Poland

ISBN 978-1-84996-848-5 e-ISBN 978-1-84882-530-7
DOI 10.1007/978-1-84882-530-7
Springer Dordrecht Heidelberg London New York

British Library Cataloguing in Publication Data
A catalogue record for this book is available from the British Library

Cover design: KuenkelLopka GmbH

Printed on acid-free paper

Springer is part of Springer Science+Business Media (www.springer.com)

Preface

Artificial Life is the science of life-like artifacts. The Artificial Life scientists are a tasty blend of computists, engineers, biologists, physicists, chemists, and mathematicians, all straying away from their mainstreams. They implement living creatures in computer programs, hardware, and robots to better understand what Life is and what other Lifes could be. Advances in virtual models of life are discussed in the sister book, *Artificial Life Models in Software*.[1] The book you are reading now is about real-world implementations of life-like artifacts, the intriguing and exciting domain of the Artificial Life research. Browsing the chapters of the book, you will see that experiments with man-made objects offer unique opportunities in studying living phenomena in physical environment and provide workable proofs of theoretical concepts based on invaluable knowledge of direct experimentation with the practical world.

In no way can the ten chapters of the book cover the whole field of real-world implementations of life-like artifacts: there are hundreds of prototypes and thousands of papers in the field. Rather, the selected pieces of work provide a representative snapshot of the state-of-art in artificial life hardware to encourage readers to enter this kingdom of curiosities and to find their way in the labyrinths of amusement.

Jim M. Conrad and Jonathan W. Mills, *The History and Future of Stiquito, a Hexapod Insectoid Robot*, put up a laymen introduction to artificial life robotics. They invite us to the "kitchen" of robot design and share their success in developing Stiquito, an insect-like hardware creature, historically first ever unconventional robot made available to and actively used by general public.

"If things walk they are living!", this thesis seems to be a key principle of hardware artificial life. There are four chapters on walking. Fumiya Iida and Simon Bovet's robots learnt to hop and then run on four legs. Hopping and walking pattern generations, adaptable walking strategies, reward-based learning, and object following are discussed in the chapter *Learning Legged Locomotion*.

[1] A. Adamatzky and M. Komosinski (eds). *Artificial Life Models in Software*, 1st edn. Springer, 2005; M. Komosinski and A. Adamatzky, 2nd edn. 2009.

Hardware designs of swimming and walking robots are presented by Alessandro Crespi and Auke Jan Ijspeer, *Salamandra Robotica: a Biologically Inspired Amphibious Robot that Swims and Walks*. The robot is controlled by onboard artificial neural network, which is crafted to be adaptable to conditions of locomotion.

Toshio Fukuda, Tadayoshi Aoyama, Yasuhisa Hasegawa, and Kosuke Sekiyama uncover designs of gorilla-like robots in *Multi-Locomotion Robot: Novel Concept, Mechanism and Control of Bio-Inspired Robot*. Transitions between different patterns of locomotion, including wall climbing, are amongst many intriguing features of the chapter.

Chapter *Self-Regulatory Hardware: Evolutionary Design for Mechanical Passivity on a Pseudo Passive Dynamic Walker* by Hiroshi Yokoi and Kojiro Matsushita is the only chapter that does not present any physical implementations. However, the models and blue-prints of passive walkers are based on extensive real-world experiments undertaken by Hiroshi Yokoi, his colleagues, and students. Their remarkable plastic-bottle walkers made a history in simplistic robotics. Yokoi and Matsushita's chapter perfectly complements Conrad and Mills designs in encouraging people to enter the field of life-like robotics.

Perceiving the environment via sensorial inputs by robots traveling without destination, that is, doing random foraging tasks, is studied by Paolo Arena, Sebastiano De Fiore, and Luca Patané in *Perception for Action in Roving Robots: A Dynamical System Approach*. Field programable gate arrays are used as the robot "brains," which learn to control chaotic information of sensorial stimuli. The ideas presented in the chapter are justified in experiments with wheeled robots and on-board learning hardware controllers.

Works on hardware implementations of life-like forms are dominated by robotic applications. Chapter *Nature-Inspired Single-Electron Computers* by Tetsuya Asai and Takahide Oya brings some diversity: Asai and Oya tell us how to build brains from networks of single-electron oscillators. Their architectures are scalable and thus offer novel ways of developing controllers for micromodel of life-like creatures. Computational power of neuromorphic processors, based on single-electron circuits, is demonstrated on the tasks of computational geometry and image processing.

All pieces of work in this book are about unconventional, novel, and emerging robotic and hardware designs. The last three chapters push boundaries of "unconventional" even further. These chapters present results of experimentations with novel substrates and interfaces between hardware and wetware.

Shuhei Miyashita, Max Lungarella, and Rolf Pfeifer, chapter *Tribolon: Water Based Self-Assembly Robots*, take on meaning of "wetware" literally. They design robots self-assembling on the water surface. Miyashita, Lungarella, and Pfeifer built a modular autonomous system of self-propelling plastic tiles capable of aggregation. The experiments shed light onto hierarchical mechanism of self-assembly and morphology as a guiding factor of assembly.

Most designers of robots and hardware take power as granted and focus on motor activities, sensors, information transducers, and control functions. When a battery dies and when main power supply is cut off, even the most sophisticated gadgets become worthless than a box of matches. Ioannis A. Ieropoulos, John Greenman, Chris

Melhuish, and Ian Horsfield take care for this not happen in future. They made energetically autonomous robots which are powered directly by bacterial metabolism. Detailed know-hows are presented in their chapter *Artificial Symbiosis in EcoBots*: the robots have on board fuel cells, where electricity is produced by live microorganisms.

In less than 10 years, the plasmodium of *Physarum polycephalum* raised from a humble slime mould to the amorphous intelligent creature proficient in computational geometry and optimization on graphs, and capable of learning. This would be a crime not to employ the plasmodium as a controller for robots. This was done by Soichiro Tsuda, Stefan Artmann, and Klaus-Peter Zauner who placed the plasmodium on board of a hexapod robot and interfaced the mould with motor controllers. Their experiments are described in chapter *The Phi-Bot: A Robot Controlled by a Slime Mould*.

The book concludes with *Chemical Controllers for Robots* by Andy Adamatzky, Ben De Lacy Costello, and Hiroshi Yokoi. The chapter overviews results of laboratory experiments on wheeled robot navigation with on-board excitable chemical medium, Belousov-Zhabotinsky reaction, and interaction between the chemical medium and robotic hand. Also discussed are possible implementations of distributed manipulators made of reaction-diffusion chemical media and plasmodium of true slime mould.

The volume is not exhaustive but striking in variety and interest selection of works. The distribution of topics reflects current dynamics of pursuits at the edge of artificial life, robotics and computer sciences: interests are oscillating, focus is shifting, new sub-fields are emerging. The main goal of the book is not to feed readers with frozen knowledge of dead facts but, to tease them with new discoveries. The book is a good reading indeed at all level of technical expertise: students, scientists, engineers, academicians, laymen – everyone will find something appealing for themselves.

We have read all chapters with great pleasure, and hope you will do the same.

May 2009

Andrew Adamatzky
Maciej Komosinski

Contents

List of Contributors

Andrew Adamatzky
Department of Computer Science, University of the West of England, Bristol BS16 1QY, UK
e-mail: andrew.adamatzky@uwe.ac.uk

Paolo Arena
Department of Electrical, Electronic and System Engineering, University of Catania, I-95125 Catania, Italy
e-mail: parena@diees.unict.it

Stefan Artmann
Frege Centre for Structural Sciences, Friedrich-Schiller-University, Zwätzengasse 9, 07737 Jena, Germany
e-mail: stefan.artmann@uni-jena.de

Tetsuya Asai
Graduate School of Information Science and Technology, Hokkaido University, Kita 14, Nisi 9, Kita-ku, Sapporo, 060-0814, Japan
e-mail: asai@ist.hokudai.ac.jp

Tadayoshi Aoyama
Department of Mechanical Science and Engineering, Nagoya University, 1 Furo-cho, Chikusa-ku, Nagoya, 464-8603, Japan
e-mail: aoyama@robo.mein.nagoya-u.ac.jp

Simon Bovet
Artificial Intelligence Laboratory, Department of Informatics, University of Zurich, Andreasstrasse 15, CH-8050 Zurich, Switzerland
e-mail: bovet@ifi.uzh.ch

James M. Conrad
Department of Electrical and Computer Engineering, University of North Carolina at Charlotte, Charlotte, NC, USA
e-mail: jmconrad@uncc.edu

Alessandro Crespi
Biologically Inspired Robotics Group, School of Computer and Communication Sciences, Ecole Polytechnique Fédérale de Lausanne, Station 14, CH-1015 Lausanne
e-mail: alessandro.crespi@epfl.ch

Sebastiano De Fiore
Department of Electrical, Electronic and System Engineering, University of Catania, I-95125 Catania, Italy
e-mail: sdefiore@diees.unic

Benjamin De Lacy Costello
Centre for Analytical Chemistry and Smart Materials, University of the West of England, Bristol, UK
e-mail: ben.delacycostello@uwe.ac.uk

Toshio Fukuda
Department of Micro-Nano Systems Engineering, Nagoya University, 1 Furo-cho, Chikusa-ku, Nagoya, 464-8603, Japan
e-mail: fukuda@mein.nagoya-u.ac.jp

John Greenman
Bristol Robotics Laboratory, University of Bristol and University of the West of England, Bristol, UK
e-mail: John.Greenman@uwe.ac.uk

Yasuhisa Hasegawa
Department of Intelligent Interaction Technologies, University of Tsukuba, 1-1-1 Tenodai, Tsukuba, 305-8573, Japan
e-mail: hase@esys.tsukuba.ac.jp

Ian Horsfield
Bristol Robotics Laboratory, University of Bristol and University of the West of England, Bristol, UK
e-mail: Ian.Horsfield@brl.ac.uk

Ioannis A. Ieropoulos
Bristol Robotics Laboratory, University of Bristol and University of the West of England, Bristol, UK
e-mail: Ioannis.Ieropoulos@brl.ac.uk

Fumiya Iida
Computer Science and Artificial Intelligence Laboratory, Massachusetts Institute of Technology, 32 Vassar Street, Cambridge, MA 02139, USA
e-mail: iida@csail.mit.edu

Auke Jan Ijspeert
Biologically Inspired Robotics Group, School of Computer and Communication Sciences, Ecole Polytechnique Fédérale de Lausanne, Station 14, CH-1015 Lausanne, Switzerland
e-mail: auke.ijspeert@epfl.ch

Maciej Komosinski
Institute of Computing Science, Poznan University of Technology, Piotrowo 2, 60-965 Poznan, Poland
e-mail: maciej.komosinski@cs.put.poznan.pl

Max Lungarella
Artificial Intelligence Laboratory, Department of Informatics, University of Zurich, Andreasstrasse 15, 8050 Zurich, Switzerland
e-mail: lunga@ifi.uzh.ch

Kojiro Matsushita
Department of Precision Engineering, University of Tokyo, Tokyo
e-mail: matsushita@robot.t.u-tokyo.ac.jp

Chris Melhuish
Bristol Robotics Laboratory, University of Bristol and University of the West of England, Bristol, UK
e-mail: Chris.Melhuish@brl.ac.uk

Jonathan W. Mills
Department of Computer Science, Indiana University, Bloomington, USA
e-mail: jwmills@cs.indiana.edu

Shuhei Miyashita
Artificial Intelligence Laboratory, Department of Informatics, University of Zurich, Andreasstrasse 15, 8050 Zurich, Switzerland
e-mail: miya@ifi.uzh.ch

Takahide Oya
Graduate School of Engineering, Yokohama National University, 79-5, Tokiwadai, Hodogaya-ku, Yokohama 240-8501, Japan
e-mail: t-oya@ynu.ac.jp

Luca Patané
Department of Electrical, Electronic and System Engineering, University of Catania, I-95125 Catania, Italy
e-mail: lpatane@diees.unict.it

Rolf Pfeifer
Artificial Intelligence Laboratory, Department of Informatics, University of Zurich, Andreasstrasse 15, 8050 Zurich, Switzerland
e-mail: pfeifer@ifi.uzh.ch

Kosuke Sekiyama
Department of Micro-Nano Systems Engineering, Nagoya University, 1 Furo-cho, Chikusa-ku, Nagoya, 464-8603, Japan
e-mail: sekiyama@mein.nagoya-u.ac.jp

Soichiro Tsuda
School of Electronics and Computer Science, University of Southampton, SO17 1BJ, UK
e-mail: soichiro.tsuda@gmail.com

Hiroshi Yokoi
Department of Precision Engineering, University of Tokyo, Tokyo, Japan
e-mail: hyokoi@robot.t.u-tokyo.ac.jp

Klaus-Peter Zauner
School of Electronics and Computer Science, University of Southampton, SO17 1BJ, UK
e-mail: kpz@ecs.soton.ac.uk

Chapter 1
The History and Future of Stiquito: A Hexapod Insectoid Robot

James M. Conrad and Jonathan W. Mills

1.1 Introduction

Although the interest in walking robots is increasing, their use in industry is very limited. Still, walking robots have been popular in the "entertainment" market. Walking robots have advantages over wheeled robots when traversing uneven terrain. Two recent entertainment robotic dogs were the Tiger/Silverlit i-Cybie™ and the Sony Aibo™. WowWee Toys now designs and sells several robotic toys, including the Robosapien™ and Bugbots™. Boston Dynamics has created a robotic prototype of a pack mule called "Big Dog" that can carry hundreds of pounds of equipments on its four legs.

These examples of large walking robots have the advantage of large power supplies and heavy electronics. Smaller, insect-like robots are more difficult to implement. However, in the 1990s, Drs. Mills and Conrad developed a kit that allows a user to make a very inexpensive insect robot, "Stiquito" (Fig. 1.1), from nitinol and music wire. With additional electronics, one can actually create a "walking stick-like" insect robot to traverse a small area and sense its surroundings.

Although the Stiquito books and kit were originally developed for educational uses, the robot has been adapted for use in research and possible space exploration. Stiquito is large by insect standards – we are building smaller and more power-efficient versions. Current research includes its use in a Martian environment as a "colony" of insects discovering its new home.

This chapter will survey the history of Stiquito in education and research, and also describes current and future research and applications.

1.2 The Origins of Stiquito

Most walking robots do not take on a true biological means of propulsion, defined as the use of contracting and relaxing muscle fiber bundles. The means of propulsion for most walking robots is either pneumatic air or motors. True muscle-like

A. Adamatzky and M. Komosinski (eds.), *Artificial Life Models in Hardware*,

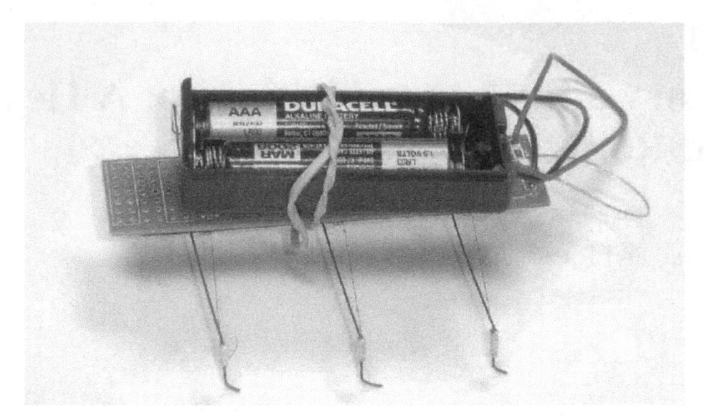

Fig. 1.1 Stiquito controlled, an autonomous walking hexapod robot

propulsion did not exist until recently. A new material, nitinol, is used to emulate the operation of a muscle. Nitinol has the properties of contracting when heated and returning to its original size when cooled. An opposable force is needed to stretch the nitinol back to its original size. This new material has spawned a plethora of new small walking robots that originally could not be built with motors. Although several of these robots have been designed in the early 1990s, one of them has gained international prominence because of its low cost. This robot is called Stiquito.

In the early 1990s, Dr. Jonathan Mills was looking for a robotic platform to test his research on analog logic. Most platforms were prohibitively expensive, especially for a young assistant professor with limited research money. As "necessity is the mother of invention," Dr. Mills set out to design his own inexpensive robot. For propulsion, he selected nitinol (specifically, Flexinol^{TM} from Dynalloy, Inc.). This material would provide a "muscle-like" reaction for his circuitry, and would closely mimic biological actions. For a counter-force to the nitinol, he selected music wire. The wire could serve as a force to stretch the nitinol back to its original length and provide support for the robot. For the body of the robot, he selected one-eighth inch square plastic rod from Plastruct, Inc. The plastic is easy to cut, drill, and glue. It also has relatively good heat-resistive properties. For leg support, body support, and attachment of nitinol to plastic, he chose aluminum tubing from K & S Engineering.

Dr. Mills experimented with various designs, from a tiny four-legged robot 2 in. long, to a six-floppy-legged, 4-in. long robot. Through this experimentation he found that the best movement of the robots was realized when the nitinol was parallel to the ground, and the leg part touching the ground was perpendicular to the ground.

The immediate predecessor to Stiquito was Sticky, a large hexapod robot. Sticky is 9 in. long by 5 in. wide by 3 in. high. It contains nitinol wires inside aluminum tubes, which are used primarily for support. Sticky can take 1.5 cm steps, and each leg had two degrees of freedom. Two degrees of freedom mean that nitinol wire is used to pull the legs back (first degree) as well as raise the legs (second degree).

Sticky was not cost effective, and so Dr. Mills used the concepts of earlier robots with the hexapod design of Sticky to create Stiquito (which means "little Sticky"). Stiquito was originally designed for only one degree of freedom, but has a very low cost. It was first described in a University of Indiana Technical report [9] and

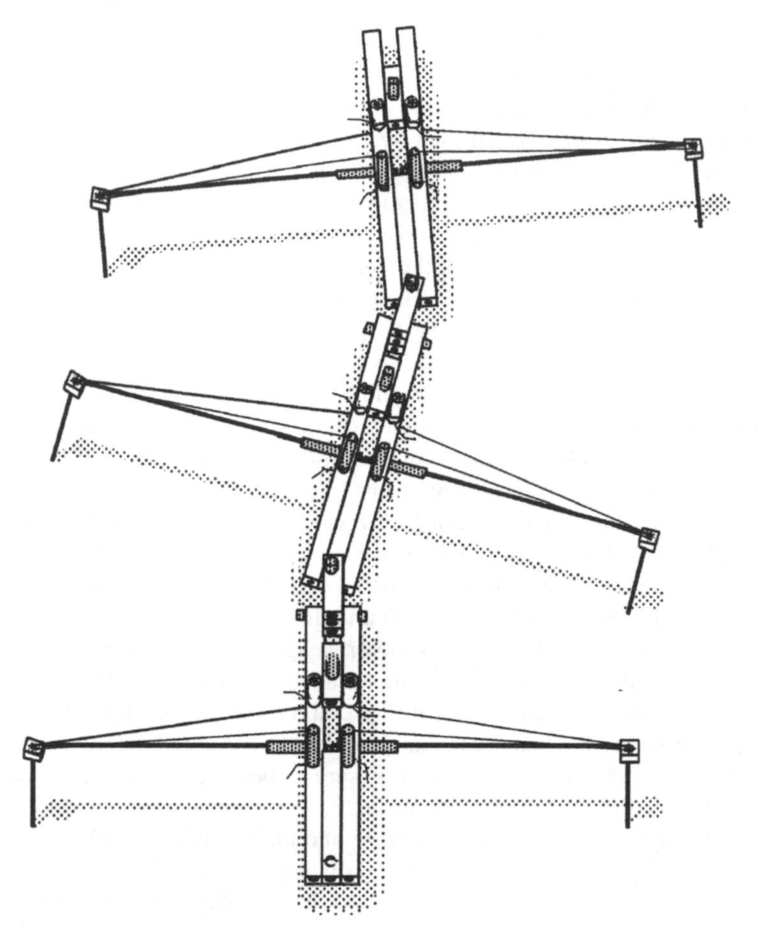

Fig. 1.2 The Stiquito II robot

announced to world in a short artificial intelligence bulletin article in 1993 [10]. Two years later, Dr. Mills designed a larger version of Stiquito, called Stiquito II, which had two degrees of freedom [6]. A drawing of Stiquito II is shown in Fig. 1.2.

At about the same time that Dr. Mills was experimenting with these legged robots, Roger Gilbertson and Mark Tilden were also experimenting with nitinol. Gilbertson and Tilden's robots are described in the first Stiquito book [6].

1.3 Engineering a Commercial Stiquito

The original Stiquito [9] can be considered as the prototype for the currently commercially available Stiquito and kit. The original Stiquito and the original Stiquito kit had several shortcomings with respect to being a commercial product. Users told us that several hard-to-get tools were used to build the robot, drilling very small holes precisely was difficult, and many metal parts touched each other.

The original Stiquito body required two pieces of plastic to be cut and glued together. The glued pieces were drilled by hand, six holes with one-eighth inch in diameter and six holes with one-sixty fourth inch in diameter were needed. The drilling was required to be precise. Information gathered from our existing users indicated that the glue sometimes was not strong enough to hold the two parts together. Also, the objects inserted into the body had very little room for error, as the body was only one-quarter inch wide. Electrical shorts were often made.

To make a commercial product, we proposed that a single piece of plastic be provided so that the user would not need to do any cutting, drilling, or gluing in the plastic body. By making the body one-eighth inch wider, the builder of the Stiquito has more room for error. It was determined that the best way to create this plastic body was to create a mold and make a plastic injection molding of the body as well as the plastic used for the manual controller.

A prototype was created to examine how this would work, and if it were possible. A mold maker at the University of Arkansas made a quick, rough mold and produced about 50 pieces to test. The mold was made more for speed than for durability.

Further iteration of the design process found that sometimes people had difficulty crimping the nitinol in the aluminum. Another design was created to allow the use of screws in the manufacture of Stiquito, rather than aluminum clips. The designer saw that to wrap screws, one needs to take into account the direction of the threads. The new Stiquito body design had holes that would ensure that the nitinol would wrap in the direction of the threads. This use of screws became the origins of Stiquito Controlled [1], described later in this chapter.

A drawing (Fig. 1.3) of the Stiquito Body and the manual controller was made. A mold was made and tested. The molded parts are used for the Stiquito kits in all the books [1, 5, 6]. The final Stiquito robot built from the kit is shown in Fig. 1.4.

1.4 How the Stiquito Insect Walks

The Stiquito robot walks when heat-activated nitinol actuator wires attached to the legs contract. The heat is generated by passing an electric current through the nitinol wire (Fig. 1.5). The legs can be actuated individually or in groups to yield tripod, caterpillar, or other gaits.

Stiquito is small and simple because it uses these nitinol actuator wires. The nitinol wire translates the heat induced by an electric current into mechanical motion, replacing stepping motors, screws, and other components otherwise needed to make a leg move. The mechanical motion results from changes in the crystalline structure of nitinol. The crystalline structure is in a deformable state (the martensite) below the martensite transformation temperature, M_t. In this state, the wire may change its length by as much as 10%. The nitinol wire used is an expanded martensite (i.e., a trained wire).

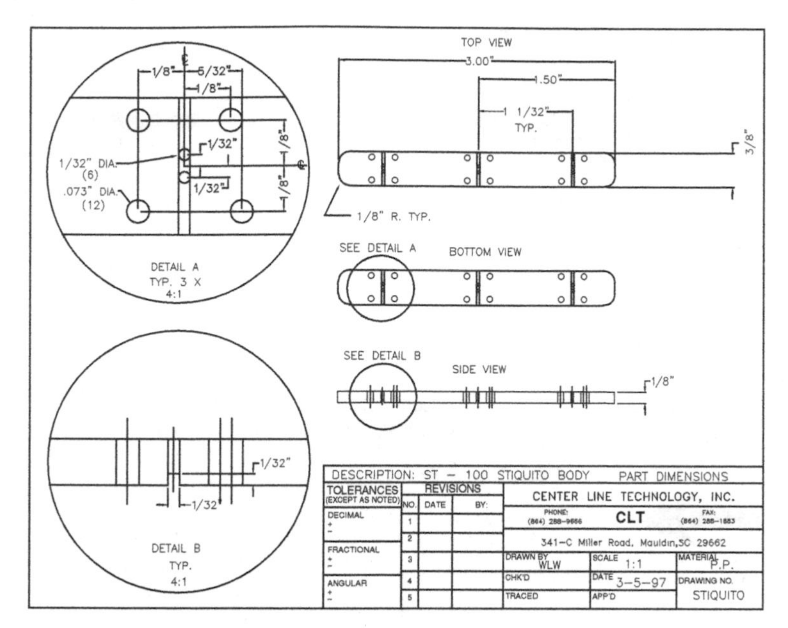

Fig. 1.3 The Stiquito Design Document

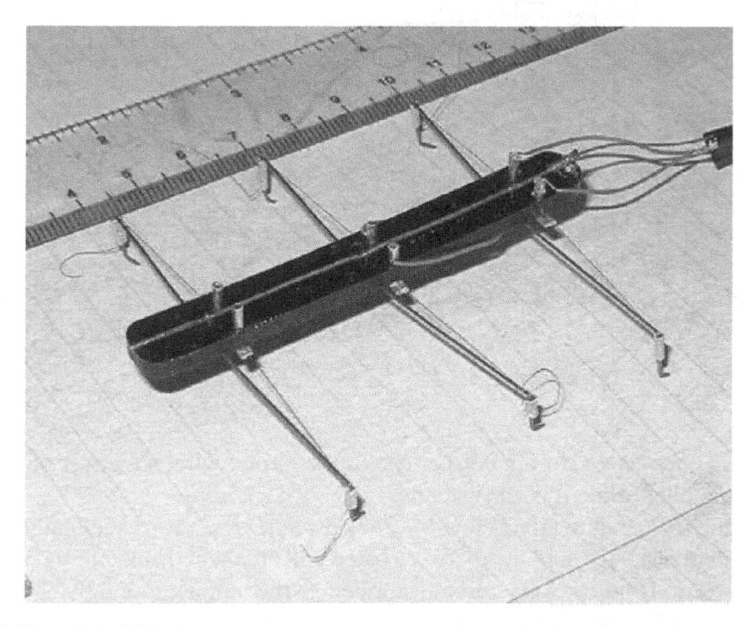

Fig. 1.4 The original Stiquito

When the wire is heated above the austenite transformation temperature A_t (1 in Fig. 1.6), the crystalline structure changes to a strong and undeformable state (the austenite). As long as the temperature of the wire is kept slightly above A_t, the wire

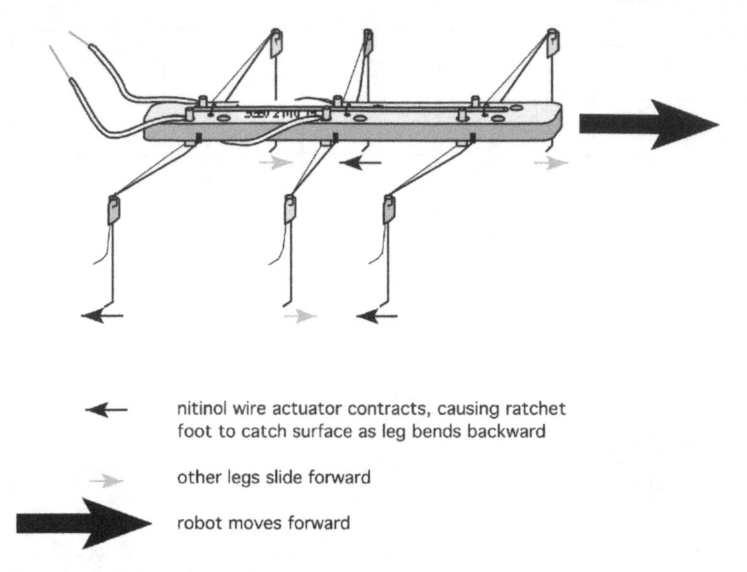

Fig. 1.5 How the Stiquito robot walks

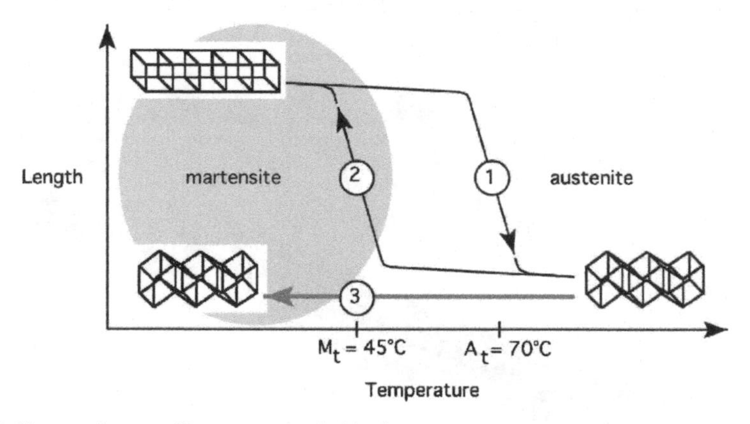

Fig. 1.6 Changes in crystalline structure of nitinol

will remain contracted. During normal use of the nitinol wire, a recovery force, or tension, is applied while it is an austenite.

When the temperature falls below M_t, the austenite transforms back into the deformable martensite (2), and the recovery force pulls the wire back into its original, expanded form. If no recovery force is applied as the temperature falls below M_t, then the wire will remain short as it returns to the martensite (3), although it can recover its original length by cycling again while applying a recovery force. If the wire is heated too far above A_t, then a new, shorter length results upon transformation to the martensite; the"memory" of the original, longer length cannot be restored.

Nitinol wire will operate for millions of cycles if it is not overheated and if a suitable recovery force is applied during each transformation. Stiquito's controller prevents overheating if used as directed. Autonomous controllers and software must limit the current supplied to the nitinol actuator wires to avoid overheating them. The music wire legs provide the correct recovery force.

1.5 Microprocessor Control and Stiquito Controlled

The roots of the Stiquito Controller Board are based on the board developed by Nanjundan Mohon in 1993 and described in the first Stiquito book [6]. This board contained a Motorola 68HC11 microcontroller (with EPROM memory), transistors to drive current to each nitinol leg independently, and an infrared sensor. The software controlled the legs and accepted signals on the infrared sensor to change the gait. This implementation was only a one degree-of-freedom controller.

Jonathan Mills observed a two degrees-of-freedom Stiquito in 1995 and realized that this form of locomotion was needed to make a robot walk quickly. Students have worked on two degrees-of-freedom robots, but none was based on the Stiquito body until 2001, when two student groups successfully built robots for a class and a race [1]. These robots, while successful, were based on the Parallax Basic Stamp 2 microcontroller, an easy-to-use but expensive platform (Fig. 1.7a). What was needed was a low-cost yet easy-to-use implementation.

In the spring semester of 2003, a senior design team at North Carolina State University was given the requirements for a Stiquito Controller Board. After investigating low-cost microcontrollers by Microchip (PIC), Renesas, Texas Instruments, and others, they decided that the TI MSP430F1101 was an appropriate device. Through hard work they created a printed circuit board of the design (Fig. 1.7b).

While the design had most of the required features, it did not have the ability to be programmed by users once the microcontroller was soldered to the board. An expert circuit designer at the University of North Carolina at Charlotte took the existing design, added a JTAG interface, made the board smaller, and had a low-cost PCB manufacturer to create several boards. Several graduate and undergraduate students tested this second prototype. Through their investigations, they found that the original microcontroller selected did not have enough features.

A new microcontroller, the MSP430F1122, was selected to expand the size of flash memory, to add a true and simple A/D converter, and to add brown-out control. Other electronics on the board included the following:

- A ULN2803 transistor array (Darlington Sink Driver) to drive the nitinol legs
- LEDs to indicate the movement of Stiquito with the current output
- A potentiometer to control the speed of the Stiquito movement

Several features were added to the board so that users could solder low-cost connectors to the board and be able to expand its capabilities. These features include a JTAG connection, a reset switch connection, a watch crystal connection, and a prototyping area. The final robot is as shown in Fig. 1.7c. The robot comes preprogrammed, but the board can be reprogrammed via the JTAG port [1, 7].

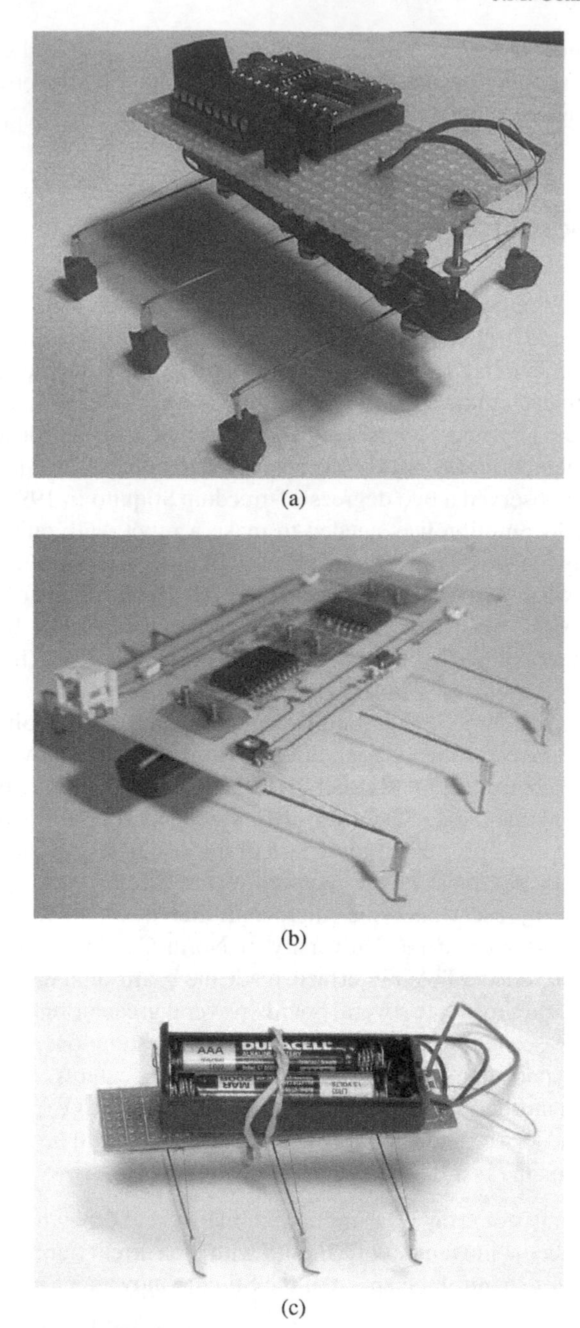

Fig. 1.7 The evolution of Stiquito Controlled: (**a**) basic stamp-based robot, (**b**) a senior design version, and (**c**) the final version

1.6 The Extended Analog Computer as a Biologically Based Stiquito Controller

1.6.1 Description of the Extended Analog Computer (EAC)

The EAC is constructed with a two-sided motherboard. On the top are a five-by-five input/output array of LEDs, programmable current sources and sinks (Fig. 1.8a), the analog-to-digital and digital-to-analog converters used to interface it to the onboard controller, a microprocessor that controls the array, and the USB interface to the host computer [8–17].

On the bottom, underneath the input/output array, is the primary computing element of the EAC, a sheet of electrically conductive plastic foam ((Fig. 1.8b)), which may be replaced with a VLSI *p*-well conductive sheet. It computes partial differential equations of orders of magnitude faster than the finite-element mesh methods used on a digital computer.

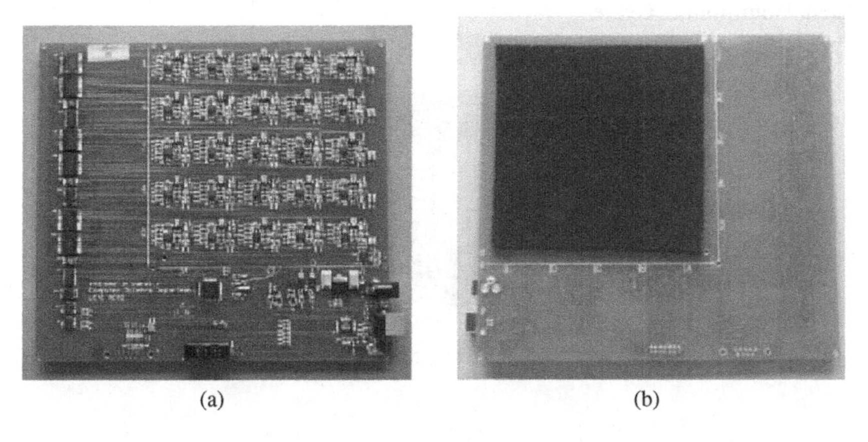

(a) (b)

Fig. 1.8 Extended analog computer: (**a**) array of probes (sources and sinks), (**b**) array of probes embedded into conductive foam

EAC is a generally applicable computer that trades a degree of precision for speed and the solution to problems that have plagued von Neumann architectures for decades: the von Neumann bottleneck (the EAC does not have one), the "memory wall" (the EAC does not use memory to compute), scalable parallelism (no memory model to contend with), and Moore's Law that limits transistor/gate scaling (the EAC uses bulk matter to compute). When the EAC implements large-scale neural aggregates by analogy, the power consumption is Femto-Joules per "device," with speeds in the sub-nanosecond range. The EAC is computationally robust, possessing the "extremely well-posed" property that prevents small input variations from causing large output errors.

The EAC is physically robust, tolerating extensive damage (loss of up to 80% of the machine) before the computation fails. In silicon, this means that fabrication defects do not affect device yield. All VLSI EACs fabricated in 1996 functioned correctly and continue to operate without errors today.

1.6.1.1 Configuring the EAC

The EAC is configured, not programmed. The computation consists of two parts: (1) the EAC configuration, which constrains the physical properties of its operation, and (2) the meaning ascribed to the evolution of those physical processes. Together, these two parts define computation by analogy. One EAC configuration (Fig. 1.9) may have many meanings, possibly very different. For example, the same EAC configuration, unchanged, can represent computation as a spike-time dependent neuron, a neuron cluster computing exclusive-OR, a tissue-level retina, an array of laminar neural sheets (such are found in visual columns), a cognitive behavioral model such as the RETIC, and even non-neuromorphic computations such as the NP-complete problem Hamiltonian Cycle.

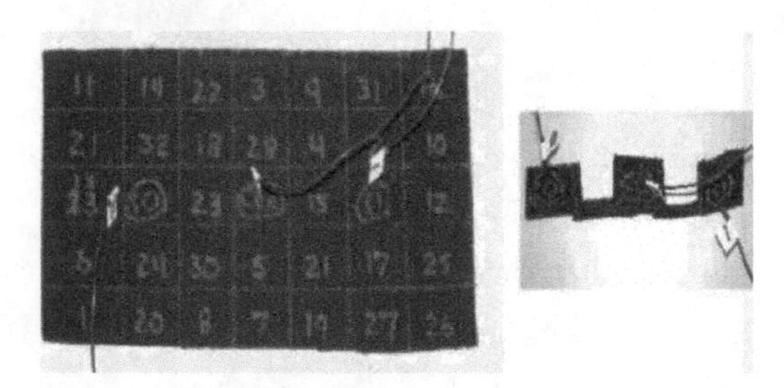

Fig. 1.9 Configuration of extended analog computer persists, even when damaged

1.6.2 The EAC as Neuromorphic Hybrid Device

The EAC is a neuromorphic hybrid device (Fig. 1.10a). The EAC can model single neurons (Fig. 1.10b). Spike trains arrive at the dendritic field (a), shown in Fig. 1.10b, and trigger a threshold response (b), which is output on the synaptic field (c). Eventually, the inputs polarize the neuron and so it stops firing (d). The transition to larger groups of neurons is smooth, implemented by analogy as in binaural location (the exclusive-OR shown here is a simple model), a tissue-level retina, reticular and thalamic behavior generation, etc.

The EAC also implements visceral systems. An electronic barn owl, *Tyto computatrix*, was modeled in 1998. The problem of motivation arose as the owl model had no motivation to hunt. This was solved by adding a model of the owl's crop using

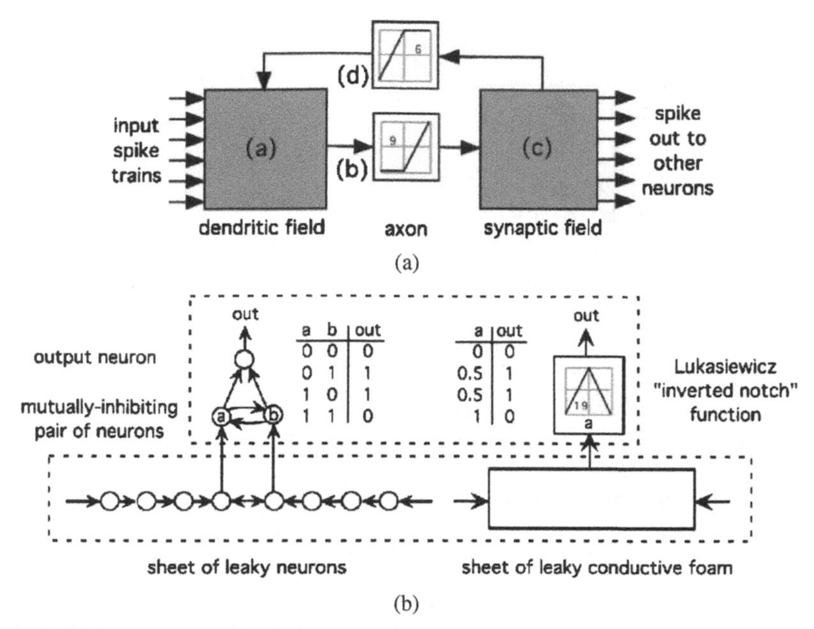

Fig. 1.10 EAC as neuromorphic device

an EAC sheet and Lukasiewicz logic elements that represented the crop's emptying, which in turn activated the hunting behavior. This model of visceral organs and their internal stimuli is needed for any mouse or cat model to motivate behavior that the neuromorphic circuits control.

1.6.3 A Proprioceptic Nervous System Model

Multiple levels of behavior can be integrated using the EAC. By integrating low-level neuromorphic circuits (Fig. 1.11) as inputs to EAC neurons and by using EAC sheets to model large scale neural aggregates for biological systems, the problem of scaling to larger systems can be avoided. Shown here is the architecture for a proprioceptic nervous system that learns by association.

1.6.4 EAC as an Analog Nervous System for Stiquito

EAC can be used as an analog nervous system for Stiquito. Buy building nervous systems that learn and exhibit adaptive behavior at multiple levels, simply by watching the robot itself move within an environment, the classic "Umwelt," or worldview-advanced Stiquito controllers can be built out of one basic EAC: the retina.

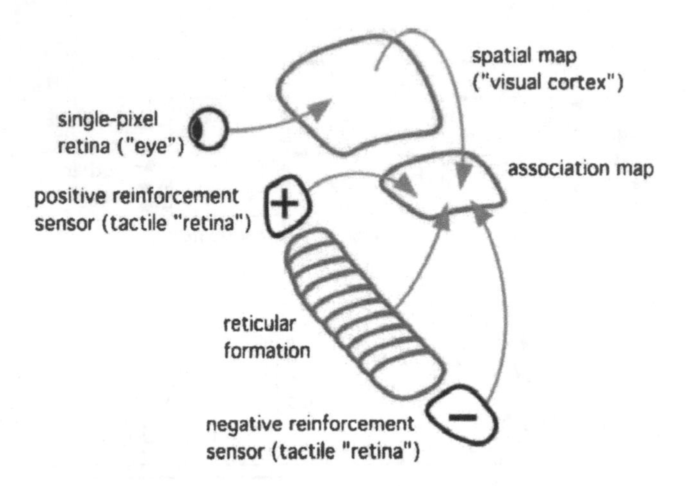

Fig. 1.11 Proprioceptic nervous system

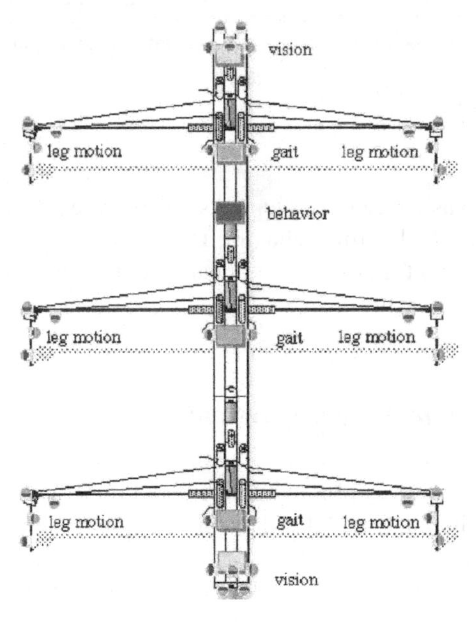

Fig. 1.12 Analog controller for Stiquito

An EAC retina has already been built and demonstrated plasticity, and it is so efficient in that role that the machine defines a new analog computing method: generate-and-recognize (Fig. 1.12). As is shown in the architecture, only one kind of EAC chip is necessary (although the retina chips will need lenses to focus local images, these are the baby-blue single-pixel "eyes").

Thus, both neurally precise behavior and large-scale aggregate behavior can be implemented in silicon devices [8–17]. This is the Stiquito controller first proposed in 1995, which is waiting for the engineer who is conversant with VLSI, fuzzy logic, noisy circuits, neural mass-action behavior, and the EAC to build. It is possible to free VLSI "intelligence" from the Carver Mead model and move to the universal model of the brain-any brain, given enough scale! – first proposed by Lee A. Rubel [18, 19].

1.7 The Sessile Stiquito Colony

From 1992 until 1998, research into colony robotics using the Stiquito robot was a primary goal at the Indiana University. Eventually, the research was abandoned, but not before it had realized a unique achievement in robotics. This section of the chapter describes a never-before-published robot design and includes the photographs that document its behavior.

"It" was a colony of 16 sessile robots that combined electronic and mechanical behavior, and took advantage of the different operating time intervals to create chaotic behavior during which the colony implemented an electromechanical form of evolutionary algorithm. Put another way, the individuals in the sessile colony were constantly engaged in robotic sexual intercourse to the tune of their own mating calls, the chiming caused by their motion, that created new behavior in each generation of successfully mated individuals.

1.7.1 The Failure of Indiana University's Hexapod Stiquito Colony

Before going further, let us briefly discuss why colony robot research with hexapod Stiquito robots was eventually abandoned. The answer lies in three words: fragility, complexity, and power.

1.7.1.1 Fragility

The Stiquito robot, robust as it is, had a weak point in its design, which after weeks and sometimes months of use caused it to fail. The weak point was the fragility of the nitinol wire if it had even a slight bend in the attachment. Under linear strain, nitinol is robust. The problem is to ensure that at the points of attachment to the robotic leg, the nitinol is not only attached linearly, but its contraction and expansion also retain that linearity. Therein lies the problem. The nitinol "muscle" of the Stiquito's leg makes a very slightly conical orbit, with the tip of the cone being the body attachment. After several thousand leg movements, even when great care is taken to keep the nitinol wire linear, it breaks. Examinations of the breaks with a microscope revealed that the breaks were not sharply sheared off nor jagged. They were bowl-like, polished depressions with the longer length of wire terminating in an equally polished, corresponding rounded end. It was as if the nitinol had formed a ball-and-socket joint before the wire broke. Given the immense contractile strength of nitinol, even though the contraction looked linear, there was a point, often inside a crimp, where the wire's strength opened sufficient room for the conical orbit to create a break.

1.7.1.2 Complexity

The first Stiquito robot described in the first two books [5, 6] was a simple design. The current Stiquito described in the third book [1] carries a sufficiently complex electronics and power "backpack" to operate for up to 1 h. The final designs at the Indiana University consisted of complex robotic bodies that included sensors, several microprocessor controllers, and the drivers for the two degrees-of-freedom Stiquito legs. These robots were 8 in. long and 2 in. wide (20 cm long × 5 cm wide). Part of the reason for the increase in the robot's size was to gain a greater sweep of motion in the leg. It was desirable for the robot to move faster and to carry the electronics. The functions needed could not be built with surface-mount electronics at that time. However, the more complex Stiquitos needed a larger arena in which to operate, and an increased likelihood that some part of the robot would fail. As has already been mentioned, the Stiquito leg is a weak point in the design. But the electronics were also prone to fail, which was eventually traced back to the torsion placed on the circuit board as opposing legs moved in tripod and other multi-leg gaits. Eventually, after weeks of continuous operation, the soldered connections broke. In a large and relatively complex circuit board, tracking the source of the failure involved some lengthy tests, even if the failed component was easily identified.

1.7.1.3 Power

The power consumed by even a small Stiquito colony was surprising. In retrospect, it should have been obvious: four robots with six large legs, and each robot with several 1990s-era microprocessors and sensors onboard consumed at least 10 A of current in total. Making this problem worse was the need to deliver the current to the robots. All were tethered to a power supply, but the tethers were a problem themselves. One of the funniest (and, unfortunately, prophetic) movies ever to come out of the late Stiquito colony research is of a single robot winding its tether into a knot and immobilizing itself. It spelled the doom of a Stiquito colony powered by tethers. The solution was the *bumper-car Stiquito* robot. This design used an onboard transformer to convert power from a 50 A alternating current source into the 12 V direct current needed for each robot. The tips of each leg returned the current to a copper plate connected to ground. A long metal rod tipped with a copper brush obtained current from an overhead mesh, as do bumper cars in a carnival ride or an electric trolley or train, as are still used in some cities. But this final attempt to build a group of hexapod robots to study colony behavior came to an end when the force needed to keep at least one leg grounded and the rod-brush assembly pushed against the mesh caused the robots to occasionally get stuck. Then, unable to move, in one case the nitinol overheated and burned like an American Fourth-of-July sparkler. The display was awesome, but the robot literally went up in smoke. It was also obvious that the arena, powered by a 50 A alternating current, was a potential cause of death by electrocution. Even the safety restrictions in place might fail, and after the "sparkler" incident, this design was considered untenable.

1.7.2 Moving Data, Sessile Robots, Robot Sex

Having exhausted local ideas to build a colony of mobile robots, the final attempt, which succeeded too late to inspire further research in the students, was to give up the idea of moving the robots, but instead move the data. The result was a successful design that combined features of the Stiquito robot leg, cellular automata, chaos theory, genetic algorithms, inverted pendulums, and the interplay between mechanical and electronic timescales. The new Stiquito colony was a four-by-four array of specially designed Stiquito robot legs, but the "robots" were partially virtual, keeping their genomes and sensorimotor behaviors in the memory of small microprocessors (the BASIC Stamp).

1.7.3 The Evolution of the Colony

The only records that preserve the existence and behavior of the sessile Stiquito colony are a sequence of time-lapse photographs. The five shown here come from a session that show the emergence of individuals who have evolved successful behaviors from the genomes that defined how each individual would transmit data and wave its long wand, the inverted pendulum "body" that was attached to the actuating leg. The first photograph shows that some individuals are strongly activated and make wide excursions, striking other individuals to create new trajectories and opportunities to exchange chromosomes through their wands (and mutate them, too, due to the electronically noisy interactions). Other individuals maintain smaller orbits, and are out of reach of the more "aggressive" swingers – for the moment (Fig. 1.13).

Figure 1.14 shows that more individuals are making wider excursions as their genetic material is modified. However, even some early "swingers" have been influenced in turn, and are beginning to "stake out turf" where their movements are limited.

From Fig. 1.15, some individuals have evolved short, only slightly activated orbits that are stable, even as their neighbors make wide excursions. Their problem will be survival in this environment. The colony can be programmed so that transfer of genetic material ranges from a very low probability to a very high probability. However, transfer of a "lethal" gene, say, to stop swinging, can "kill" the colony.

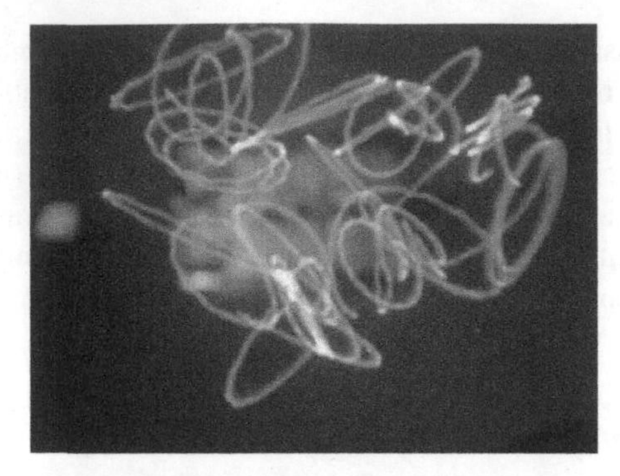

Fig. 1.13 Trajectories of "aggressive" swingers

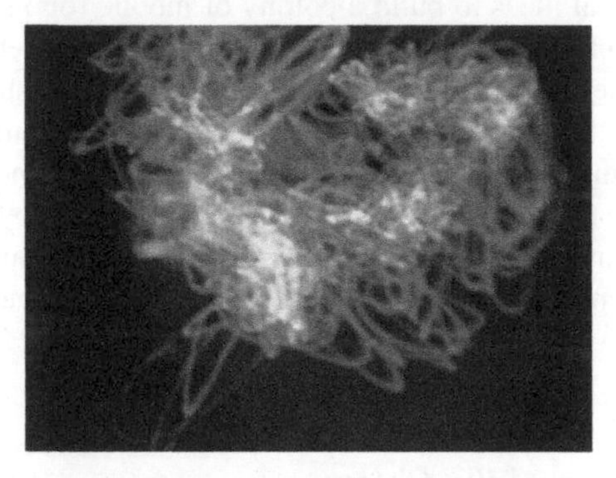

Fig. 1.14 Robots goes for a wider sex-excursions

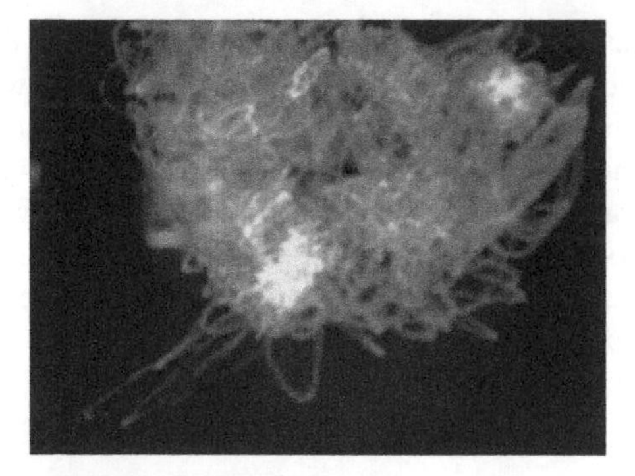

Fig. 1.15 Sub-division of the colony by types of motion

Figure 1.16 shows that a population of individuals is emerging that interact in almost a hierarchical fashion. At the lower left can be seen three "dominant" individuals that make occasional wide swings, but have otherwise short stable orbits. They are open to genetic exchange, but do not engage in it often. One individual in the upper middle right has an orbit that oscillates with different periods, as shown by its "dotted arc."

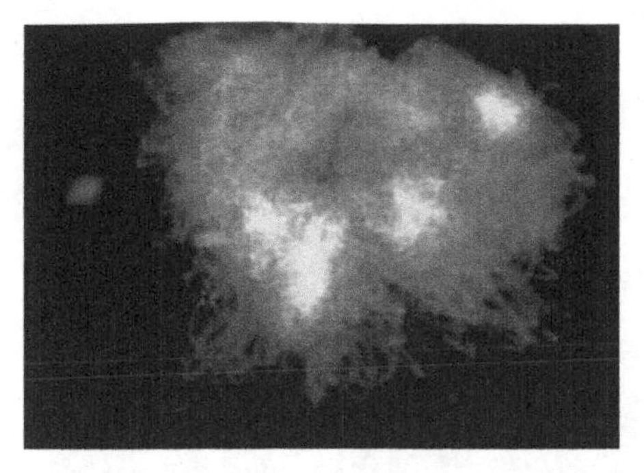

Fig. 1.16 Stabilizing the colony structure

1.7.4 Extinction?

This was the last robotic project undertaken with a Stiquito-like robot at Indiana University. After a hiatus of 10 years, robotics emerged again under the direction of other professors, such as Randall Beer, Steven Johnson, and Matthias Scheutz. Currently, there is no research in colony robotics. Individual mobile robots, such as a golf-cart with video cameras, global position sensing and laser tracking, or a Segway personal vehicle with similar equipment, are currently used successfully to study robotic navigation and control in large environments.

But before the colony of sessile Stiquitos vanished into history, leaving only a photographic trace, they showed how a low cost, simple colony of electromechanical robots could successfully model sophisticated colony behaviors. Indeed, much more could be done, especially with larger colonies. Is there a "Garden of Eden" gene? What happens when individual members are designated as "food" or "prey"? The colony chimes as it operates. Can acoustic as well as visual sensors be used to activate the individuals? What is the rate at which information is transferred between individuals? What kind of "diseases" (viral genomes) evolve to kill or weaken a colony? The future of Stiquito in research still is wide open, but it must wait for a researcher who is enamored of sessile robots and mobile data!

1.8 Educational Uses of Stiquito

Stiquito has been used in education long before the Stiquito books were published. In fact, the first Stiquito book, Stiquito: Advanced Experiments [6], contains many chapters that describe projects by student. The second book, Stiquito for Beginners [5], contains educational experiments on nitinol, electronics, and Stiquito control. The book Stiquito Controlled [1] has been used to introduce students to microcontroller technology. Stiquito has been used in high school classes [4], college

Fig. 1.17 Stiquito in education

freshmen courses [3], and even biomedical engineering classes [2]. Figure 1.17 shows a class exercise where computer engineering students taught mechanical engineering students how to reprogram the Stiquito Controlled board.

1.9 The Future of Stiquito

So what can Stiquito (or other similar insect robot) be used for in the future? Legged robots have been shown in futuristic movies as valuable for certain applications. Surely Stiquito could have some applications, even though it is small.

- Mark Tilden has suggested biodegradable insect robots for pest control on farms [6]. Instead of spraying environmentally hazardous chemicals on crops, one could release thousands of small Stiquitos in the fields. The robots would seek out and "eat" harmful pests. It would rely on sunlight for power. At the end of the growing season, the robots would be tilled into the earth and decompose, thus supplying nutrients back into the soil.
- Instead of sending one, very expensive wheeled vehicle to Mars, why not send thousands of small Stiquitos? Each robot would have sensors and a radio to transmit data to the "ant hill," or the home base on Mars. These would also be solar powered, and could walk and collect data for years. How far would it walk?
- In schools of the future, "scavenger Stiquitos" would leave their home base in the cafeteria and wander the floor in search of crumbs ··· and other bugs to eat! In fact, why limit them to schools? They could even be used in your home kitchen!

- Stiquitos could be inserted into one end of a large, clogged pipe. They could dig their way through the clog to the other side.
- More nimble variations of Stiquito, equipped with microphones and cameras, could be sent into the rubble of a building to look for survivors.
- Grass-chomping Stiquitos could cut the lawn while you sleep.

In nature, Stiquito would be a greedy, blind, crippled, dumb ant. They consume electricity voraciously. Their legs lack the numerous degrees of freedom like those of an ant. Their controller, whether a PC, an analog circuit, or a microcontroller, is overwhelmingly large and clunky as compared to the ganglia that serve an ant (at least to accomplish an ant's tasks).

So why build Stiquitos? We build them because they are simple, they are fun, and they let us investigate robotics without getting a million-dollar grant. We build Stiquito because we have yet to construct anything of the size of an ant that has the capability of an ant. But we are working on it!

Acknowledgments

Special thanks go out to all of the past students and researchers who have contributed to Stiquito's success and helped write the articles and text in the Stiquito "Body of Knowledge," the books and articles.

Some of the text in this chapter has appeared, in part, in the three Stiquito books.

Stiquito is a registered trademark of Indiana University.

References

1. Conrad, J.M.: Stiquito Controlled! Wiley, Los Alamitos, CA (2005)
2. Conrad, J.M.: Stiquito for robotics and embedded systems education. IEEE Computer Magazine 38, 77–81 (2005)
3. Conrad, J.M., Brickley, J.J.: Using Stiquito in an introduction to engineering skills and design course. Proceedings of the 1997 Frontiers in Education Conference, Pittsburgh, PA 1212–1214 (1997)
4. Conrad, J.M., Chitturi, V.R.: Success of an institute on engineering and technology institute for secondary school teachers, Proceedings of the 1995 Frontiers in Education Conference, Atlanta, GA, 45–46 (1995)
5. Conrad, J.M., Mills, J.W.: Stiquito for beginners: An introduction to robotics, IEEE Computer Society Press, Los Alamitos, (1999)
6. Conrad, J.M., Mills, J.W.: Stiquito: Advanced experiments with a simple and inexpensive robot. IEEE Computer Society Press, Los Alamitos, CA (1997)
7. McClain, A., Conrad, J.M.: Software design of the Stiquito micro robot. Proceedings of the 2005 IEEE Southeast Conference, Ft. Lauderdale, FL, 143–147 (2005)
8. Mills, J., Beavers, M.G., Daffinger, C.A.: Lukasiewicz logic arrays. Proc. 20th Int. Symp. Multiple-Valued Logic, IEEE Press (1990)
9. Mills, J.M.: Stiquito: A small, simple, inexpensive hexapod robot. Technical Report 363a, Computer Science Department, Indiana University, Bloomington, IN (1992)
10. Mills, J.M.: Stiquito: A small, simple, inexpensive hexapod robot. SIGART Bulletin, 4, (1993)

11. Mills, J.: The continuous retina: image processing with a single-sensor artificial neural field network. Proc. IEEE Int. Conf. Neural Networks 2, 886–891(1996)
12. Mills, J.: Area efficient implication circuits for very dense Lukasiewicz Logic Arrays, United States Patent No. 5, 770.966 (1998)
13. Mills, J.: Polymer processors, 2nd Int. Symp. Non-Silicon Computation (2000)
14. Mills, J., Walker, T., Himebaugh, B.: Lukasiewicz' insect: Continuous-valued robotic control after ten years. Int. Jour. Multiple-Valued Logic, IEEE Press (2003)
15. Mills, J., Himebaugh, B. et al: "Empty space" computes: The evolution of an unconventional supercomputer, Proc. ACM Computing Frontiers Conf., [on IEEE CD of conference proceedings] (2006)
16. Mills, J.: The architecture of an extended analog computer core. Fourth Workshop on Unique Chips and Systems (UCAS-4), April (2008)
17. Mills, J.: The nature of the extended analog computer. Physica D 237, 1235–1256 (2008)
18. Rubel, L.A.: The brain as an analog computer. J. Theoret. Neurobiol. 4, 73–81 (1985)
19. Rubel, L.A.: The extended analog computer. Adv. Appl. Math. 14, 39–50 (1993)

Chapter 2
Learning Legged Locomotion

Fumiya Iida and Simon Bovet

2.1 Introduction

Legged locomotion of biological systems can be viewed as a self-organizing process of highly complex system–environment interactions. Walking behavior is, for example, generated from the interactions between many mechanical components (e.g., physical interactions between feet and ground, skeletons and muscle-tendon systems), and distributed informational processes (e.g., sensory information processing, sensory-motor control in central nervous system, and reflexes) [21]. An interesting aspect of legged locomotion study lies in the fact that there are multiple levels of self-organization processes (at the levels of mechanical dynamics, sensory-motor control, and learning).

Previously, the self-organization of mechanical dynamics was nicely demonstrated by the so-called Passive Dynamic Walkers (PDWs; [18]). The PDW is a purely mechanical structure consisting of body, thigh, and shank limbs that are connected by passive joints. When placed on a shallow slope, it exhibits natural bipedal walking dynamics by converting potential to kinetic energy without any actuation. An important contribution of these case studies is that, if designed properly, mechanical dynamics can generate a relatively complex locomotion dynamics, on the one hand, and the mechanical dynamics induces self-stability against small disturbances without any explicit control of motors, on the other. The basic principle of the mechanical self-stability appears to be fairly general that there are several different physics models that exhibit similar characteristics in different kinds of behaviors (e.g., hopping, running, and swimming; [2, 4, 9, 16, 19]), and a number of robotic platforms have been developed based on them [1, 8, 13, 22].

Dynamic interactions of distributed information processing also play an important role in stable and robust legged locomotion, which has previously been shown in the locomotion studies of biologically inspired motor control architectures, the so-called central pattern generator models (CPGs; [14]). This approach typically

A. Adamatzky and M. Komosinski (eds.), *Artificial Life Models in Hardware*,

simulates the dynamic interactions of neurons, and the periodic oscillatory signal output of the neural network is connected to the motors of legged robots. Because of the dynamic stability in the signal output, the locomotion processes using this architecture generally exhibit robust locomotion of complex musculo-skeletal structures [20, 25], and it has been shown that the legged robots are capable of legged locomotion in relatively complex environment [7, 10, 15, 17].

As exemplified in these case studies, one of the most challenging problems in the studies of legged locomotion is to identify the underlying mechanisms of self-organization that induces physically meaningful behavior patterns in terms of stability, energy efficiency, and controllability, for example, [3]. From this perspective, the goal of this article is to explore how the self-organization processes in the physical system–environment interactions can be scaled up to more "intellignet" behaviors such as goal-directed locomotion by discussing two case studies of learning legged robots. More specifically, while the dynamic legged locomotion research were limited to only periodic behavior patterns, we will explore the mechanisms in which the rules of motor control can be generated from the physical interactions in the legged robotic systems. Note that this article shows only the important aspect of the case studies in order to discuss conceptual issues. More technical details can be found in the corresponding publications [6, 11].

2.2 Learning from Delayed Reward

Physical dynamic interactions play an important role not only for the repetitive behavior patterns such as walking and running on a flat terrain, but also for the resilient behaviors such as high jumps and kicking a ball. Generating such resilient behaviors generally involves nonlinear control that requires a certain form of planning. For example, a high jump requires a preparation phase of several preceding steps; ball-kicking requires a swing back of the leg in a specific way to gain the maximum momentum at impact. The optimization of such behavior control can be characterized as a "delayed reward" learning problem [24], which means, for example, that a system can realize it was a bad step only after falling over. To deal with such nonlinear control of body dynamics, this section explores a case study of a one-legged hopping robot that learns to generate a series of high-jumps to traverse a rough terrain [11].

2.2.1 One-legged Hopping Robot

Figure 2.1 shows one of the simplest legged robot models. This robot consists of one motor at the hip joint and two limb segments connected through an elastic passive joint. This system requires only a simple motor oscillation to stabilize itself into

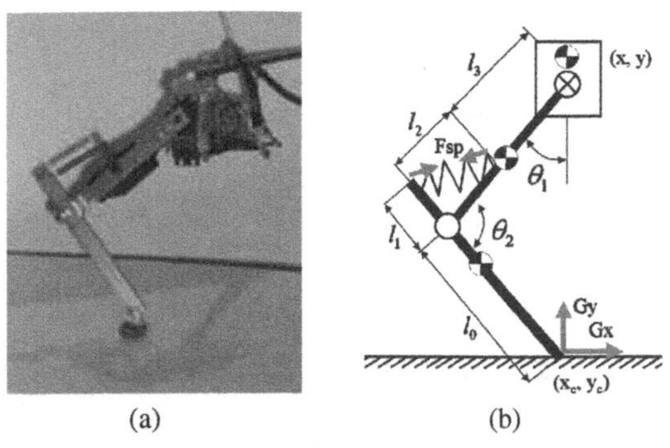

Fig. 2.1 (**a**) Photograph and (**b**) schematic of the one-legged hopping robot. It consists of one servomotor at the hip joint (represented by a *circle with a cross*) and two limb segments connected through a compliant passive joint (marked by an *open circle*)

a periodic hopping behavior [23]. The hip motor uses a position feedback control, in which the angle of hip joint is determined by three parameters: amplitude A, frequency f, and offset of oscillation B.

$$P(t) = A\sin(2\pi ft) + B. \tag{2.1}$$

When these parameters are set properly, the robot shows stable periodic hopping behaviors (Fig. 2.2), and behavioral characteristics resulting from its particular morphology can be summarized as follows. First, locomotion can only be achieved dynamically. As the leg has one single actuated degree of freedom, the only way the robot can lift its legs off the ground is by delivering enough energy through the motors to make the whole body jump. Second, stability is achieved through the material properties of the legs (especially the compliance of the passive joints) rather than by actively controlling the positions of all joints. For instance, an inadequate position of the lower limb (which is only passively attached to the upper limb) during the flight phase will automatically be corrected by the spring on contact with the ground. In particular, this characteristic allows the robot to be controlled in an open-loop manner (i.e., without any sensory feedback) over a continuous range of control parameters. By simply actuating periodically the motors back and forth, the robot put on the ground will automatically settle after a few steps into a natural and stable running rhythm. Third, the elasticity of the legs, partially storing and releasing energy during contact with the ground, allows to achieve not only stable, but also rapid and energy efficient locomotion. The importance of such elastic properties in muscle–tendon systems has been long recognized in biomechanics, where it has a particular significance in theoretical models for the locomotion of legged animals [2, 19].

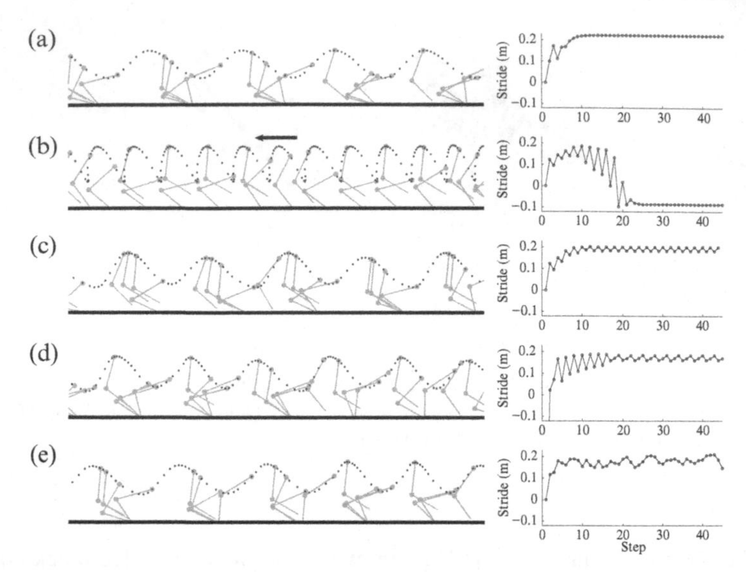

Fig. 2.2 Self-stability and variations of locomotion processes: (**a**) stable forward locomotion with a constant stride length, (**b**) backward locomotion, (**c**) and (**d**) forward locomotion with two and three step cycles, (**e**) stable locomotion with chaotic stride lengths. The oscillation frequencies of the hip motor are $f = 2.78$, 2.72, 2.85, 2.78, and 2.73 Hz (from top to bottom)

2.2.2 Learning to Hop Over Rough Terrain

Although the periodic dynamic locomotion can be mechanically stabilized against small disturbances, the hopping robot needs to actively manipulate the motor control parameters to deal with more complex environment. In this experiment, we applied a machine learning method, the so-called Q-learning algorithm [24], for optimizing the oscillation frequency of the actuated joint.

The learning process repeats locomotion experiments in a given environment until it reaches to a certain number of leg steps. In a learning step n, the system tests a motor control policy consisting of a series of motor oscillation frequencies $f_{i=1,2,\ldots}$, which is determined from a probability matrix $Q^n(i,f)$ (i is the number of leg steps). After each trial, the learning process receives a positive reward signal proportional to the traveling distance and negative reward in case the robot falls over.

$$R^n(i) = \begin{cases} -5.0 & : i = \text{FailedStep} \\ \text{FinalDistance} & : i \neq \text{FailedStep} \end{cases} \tag{2.2}$$

The learning process then updates the probability matrix with a certain learning rate α as follows:

$$\begin{aligned} Q^{n+1}(i,f_i) = (1.0-\alpha)Q^n(i,f_i) + \alpha(R^n(i) \\ +\gamma\max(Q^n(i+1,f))) \end{aligned} \tag{2.3}$$

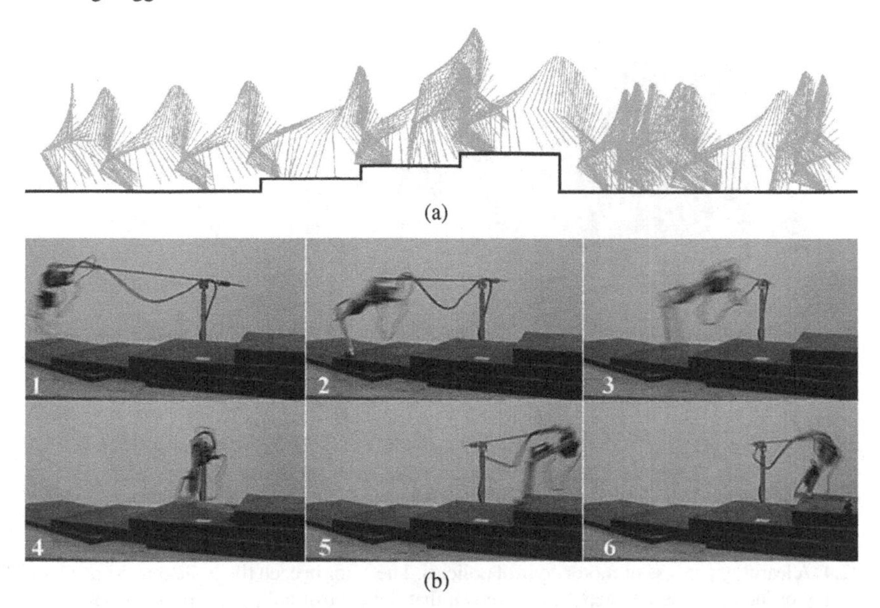

Fig. 2.3 (**a**) Learning results of motor control in simulation. The optimized sequence of motor frequencies exhibits 12 leg steps successfully traveling through a rough terrain. (**b**) Time-series photographs of the robot hopping over the steps. The motor control parameter was first optimized in simulation and transferred to the robot for the real-world experiment

As in the typical reinforcement learning, this learning process utilizes a discount factor γ, which influences the selection of action with respect to the prior action. For example, when the action f_i at a leg step i resulted in a successful continuation of locomotion, the learning process reinforces the probability of choosing f_{i-1} with a discount factor γ as well as that of f_i.

The hopping robot was implemented in a physically realistic simulator to facilitate a number of trials and errors in the learning process, and the learned parameters were transferred to the real-world robotic platform. After a few hundred iterations in the learning phase, the system is able to find a sequence of frequency parameters that generates a hopping gait of several leg steps for the locomotion of the given rough terrain (Fig. 2.3).

Searching for a specific series of frequency parameters is not a trivial problem, because the choice of parameter not only influences behavior of the corresponding leg step, but also those of subsequent leg steps. For example, if the system changes a control parameter at the leg step i, the exactly same motor output of the leg step $i+1$ often results in completely different behaviors. It is, therefore, necessary to utilize the delayed-reward learning such as the Q-learning algorithm explained above, and the typical characteristics in the learning process is illustrated in Fig. 2.4. At the earlier learning steps, the robot attempts mostly random sequences of the control parameters, which are more structured at the later stage. The search process is, however, not straight forward in a way that, at a certain learning step, the control parameters at earlier leg step is modified to achieve breakthroughs. For example,

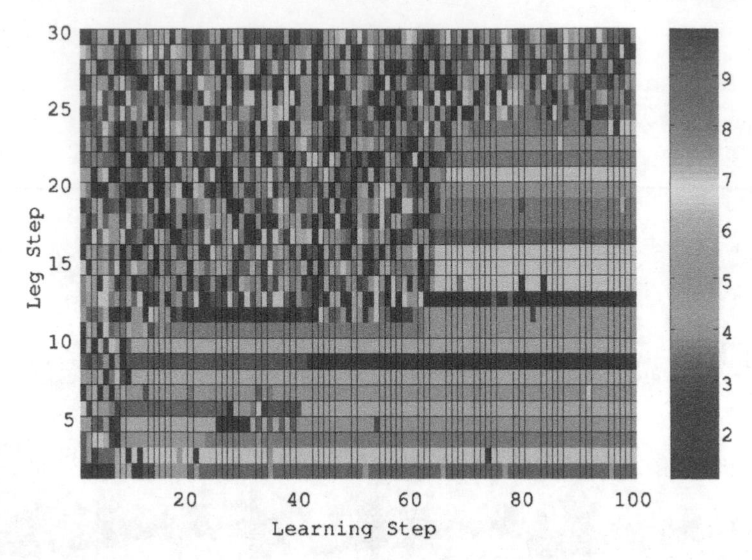

Fig. 2.4 A learning process of motor control policies. The color in each tile indicates the oscillation frequency of motor at the leg step N. It is shown that the control policy is structured towards the end of the learning process

while the learning process in Fig. 2.4 could not find the adequate parameter at the leg step 12 (after the learning step 17), it had to explore the parameter space until the parameter change at the leg step 9 (at the learning step 42), which eventually resulted in a breakthrough to continue the locomotion thereafter.

In summary, this case study explored a learning architecture that exploits dynamics of a compliant leg for goal-directed locomotion in rough terrain. To achieve highly dynamic locomotion such as a series of high jumps over large steps, the learning architecture requires a self-organization process that explores time-series motor output: because behavior of the robot is not only dependent on an immediate motor output but also on the prior ones, the delayed-reward mechanism (the propagation of reward signals over multiple leg steps) is necessary in the learning architecture. It is important to note that the goal-directed behavior shown in this case study was a result of the two levels of self-organization processes (i.e., in mechanical and informational dynamics): because the learning process exploited the underlying mechanical self-stability, the basic forward locomotion dynamics do not require parameter optimization, on the one hand, and the rich behavioral diversity of various hopping heights can be generated only by manipulating frequency parameters.

2.3 Learning from Implicit Reward

The previous case study employed a rather simple setup of learning experiments to emphasize the roles of delayed-reward signals in a learning process of legged locomotion. In contrast, this section discusses how the complexity of self-organization

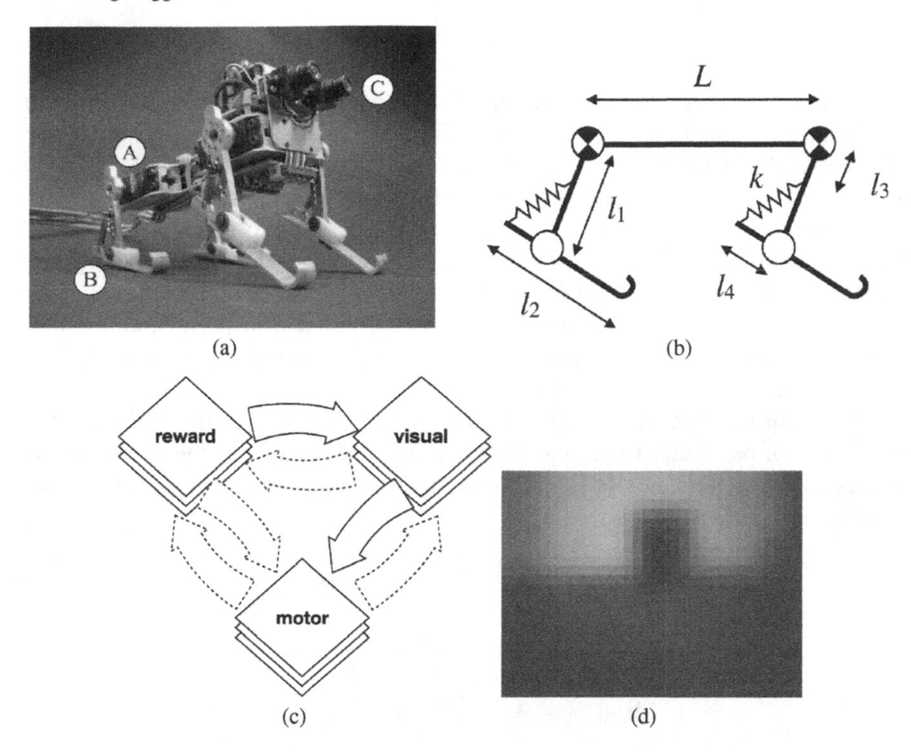

Fig. 2.5 (**a**) Four-legged running robot with two cameras (only one of them was used in the experiments). (**b**) Schematic of the robot, illustrating the locations of motors (*circles with crosses*) and passive joints (*open circles*). (**c**) Neural network architecture that illustrates synaptic weights. *Solid arrows* represents nonzero synaptic connections after the learning phase through which neural activities can propagate, while *dotted arrows* represent synaptic connections with essentially zero weights. (**d**) Snapshot image correlated to the reward signals

processes can be scaled up such that nontrivial signal pathways can be developed between sensory input and motor output. Here, we introduce another learning architecture that extracts correlation between signals to propagate implicit reward signals for a visually mediated target following behavior.

2.3.1 Four-legged Running Robot

The robotic platform used in this case study is a running robotic dog [12] shown in Fig. 2.5. This robot has four identical legs, each of which consists of one servomotor actuating a series of two limbs connected through a passive elastic joint as in the previous case study. Dynamic locomotion is also achieved by periodically moving back and forth the servomotors actuating the legs of the robot, and the target angular position $P_i(t)$ of motor i at time t is given by

$$P_i(t) = A_i \sin(2\pi ft + \phi_i) + B_i, \tag{2.4}$$

where A_i is the amplitude and B_i the set point of the oscillation. Compared with the one-legged hopping robot, this robot has a few additional parameters ϕ_i, the phase offsets, which determine the phase delay of oscillation between the legs.

The learning architecture of this robot has a form of neural network, which receives signals online from a visual sensor and provides output signals to the control parameters of (2.4). The neural network is specifically designed for extracting correlations in sensory-motor signals by using a modified Hebbian learning rule (see [6] for more details). We have modeled three groups of neurons, that is, motor neurons, sensor neurons, and "reward neurons," which are fully connected internally as shown in Fig. 2.5c.

The motor neurons are connected to a set of motor variables that represent the *differences* of parameter values between left-side and right-side motors, as well as between fore and hind motors. For instance, the oscillation amplitudes A_i of the four motors are defined as follow:

$$A_{\text{fore,left}} = A_0 - \frac{1}{2}\Delta A_{\text{lat}} - \frac{1}{2}\Delta A_{\text{long}} \tag{2.5}$$

$$A_{\text{fore,right}} = A_0 + \frac{1}{2}\Delta A_{\text{lat}} - \frac{1}{2}\Delta A_{\text{long}} \tag{2.6}$$

$$A_{\text{hind,left}} = A_0 - \frac{1}{2}\Delta A_{\text{lat}} + \frac{1}{2}\Delta A_{\text{long}} \tag{2.7}$$

$$A_{\text{hind,right}} = A_0 + \frac{1}{2}\Delta A_{\text{lat}} + \frac{1}{2}\Delta A_{\text{long}} \tag{2.8}$$

ΔA_{lat} and ΔA_{long} are the lateral and longitudinal differences of amplitude, and A_0 is the average amplitude. The other motor parameters (i.e., the set points B_i and the phase offsets ϕ_i) are defined accordingly. Eventually, we have the eight state components (i.e., A_0, ΔA_{lat}, ΔA_{long}, B_0, ΔB_{lat}, ΔB_{long}, $\Delta\phi_{\text{lat}}$, and $\Delta\phi_{\text{long}}$) whose values are represented by the activity of eight motor neurons. Note that the frequency of oscillation f is constant for all motors, which provides a basic setup of the robot running forward.

The robot is equipped with a vision system consisting of a camera attached to the body and pointing in the forward direction (see Fig. 2.5). The sensor neurons are receiving both intensity and estimated optical flow extracted from the gray-scale visual input of the 32×24 pixel values. For enabling reinforcement learning, we also include a set of "reward neurons" as described later.

2.3.2 Learning to Follow an Object

The experiment of this case study consists of two phases. In the initial phase, the motor neurons are randomly activated, thus producing arbitrary motions of the robot. This initial phase allows the neural network to learn the basic cross-modal

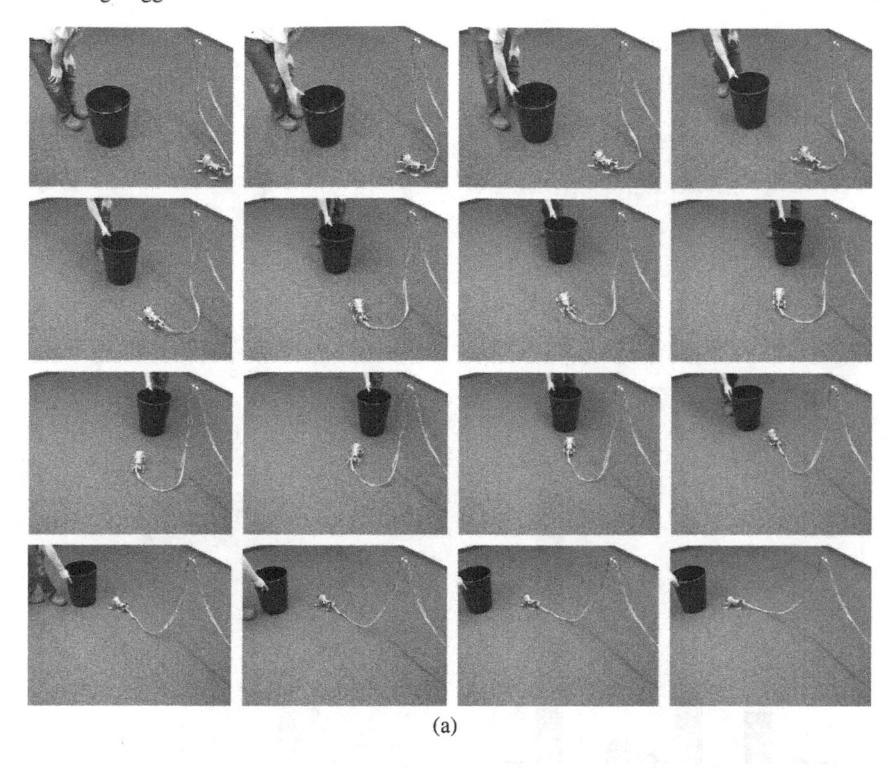

(a)

Fig. 2.6 Four-legged running robot following a black object in an unstructured environment

correlations as follows. The reward is delivered when the robot is facing a large black bin placed in the environment (as shown in Fig. 2.6). The synaptic connections between the reward and the sensor neurons therefore learn a correlation between reward signals and a set of visual input signals that correspond to the image of the black bin in the center of the visual field (as shown in Fig. 2.5d). At the same time, because of the Hebbian-like learning rule, the synaptic connections between the visual and motor groups of neurons capture another significant correlation. This correlation, which we will elaborate later, involves the visual neurons that receive optical flow signals and a particular pattern of activity in the motor neurons. In the second phase, the robot is let to move on its own while activating the sensory neurons receiving reward signals. Because of the particular synaptic connections that have been strengthened during the initial phase of the experiment, the reward signals are propagating through visual neurons to motor neurons, which eventually activate the oscillation of the legs such that the robot follows the object.

The observed behavior, generated from the propagation of neural activity across the network, is illustrated in Fig. 2.6, where the robot turns towards any black object that is placed in the center of its field of view and follows the object as it is moved around. The key aspect of the network connectivity is the correlation between perceived visual flow and motor activity, which is captured by the synaptic

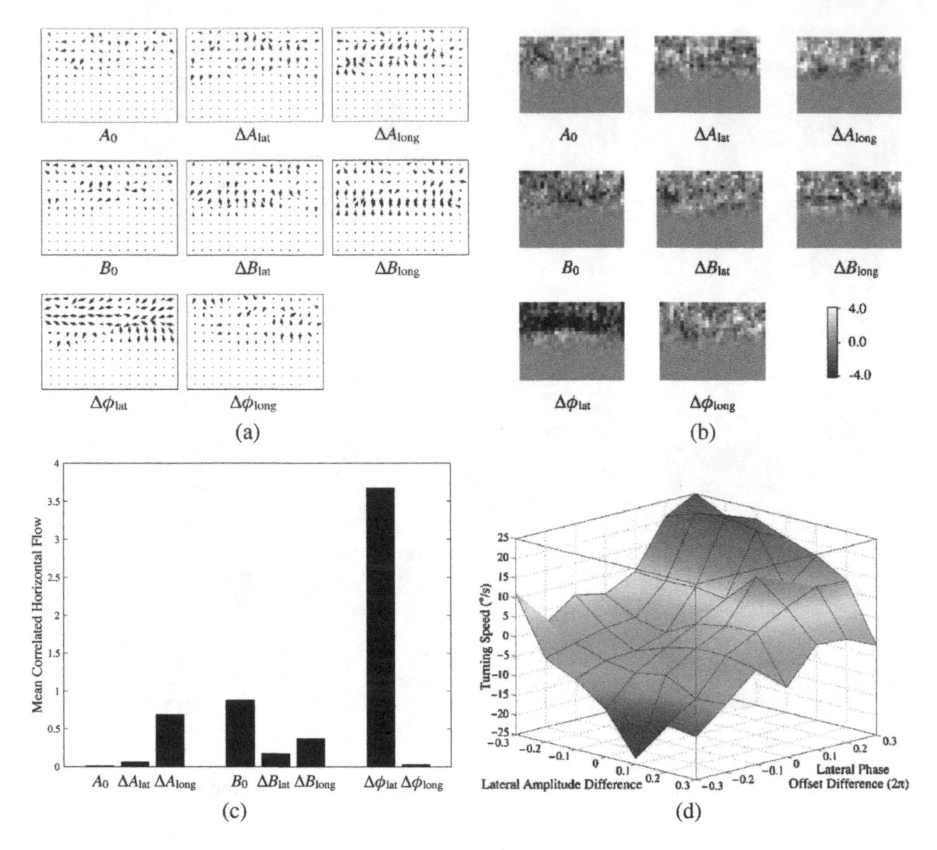

Fig. 2.7 Graphical representation of the synaptic weights coupling the visual modality to the motor modality, showing (**a**) the visual flow field, and (**b**) only the horizontal component thereof, correlated to each of the eight components of the motor state. (**c**) Average horizontal component of the visual flow correlated to each motor component (absolute value). (**d**) Turning speed of the robot as a function of both lateral amplitude difference ΔA_{lat} and later phase offset difference $\Delta \phi_{\text{lat}}$

weights coupling the visual modality to the motor modality. Figure 2.7 shows a graphical representation of these weights, illustrating the visual flow correlated to each motor control parameter. Clearly, the neural architecture captures a significant correlation only between visual flow and the motor parameter corresponding to lateral phase offset difference ($\Delta \phi_{lat}$). This means that the quadruped robot learns a control strategy for turning that modifies essentially the phase difference between the oscillations of the left and the right legs.

To better understand this result, we systematically quantify the turning rate of the robot as a function of various motor control parameters. Figure 2.7d shows that the turning speed is most easily and robustly controlled with the lateral phase difference, the relation between the two quantities being almost linear in the considered range. In contrast, when the other motor control parameters are varied, the turning speed of the robot either does not change significantly or displays no linear relation: for instance, as the lateral amplitude difference is steadily increased, the robot does not always change the turning rate monotonously.

In summary, the modified Hebbian learning rule, which captured the correlation patterns of sensory-motor activity in the neural network, developed a nontrivial synaptic structure that produces an object following behavior based on the visual sensory input. To achieve this task, the network has to find a nontrivial correlation between visual sensory input, reward signals, and motor signals. This experiment shows how self-organization processes that capture correlation of sensory-motor signals can generate sensible behavior patterns.

2.4 Conclusion

This article discussed the issues of legged locomotion from the perspective of artificial life in the real world. By treating legged locomotion as a self-organizing process resulting from complex physical and informational dynamics, we argue that one of the most significant challenges lies in the grounding of self-organization processes for physically meaningful behaviors. While our exploration is still at a nascent stage, we extracted a few important principles from the case studies presented in this article. In particular, we have shown that a learning architecture requires, on the one hand, reward signals evaluating a series of motor actions to make full use of nonlinear mechanical dynamics, and on the other, a specific form of signal propagation to capture the patterns of sensible physical system–environment interactions for goal-directed behaviors.

There are still a number of open questions that we have not explicitly discussed in this article so far. One of the fundamental questions is how we could extend the complexity of self-organizing processes further with less "hand-coded" elements in the embodied systems. For example, in the case studies presented in this article, we predefined a number of elements such as the basic controllers that generate sinusoidal oscillation, basic sensory information processing (e.g., optical flow estimation), mechanical dynamics with fixed viscous-elasticity in passive joints, and the basic reward signals, to mention but a few. Although we found these predefined elements essential to maintain the learning phase within a reasonable amount of time, it requires further studies to discuss how the self-organizing processes should be structured. We are expecting that the comparative study with some of the related work (e.g., [5, 17, 26]) will clarify more general rules to manage the higher dimension of parameter space in self-organizing processes of embodied systems.

Acknowledgment

This work is supported by the Swiss National Science Foundation (Grant No. 200021-109210/1) and the Swiss National Science Foundation Fellowship for Prospective Researchers (Grant No. PBZH2-114461).

References

1. Ahmadi, M., Buehler, M.: Controlled passive dynamic running experiments with ARL monopod II. IEEE Transactions on Robotics, 22, 974–986 (2006)
2. Alexander, R.McN.: Three uses for springs in legged locomotion. International Journal of Robotics Research, 9, 53–61 (1990)
3. Bedau, M.A., McCaskill, J.S., Packard, N.H., Rasmussen, S., Adami, C., Green, D.G., Ikegami, T., Kaneko, K., Ray, T.S.: Open problems in artificial life. Artificial Life 6, 363–376 (2000)
4. Blickhan, R., Seyfarth, A., Geyer, H., Grimmer, S., Wagner, H.: Intelligence by mechanics. Philosphical Transactions of the Royal Society of London Series A: Mathematical and Physical Sciences 365, 199–220 (2007)
5. Bongard, J., Zykov, V., Lipson, H.: Resilient machines through continuous self-modeling. Science 314, 1118–1121 (2006)
6. Bovet, S.: Robots with self-developing brains. Dissertation, University of Zurich (2007)
7. Buchli, J., Ijspeert, A.J.: Self-organized adaptive legged locomotion in a compliant quadruped robot. Autonomous Robots 25, 331–347 (2008)
8. Collins, S., Ruina, A., Tedrake, R., Wisse, M.: Efficient bipedal robots based on passive dynamic walkers. Science 307, 1082–1085 (2005)
9. Dickinson, M.H., Farley, C.T., Full, R.J., Koehl, M.A.R., Kram, R., Lehman, S.: How animals move: An integrative view. Science 288, 100–106 (2000)
10. Geng, T., Porr, B., Wörgötter, F.: A reflexive neural network for dynamic biped walking control. Neural Computation 18, 1156–1196 (2006)
11. Iida, F., Tedrake, R.: Optimization of motor control in underactuated one-legged locomotion. International Conference on Robotics and Systems (IROS 07), 2230–2235 (2007)
12. Iida, F., Gomez, G.J., Pfeifer, R.: Exploiting body dynamics for controlling a running quadruped robot. Proceedings of International Conference on Advanced Robotics (ICAR 2005), 229–235 (2005)
13. Iida, F., Rummel, J., Seyfarth, A.: Bipedal walking and running with spring-like biarticular muscles. Journal of Biomechanics 41, 656–667 (2008)
14. Ijspeert, A.J.: Central pattern generators for locomotion control in animals and robots: A review. Neural Networks 21, 642–653 (2008)
15. Kimura, H., Fukuoka, Y., Cohen, A.-H.: Biologically inspired adaptive walking of a quadruped robot. Philosophical Transactions of the Royal Society of London Series A: Mathematical and Physical Sciences 365, 153–170 (2007)
16. Kubow, T.M., Full, R.J.: The role of the mechanical system in control: A hypothesis of self-stabilization in hexapedal runners. Philosophical Transactions of the Royal Society of London Series B: Biological Sciences 354, 849–861 (1999)
17. Matsubara, T., Morimoto, J., Nakanishi, J., Sato, M., Doya, K.: Learning CPG-based biped locomotion with a policy gradient method. Proceedings of 2005 5th IEEE-RAS International Conference on Humanoid Robots, 208–213 (2005)
18. McGeer, T.: Passive dynamic walking. The International Journal of Robotics Research 9, 62–82 (1990)
19. McMahon, T.A.: Muscles reflexes and locomotion. Princeton University Press, Princeton, NJ (1984)
20. Ogihara, N., Yamazaki, N. Generation of human bipedal locomotion by a bio-mimetic neuro-musculo-skeletal model. Biological Cybernetics 84, 1–11 (2001)
21. Pfeifer, R., Lungarella, M., Iida, F.: Self-organization, embodiment, and biologically inspired robotics. Science 318, 1088–1093 (2007)
22. Raibert, H.M.: Legged robots that balance. MIT Press, Cambridge, MA (1986)
23. Rummel J., Seyfarth A.: Stable running with segmented legs. International Journal of Robotics Research 27, 919–934 (2008)
24. Sutton, R., Barto, A.: Reinforcement learning. MIT Press, Cambridge, MA (2000)

25. Taga, G., Yamaguchi, Y., Shimizu, H.: Self-organized control of bipedal locomotion by neural oscillators in unpredictable environment. Biological Cybernetics 65, 147–159 (1991)
26. Tedrake, R., Zhang, T.W., Fong, M., Seung, H.S.: Actuating a simple 3D passive dynamic walker. Proceedings of the IEEE International Conference on Robotics and Automation (ICRA 2004), 4656–4661 (2004)

Chapter 3
Salamandra Robotica: A Biologically Inspired Amphibious Robot that Swims and Walks

Alessandro Crespi and Auke Jan Ijspeert

3.1 Introduction

One of the key characteristics of animals, perhaps the most impressive, is their ability to move. It is the result of millions of years of evolution, and its complexity, flexibility, and energy efficiency are yet to be approached by robots. The control and coordination of many degrees of freedom in a robot is complex, and there is no well-established technique to deal with this: on their side, animals often have hundred degrees of freedom and use them with a surprising ability.

The purpose of this project is to develop an amphibious robot inspired by the salamander in two aspects: the biomechanical structure and the locomotion control. The first purpose of the project is to explore and develop new technologies inspired by the salamander. In particular, we aim at developing a robot that can robustly swim, crawl, and walk. The second purpose is to use the developed robot as a test-bed for neurobiological models in a real (as opposed to simulated) embodiment. Finally, with its multiple gaits, such a robot would be useful for inspection or exploration purposes in difficult environments (e.g., flooded zones, under collapsed buildings, etc.).

The salamander, a tetrapod capable of both swimming and walking, offers a remarkable opportunity to investigate vertebrate locomotion. It is considered as closely resembling the first terrestrial vertebrates and represents, among vertebrates, a key animal in the evolution from the aquatic to the terrestrial environment [6, 18]. It also has orders of magnitudes fewer neurons than mammals have [29, 30] and is therefore at a level of complexity that is more tractable from the understanding and modeling point of views. Finally, the central nervous system of the salamander shares many similarities with that of the lamprey, an extensively studied primitive fish, and many data and models of the lamprey's swimming circuitry are therefore available to guide the understanding of the salamander's locomotor circuitry. From

A. Adamatzky and M. Komosinski (eds.), *Artificial Life Models in Hardware*,

a robotics point of view, it is very attractive to develop an amphibious robot capable of swimming, crawling, and walking. To the best of our knowledge, such a robot has never been developed before.

3.1.1 Robots as Tools for Biology

Robots are increasingly being used as tools to verify biological hypotheses or as models of biological sensorimotor systems [37]. Examples include lamprey locomotion [33, 41], lobster locomotion [1], cricket phonotaxis [40], and cat locomotion [17]. For a more detailed review, see [38, 39].

Compared to computer simulations, the use of real robots is interesting as it provides several advantages:

- The model is completely interacting with a real environment, using real sensors and real actuators. This therefore eliminates the need to simulate the sensors and the actuators (which can be generally simulated only with approximate models). The absence of simplified models or biased results owing to the simulation is a great advantage, as some aspects could strongly depend on the interaction with the environment.
- There is no need to simulate complex environments or complicated force models. A correct simulation of some phenomena (e.g., friction forces, hydrodynamical forces, etc.) is extremely difficult (especially if associated with articulated moving bodies, whose shape is not constant); simulations are therefore generally limited to a simplified model and could introduce artifacts that cause the model to behave differently than in the real world.

However, it is also important to notice that the use of real robots has some drawbacks compared to simulations:

- Reproducing the mechanical properties of real animal bodies on a robot is very difficult. A robot will only be an approximation of the real animal, as generally it is technically not feasible to build a robot having the same properties (especially for the number of degrees of freedom: a robot with hundreds of degrees of freedom like a real snake would be much larger than the animal with the current technology). The visco-elastic properties of animal muscles are also difficult to implement in robots.
- Almost everything can be designed in a simulation, including systems using components (e.g., sensors or actuators) that are expensive, hard to use, or even not existing with the current technology. For example, some animal sensory systems (like touch) are difficult to replicate with currently existing sensing devices.
- Building a robot generally requires much more work than implementing a simulation, and robots sometimes have to be repaired or maintained. Moreover, robots only run in real time (simulations can be faster).

3.1.2 Related Work

This section briefly presents existing snake and salamander robots.

3.1.2.1 Snake-like Robots

Snake-like robots can be classified into two main groups:

- Robots that move using powered wheels or caterpillars (i.e., a torque is applied on the axis of the wheels, which are in contact with the ground, producing a rotation and consequently a movement).
- Robots that move by applying torques on the joints between the segments. Among these robots, some have passive wheels.

Robots using powered wheels are simpler to control: the design techniques are well known and standard algorithms for the control of mobile robots can be used; however, the resulting locomotion is completely artificial and the wheels may not be adequate in every environment. Robots of this type are often developed for inspection tasks in difficultly accessible zones [23, 27] and are sometimes used, for example, for the inspection of pipes [5]. On the other side, robots that use powered joints instead of powered wheels are more complicated to design, and the control algorithms that can be used are partially unexplored.

One of the first known snake robots was built by Hirose and colleagues at the end of 1972 [36]. He generically named this kind of robot an *active cord mechanism* (ACM). After this first prototype, he built some other snake robots [20]. A huge snake robot has been developed in 1992 at Caltech [4]. The Jet Propulsion Laboratory of the NASA presented in 1994 a serpentine robot [24]. Miller developed several prototypes of snake robots; among them the last one, S5 [26], has a very realistic lateral undulatory gait (its locomotion is probably the most similar to a biological snake, compared to other snake robots). Saito and colleagues presented in 2002 a simple snake robot used to validate some theoretical results [31]. Conradt developed WormBot [7], a snake-like robot controlled by local central pattern generators (CPGs). For a more detailed review of snake-like robots, see [14, 42].

Swimming snake robots (also referred to as lamprey robots or eel robots) are rarer. They are generally designed to imitate the anguilliform swimming of the eel (or the very similar one of the lamprey). Several theoretical papers have been written on this subject, but there are only a few real robotic realizations. The robots in this category that are the most interesting are the eel robot REEL II [25], the lamprey robot built at Northeastern University [41], and the lamprey-like robot built at SSSA [33]. In principle, these eel and lamprey robots could be adapted to terrestrial locomotion, but such experiments have not been reported. To the best of our knowledge, there are currently only a few amphibious snake-like robots, the HELIX-I ([34] as cited in [21, 35]) and its successor ACM-R5 [43], that can both swim in water and crawl on the ground (although ground locomotion is not described in the papers).

3.1.2.2 Salamander-like Robots

Currently, only a few prototypes of salamander-like robots (i.e., quadruped robots with several degrees of freedom in the spine) have been the object of scientific publications:

- A salamander robot with six segments and an on-board FPGA-based control system has been presented in [19]. It is not amphibious and can only walk.
- Robo-Salamander, a salamander robot with two degrees of freedom for the spine and two for each leg has been presented in [2]; no experiments seem to have been done with it, and no other publications followed. This robot was not autonomous and was powered and controlled using a cable. It is only capable of walking.

There are also some legged robots with flexible spine built by hobbyists, whose descriptions can be found on Internet, but none of them has been designed or used for scientific experiments.

None of the robots listed here is capable of swimming, and none is fully autonomous or amphibious.

3.1.3 Central Pattern Generator Model

The swimming motion of salamanders is similar to that of lampreys, using axial undulations which propagate as traveling waves from head to tail. The walking motion has a different pattern: the salamander moves the diagonally opposed limbs together, generating at the same time an S-shaped standing wave (which has nodes at the girdles) with the body.

Using the salamander as model, we address three fundamental issues related to vertebrate locomotion: (1) the modifications of the spinal locomotor circuits during the evolutionary transition from aquatic to terrestrial locomotion, (2) the mechanisms needed for the coordination of limb and body (i.e., axial) movements, and (3) the mechanisms that underlie gait transitions induced by simple electrical stimulation of the brainstem.

Our model is based on the following hypotheses:

1. The body CPG is as that of the lamprey and spontaneously produces traveling waves when activated with a tonic drive. The limb CPG, when activated, forces the whole CPG into the walking mode, as previously proposed in [6].
2. The couplings from limb to body oscillators are stronger than those from body to body oscillators and from body to limb oscillators. This allows the limb CPG to "override" the natural tendency of the body CPG to produce traveling waves and forces it to produce standing waves.
3. Limb oscillators cannot oscillate at high frequencies, that is, they saturate and stop oscillating at high levels of drive. This provides a mechanism for automatically switching between walking and swimming when the drive is varied [3],

and explains why swimming frequencies are systematically higher than walking frequencies [13, 16].

4. For the same drive level, the intrinsic frequencies of the limb oscillators are lower than those of the body oscillators. This explains the rapid increase in frequency during the switching from walking to swimming and the gap between the walking and swimming frequency ranges [13, 16].

More details about the underlying biological hypotheses can be found in [22].

The body CPG model is a double chain of oscillators with nearest neighbor coupling (Fig. 3.1). An oscillator models the activity of an oscillatory center in the spinal cord (a group of about 50,000 neurons that produce rhythmic activity). The chain is designed to generate a traveling wave from the head to the tail of the robot. This wave is used to achieve anguilliform swimming in water. In addition to this body CPG, limb oscillators have been added to the model (one per limb); they are bidirectionally coupled together and unidirectionally coupled to all body oscillators (see Fig. 3.1). During swimming, these oscillators are stopped (they do not oscillate), and thus do not influence the behavior of the body CPG, which continues to produce a traveling wave. During walking, the oscillators are enabled and influence the body oscillators, which begin to produce an S-shaped standing wave that can be used for walking.

The total number of oscillators is $N = 20$: $N_B = 16$ oscillators (i.e., eight pairs) for the body CPG (which controls six real elements and two fictive joints placed in

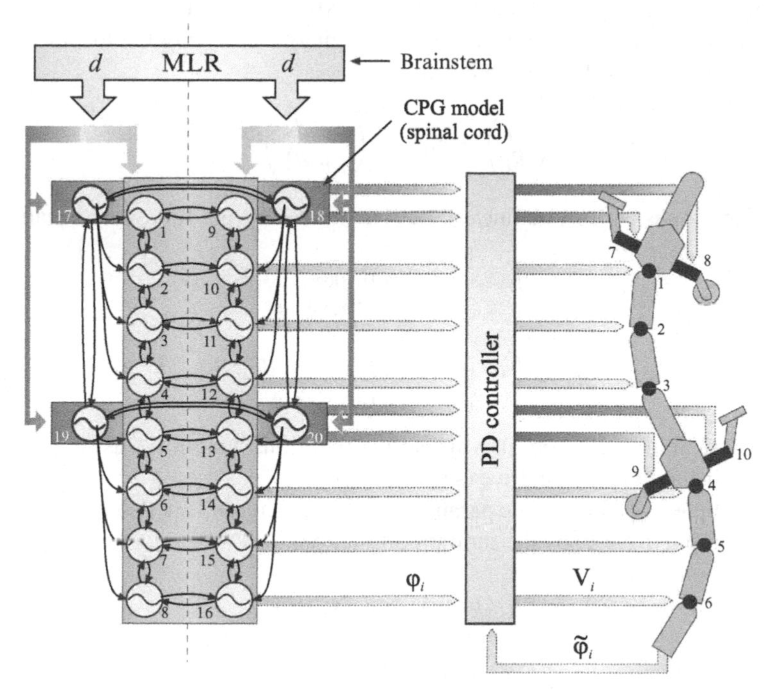

Fig. 3.1 Structure of the CPG used in the robot

the limb elements), and $N_L = 4$ oscillators for the limbs. Body joints (both real and fictive) are numbered 1–8 from head to tail. Oscillators in the left chain of the CPG are numbered 1–8 and those on the right side are numbered 9–16 from head to tail. Limb oscillators are numbered 17–20.

The oscillators that compose the CPG are implemented as amplitude-controlled phase oscillators:

$$\begin{cases} \dot{\theta}_i &= 2\pi \nu_i + \sum_j w_{ij} \sin\left(\theta_j - \theta_i - \phi_{ij}\right), \\ \ddot{r}_i &= a_i\left(\frac{a_i}{4}(R_i - r_i) - \dot{r}_i\right), \\ x_i &= r_i\left(1 + \cos(\theta_i)\right), \end{cases} \tag{3.1}$$

where the state variables θ_i and r_i represent, respectively, the phase and the amplitude of the ith oscillator, the parameters ν_i and R_i determine the intrinsic frequency and amplitude, and a_i is a positive constant. The coupling between the oscillators is defined by the weights w_{ij} and the phase biases ϕ_{ij}. The variable x_i is the rhythmic and positive output signal extracted out of oscillator i.

3.1.3.1 Influence of Drive: Saturation Function

To allow the control of the whole CPG with a single parameter, reproducing the output of the mesencephalic locomotor region (MLR) in the animal (i.e., the region of the brainstem that outputs the descending signals controlling the locomotion), a *saturation function* has been introduced:

$$\begin{pmatrix} \nu_i \\ R_i \end{pmatrix} = g(d) = \begin{pmatrix} g_\nu(d) \\ g_R(d) \end{pmatrix}. \tag{3.2}$$

This function is a stepwise linear function, defined by the following equations:

$$g_\nu(d) = \begin{cases} c_{\nu,1} d + c_{\nu,0} & \text{if } d_{\text{low}} \le d \le d_{\text{high}} \\ \nu_{\text{sat}} & \text{otherwise,} \end{cases} \tag{3.3}$$

$$g_R(d) = \begin{cases} c_{R,1} d + c_{R,0} & \text{if } d_{\text{low}} \le d \le d_{\text{high}} \\ R_{\text{sat}} & \text{otherwise.} \end{cases} \tag{3.4}$$

Body and limb oscillators use different saturations functions, as they have to saturate at different levels of drive (see Fig. 3.2).

The frequency and amplitude parameters of all oscillators are determined, on the base of the input drive d, by the saturation function:

$$\begin{aligned} \begin{pmatrix} \nu_{\text{body}} \\ R_{\text{body}} \end{pmatrix} &= g_{\text{body}}(d), \\ \begin{pmatrix} \nu_{\text{limb}} \\ R_{\text{limb}} \end{pmatrix} &= g_{\text{limb}}(d). \end{aligned} \tag{3.5}$$

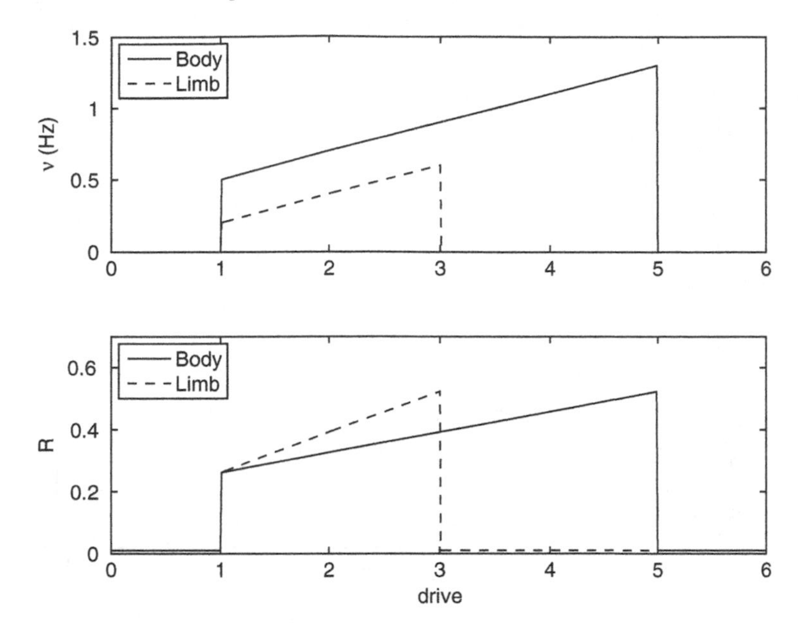

Fig. 3.2 Saturation function used on the robot

The actual saturation function is plotted in Fig. 3.2.

To achieve turning, different drives d_{L} and d_{R} can be applied to the left and right sides of the body oscillator.

3.1.3.2 Setpoints

For the body, the setpoints φ_i, that is, the desired angles for the six actuated joints, are obtained by taking the difference between the x_i signals from the left and right oscillators. A standard PD motor controller is then used to compute the voltage τ_i applied to the motor (using a PWM signal):

$$\begin{aligned} \varphi_i &= x_i - x_{N+i} \\ \tau_i &= K_{\mathrm{p}} e_i + K_{\mathrm{d}} \dot{e}_i \end{aligned} \tag{3.6}$$

where $e_i = \varphi_i - \tilde{\varphi}_i$ is the tracking error between the desired angles φ_i and the actual angles $\tilde{\varphi}_i$ measured by the motor incremental encoders, and K_{p} and K_{d} are the proportional and derivative gains.

As the limbs need to make complete rotations, their setpoints should monotonically increase, instead of having rhythmic movements like the body joints. The setpoints φ_i are therefore directly calculated from the phases θ_i of the limb oscillators

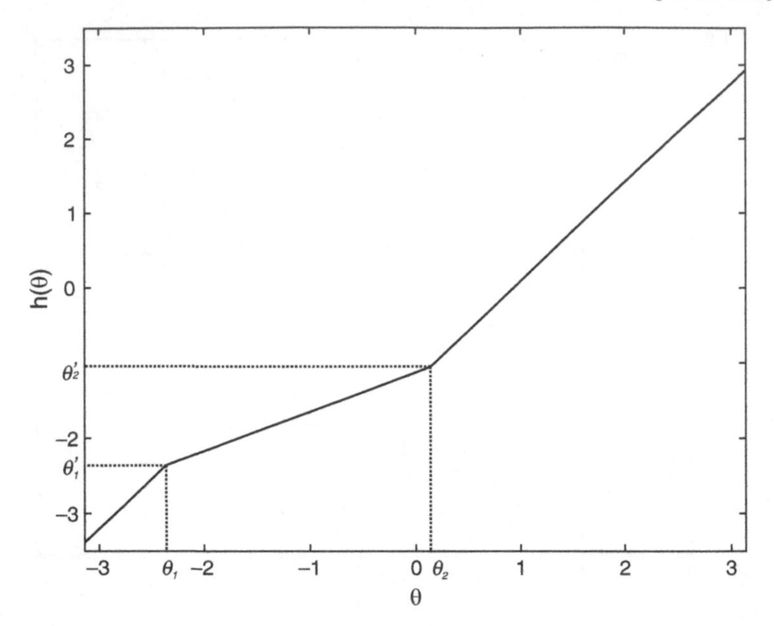

Fig. 3.3 Leg rotation function used on the robot. θ_1 and θ_2 correspond, respectively, to the begin and end of the stance phase of the leg

using a nonuniform rotation function (which accelerates the movement of the leg when it is not in contact with the ground):

$$\begin{aligned} \varphi_i &= h(\theta_i) \\ \tau_i &= K_p e_i + K_d \dot{e}_i \end{aligned} \tag{3.7}$$

The $h(\theta)$ function used on the robot is plotted in Fig. 3.3. With this function, the stance duration is approximately 40% of a whole cycle.

3.1.3.3 Inter-oscillator Couplings

In the body oscillator, the phase biases ϕ_{ij} are chosen to be equal to π between left and right oscillators (i.e., these will oscillate in antiphase). The phase biases between neighbor oscillators are set to $\frac{2\pi}{N_B/2} = \frac{2\pi}{8}$ for the descending connections and to $-\frac{2\pi}{N_B/2} = -\frac{2\pi}{8}$ for the ascending connections; this produces a complete wave of the body. In the limb oscillator and for the couplings between body and limb oscillators, $\phi_{ij} = \pi$ is used for all connections.

We use $w_{ij} = 10$ for all connections, with the exception of the couplings from limb to body oscillators, which need to be stronger, for which a value of $w_{ij} = 30$ has been used. For all oscillators, $a_i = 20$. The PD coefficients K_p and K_d are tuned manually for each element (e.g., elements in middle of the chain require larger gains than those at the extremities for good trajectory tracking).

The d parameter can be modified online by a human operator from a control PC using the wireless connection. The CPG will rapidly adapt to any parameter change

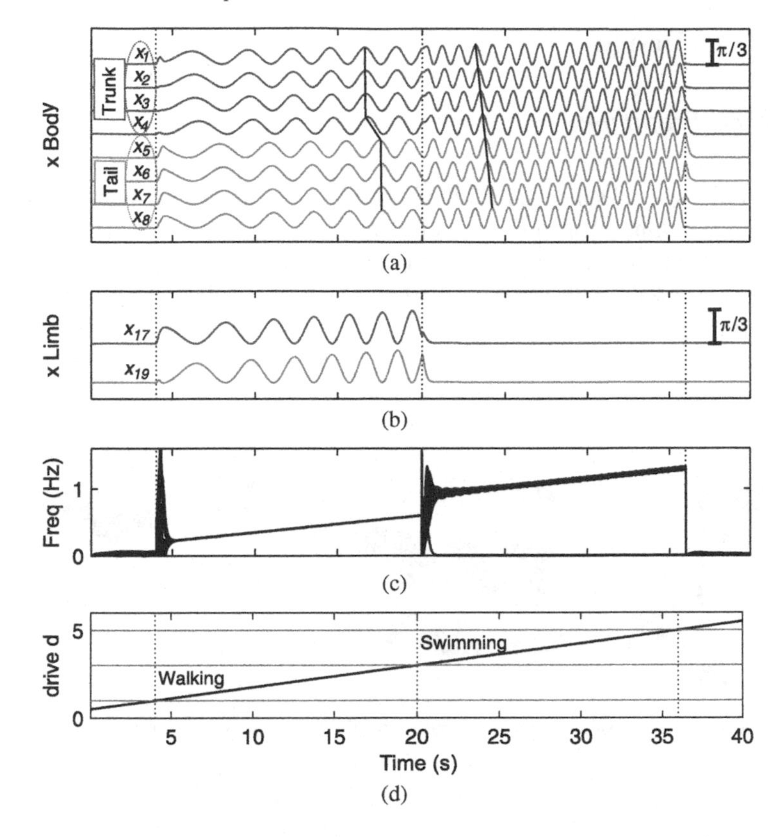

Fig. 3.4 Switching from walking to swimming; activity of the CPG model when the drive signal is progressively increased. **a** x_i signals from the left body CPG oscillators (oscillators on the right side are exactly in antiphase). Units are in radians (scale bar on the top right). Note the transition from standing waves (with synchrony in the trunk, synchrony in the tail, and an antiphase relation between the two, $4\,\text{s} < t < 20\,\text{s}$) to traveling waves ($20\,\text{s} < t < 36\,\text{s}$). **b** x_i signals from the left-limb CPG oscillators. Ipsilateral fore- and hindlimbs are in antiphase. **c** Instantaneous frequencies measured as $\dot{\theta}_i/2\pi$ in cycles s^{-1}. The variations in the instantaneous frequencies among individual oscillators at times $t = 4$ s and $t = 20$ s correspond to brief accelerations and decelerations before resynchronization. **d** Linear increase of the drive d applied to all oscillators. The horizontal lines correspond to the lower ($d_{\text{low}}^{\text{limb}} = d_{\text{low}}^{\text{body}} = 1$) and upper ($d_{\text{high}}^{\text{limb}} = d_{\text{high}}^{\text{body}} = 5$) oscillation thresholds for limb and body oscillators in arbitrary drive units

and converge to the modified traveling or standing wave after a short transient period. An example of how the CPG reacts to parameter changes can be observed in Fig. 3.4: even with a continuously changing input drive, the oscillator generates smooth trajectories without any discontinuities in the outputs.

The differential equations are integrated by the microcontroller of the head (see Sect. 3.3.3) using the Euler method, with a time step of 10 ms and using fixed point arithmetics. As the current trajectory generator has a limited computing power (10 MIPS with one 8-bit register), the code heavily uses lookup tables for calculating functions.

3.2 Robot's Design

3.2.1 First Prototype

The first prototype of salamander robot (Fig. 3.5) was built using three body elements and two limb elements. A head element was also present, but was internally empty, the trajectory generator being an offboard computer connected to the robot with a shielded cable. The body elements were those of the AmphiBot I snake robot [8–10].

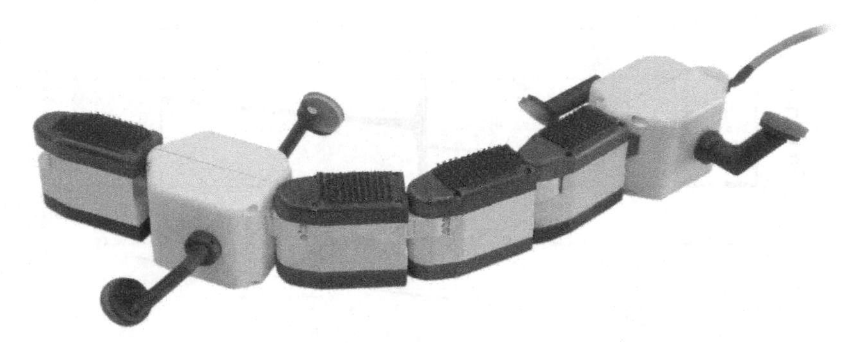

Fig. 3.5 The prototype of salamander robot built with the first generation elements

3.2.2 Body Elements

Each element has a single degree of freedom, and elements are fixed such that all axes of rotation are aligned. They consist of four structural parts: a body, two covers, and a connection piece (a drawing of two connected elements is visible in Fig. 3.6). All parts are molded using polyurethane, using molds created from positive parts in aluminum shaped with a CNC milling machine. The Li-Ion battery is directly incorporated into the bottom cover when the polyurethane is cast in the mold. To ensure the waterproofing of the robot, O-rings are placed between each cover and the body and around the output axis (the bottom O-ring has been subsequently replaced by a silicone sealant, because the complete closing of the bottom cover was generating mechanical problems to the gearbox). An element has a length of 7 cm and a section of 5.5 by 3.3 cm.

Each element contains two printed circuits (one for the power supply/battery charger and one for the motor controller, see Fig. 3.7), a DC motor, and a set of gears. Two different voltages are used inside an element: 3.6 and 5 V. The first one is the typical value of a Li-Ion battery and is only used to power the motor; the second one is used to power the electronics. When the robot is battery-powered (no

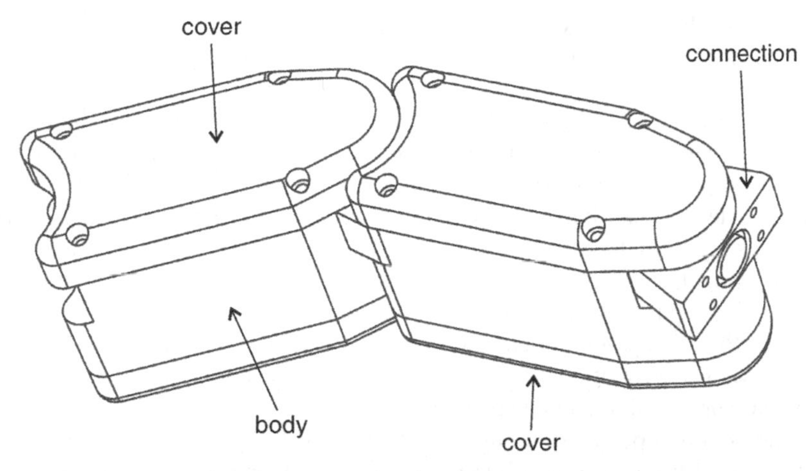

Fig. 3.6 Two first generation elements connected together

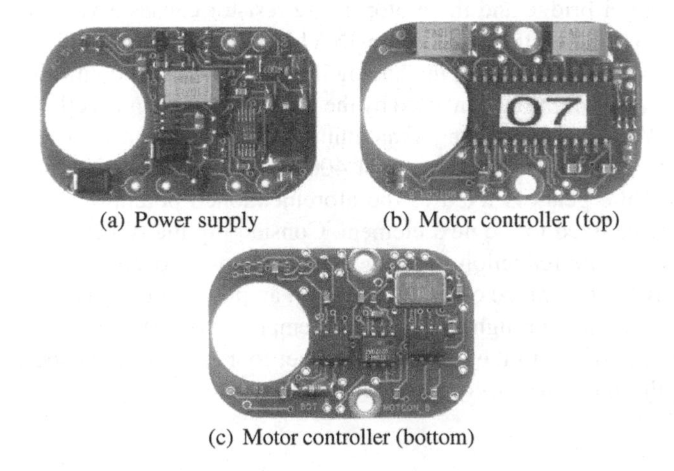

(a) Power supply (b) Motor controller (top)

(c) Motor controller (bottom)

Fig. 3.7 The printed circuit boards of the first prototype (real size)

external power source is connected), the motor is directly powered using the battery, without any intermediary regulator or converter, and the 5 V used by the electronics are generated with a capacitive charge-pump step-up converter (LTC3200). When an external (5 V) power source is connected, the 3.6 V for the motor are generated using a low-efficiency diode to create a voltage drop, and the electronics are directly powered using the external source. When the external power source is present, the battery could also be charged if this is necessary; for this reason, a small battery charger (LTC1733) is part of the power supply circuit. The charger can be enabled or disabled by the user over the I^2C bus. The battery has a capacity of 600 mAh, which is enough to power the element for an average time of approximately 2 h of continuous use (but this largely depends on the movements that the robot has to do and on the external constraints applied to it). An empty battery can be charged in approximately 1 h.

The motor controller is built with a PIC microcontroller (PIC16F876) and some external components. The motor has a magnetic encoder, which generates 16 pulses for every complete rotation of the axis. This encoder is connected to a LS7084 quadrature detector that filters and decodes the signals of the magnetic coder, generating a clock signal and a direction flag; these two signals are sent to the microcontroller, allowing it to track the current position of the motor. A 10 kΩ potentiometer is fixed to the output axis (after the reduction gears) and is connected to an analog input of the PIC; this potentiometer can be used to read the absolute position of the axis (e.g., when the robot is switched on, or to detect possible skews between the position measured with the magnetic coder and the real one).

The motor coil is powered through a SI9986 H-bridge, which supports currents up to 1 A. The H-bridge is driven by the microcontroller using a pulse-width modulation (PWM) signal, allowing the speed of the motor to be changed.

Between the H-bridge and the motor, a 1 Ω resistor causes a voltage drop. The resistor is connected to the input of an INA146 operational amplifier, the output of which is connected to one of the analog inputs of the microcontroller, therefore allowing a measure of the current used by the motor, and then indirectly of its torque.

The 0.75 W DC motor (having a maximum torque 1.2 mN m) drives a set of reduction gears with a reduction factor of 400 and an efficiency around 60%. The output axis of the gears is fixed to the aforementioned potentiometer and to the connection piece fixed to the next element. Considering the typical working speed of the motor and the reduction of the gears, a maximum oscillation frequency of approximately 0.3 Hz can be obtained if the full amplitude of 45° is used.

Five wires, passing through the (internally empty) axis, are connected to the contacts that are molded into the connection piece; four of them are used to pass the I^2C bus and the external power source all along the robot.

3.2.3 Limb Elements

The first limb elements, very similar in structure to the current ones (see Sect. 3.3.2) but with the same electronics as the old body elements, were also tested as a first prototype of a salamander robot (without its tail; see Fig. 3.5).

Most of the body elements were damaged by water leakages (which were not immediately detected) after the tests published in [10]; only three of them were still working and have been used for the salamander prototype. This was therefore only a preliminary design for testing the conception of the limb elements (no experiments have been done), and no detailed description of them will be given.

3.2.4 Design Problems

This first prototype suffered of several design problems, which have been mostly corrected in the current version of the elements (see next sections):

- The direct use of 5 V for the external power supply (mostly due to the lack of internal space for step-down converters, which require big coils) rendered the usage of the robot with external power (and the battery charging) very problematic, as only a limited amount of current can pass through the internal wires (having a section of 0.127 mm^2). For instance, a current of approximately 2 A on the wires caused a voltage drop along the robot around 2.5 V, causing part of the elements to reset (disabling battery charging).
- The torque generated by the elements was insufficient to achieve full oscillations at frequencies greater than 0.3 Hz, resulting in very slow locomotion.
- The waterproofing of the elements was very problematic and required sealing them with silicone.
- There was no possibility to detect the presence of water inside elements. Any malfunctioning supposedly owing to water leakage required the robot to be completely unmounted.
- The rigid connection between the elements combined with the small differences in the pieces caused the mounted robot to have bad contact with the ground (i.e., it was not perfectly flat).
- No battery protection mechanisms were implemented, and there was no possibility of turning off the robot; therefore, it had to remain connected to the external power all the time to preserve the batteries from being completely discharged (and thus rendered unusable).
- No connectors were on the circuits, and all the connections (including those to the motor) were realized by directly soldering the wires to the PCB. This operation was difficult (hence giving high mounting times for each element) and rather unreliable.
- The absence of any onboard trajectory generation capabilities and of radio communication required the direct control of the robot through a long shielded cable connected to a PC using a RS-232–I^{2}C converter.

3.3 Hardware

This new prototype (Fig. 3.8) address most of the problems found with the previous one, particularly in terms of mounting simplicity, electronic reliability, and waterproofing.

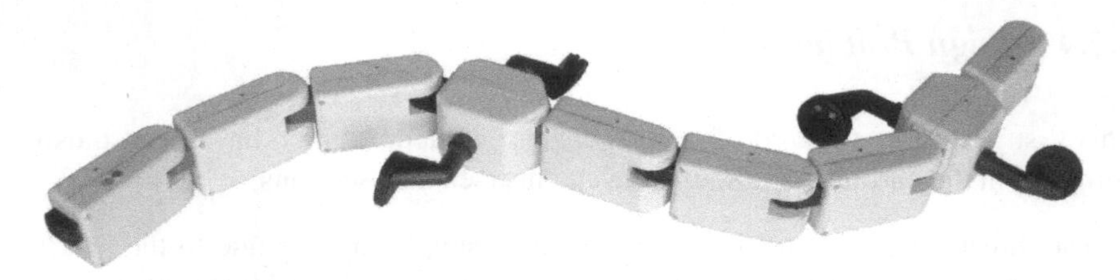

Fig. 3.8 Salamandra robotica

3.3.1 Body Elements

The same material of the first generation elements (polyurethane resin lighted with glass microballs) has been chosen for the external casing of the elements. They consist of two vertical symmetrical parts that are fixed together with screws. This is different from the first prototype, which was having a body closed with two covers (top and bottom). The elements are connected (both mechanically and electrically) using a compliant connection piece (molded with polyurethane rubber) fixed to the output axis, which contains six wires. The use of compliant connection pieces corrects the bad contact with the ground that was a serious problem of the previous generation elements, and allows the robot to better deal with irregularities of the ground. All the output axes of the elements are aligned, therefore producing planar locomotion. To ensure the waterproofing of each element, custom O-rings (placed between the two parts composing the body) are used.

Each element contains three printed circuits (a power board, a PD motor controller, and a small internal water detector) connected with a flat cable, a DC motor with an integrated incremental encoder, a set of gears (which uses two additional printed circuits as mechanical support), and a rechargeable Li-Ion battery. A view of an open element can be seen in Fig. 3.11. In opposition to the old elements, where all connections were realized by soldering the wires directly on the printed circuit boards, the new circuits use MicroMatch connectors for all the interconnections (bus, battery, motor, and inter-circuit connection); only the water detector (which was added later to the design; see below) uses directly soldered wires for space reasons. The elements are completely independent from each other (both electrically and mechanically). The density of the robot elements is slightly lower than 1 kg m^{-3} (the density of the old elements was slightly higher; therefore, the first robot was not buoyant). The battery is placed at the bottom of the elements to have the center of mass below the geometrical center, therefore, ensuring the vertical stability of the robot during both swimming and crawling.

In this description, for simplicity, we will not distinguish on which of the printed circuits each component is located. A block schematic diagram of the circuits can be found in Figs. 3.9 and 3.10. The motor controller is based on a PIC16F876A microcontroller and is basically the same of the first prototype. It is connected to the I^2C bus of the robot through a simple bidirectional repeater (built using two BSS138 MOS transistors), which is very useful to protect the microcontroller

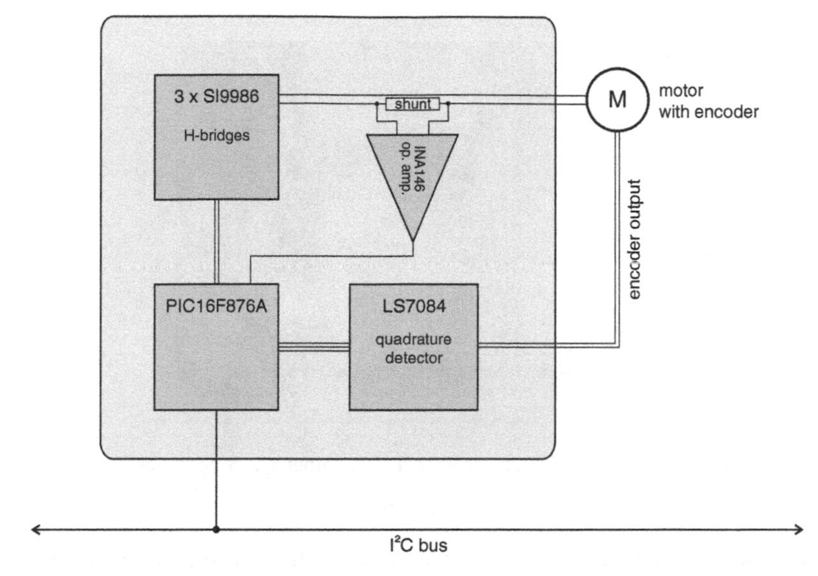

Fig. 3.9 The PD controller of the body (and legs) elements

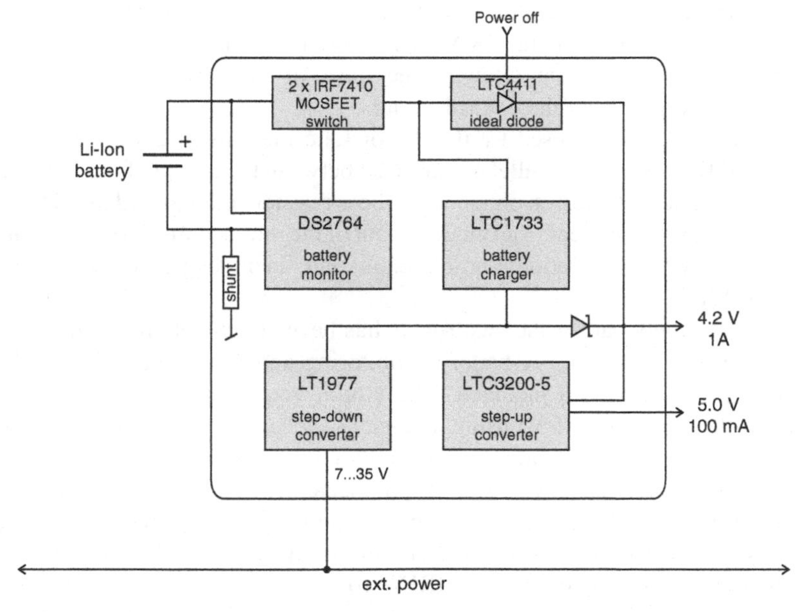

Fig. 3.10 The power circuits of the body (and legs) elements

internal drivers. The motor has an integrated magnetic incremental encoder, which generates 512 pulses for every complete rotation of the motor axis. The encoder is connected to a LS7084 quadrature detector that filters and decodes the signals coming from the encoder, generating a direction flag and a clock signal, which are

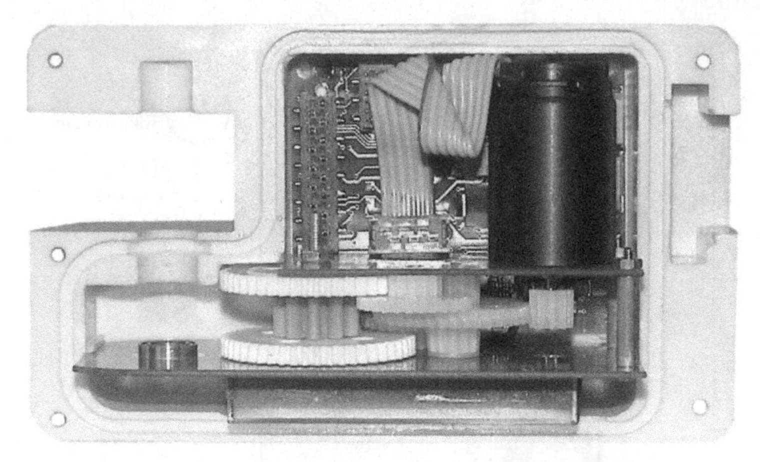

Fig. 3.11 Internal view of a body element (real size). The output axis is not mounted

connected to the microcontroller. Compared to the first generation elements, the potentiometer has been removed to simplify the mechanical structure.

The motor coil is powered through three SI9986 buffered H-bridges connected in parallel (each of which has a maximum current of 1 A; the maximal current that can be drawn by the motor is thus 3 A). These H-bridges are driven by the microcontroller with a PWM signal, allowing the speed of the motor to be changed by modifying the duty cycle of the control signal.

To measure the current used by the motor (and then, indirectly, its torque), a couple of 0.2 Ω resistors in parallel are inserted between the output of the H-bridges and the motor. The voltage drop obtained on these resistors is amplified by a INA146 operational amplifier and sent to an analog input of the microcontroller. The negative power (−5 V) for the operational amplifier is generated using a small capacitive inverter (MAX1719).

The power supply part of the electronics has been completely redesigned compared to the first prototype. A battery monitoring and protection circuit, which was missing, has also been included. The circuit generates the voltage required by most of the electronics (5 V) using a capacitive charge-pump step-up converter (LTC3200-5). All the electronics can be either powered by the internal Li-Ion battery or by an external power source (connected to the last element and distributed internally to all elements). When no external power source is connected, the battery (connected to the rest of the circuit through a DS2764 battery monitoring/protection circuit that controls two IRF7410 power MOSFETs) directly powers the motor. When an external power source is connected, an inductive step-down converter (LT1977) generates a voltage of approximately 4.6 V, which can both replace the battery voltage (to power the motor and the step-up converter) and power the LTC1733 battery charger. The circuit accepts up to 35 V (to reduce as much as possible the current on the internal wires, which have a limited section). The switch between the internally generated 4.6 V and the battery is realized with a LTC4411 "ideal diode" and a SS34 Schottky diode. The used battery is the same as that was used in the previous prototype and has a capacity of 600 mAh; it can power an element for approximately 2 h of continuous use in normal conditions. When empty, the battery can be recharged in approximately 1 h. The battery protection circuit disconnects the battery when its voltage drops below a critical threshold, thus preserving it from the often irreversible complete discharging. The circuit can also measure the instantaneous and accumulated current used by the circuit (or by the battery, during charging) and the battery voltage. This information could be read out using an I^2C bus, but these signals are currently left unconnected on the power card to limit the total number of devices on the bus (which is global to the robot).

A signal coming from a reed contact placed in one of the elements allows the user to switch off the robot by placing a magnet on it. This solution was found to be simpler than using a big waterproof switch. This signal is connected to the enable pin of the aforementioned LTC4411 (no current is drawn, the signal can therefore be directly generated using one of the batteries).

The water detector circuit (Fig. 12(e)), used internally to detect and localize any leakage, is placed at the bottom of the element. It has been introduced in the current

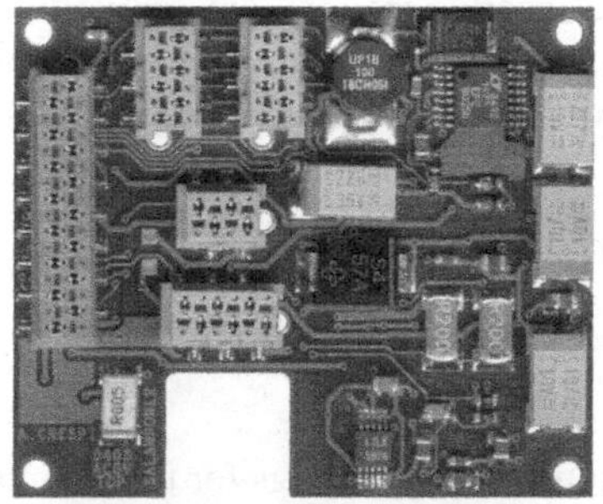

(a) Power and motor circuits (top side)

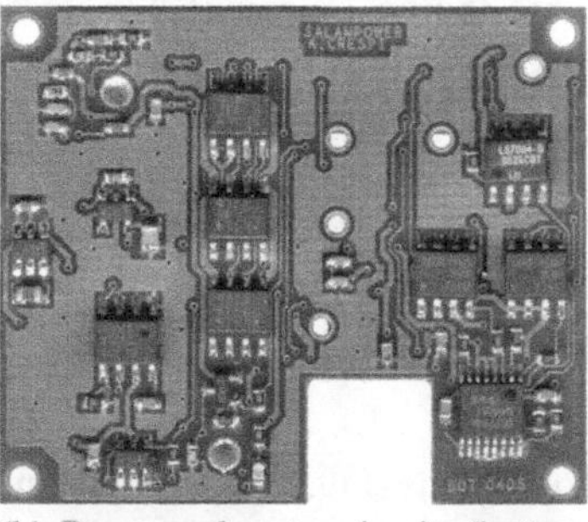

(b) Power and motor circuits (bottom side)

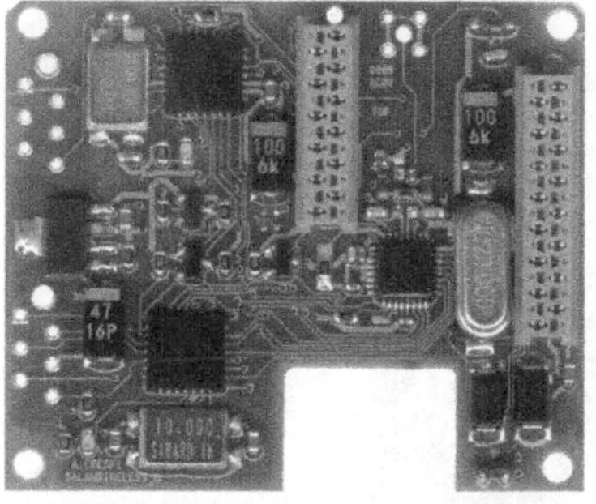

(c) Microcontroller circuit

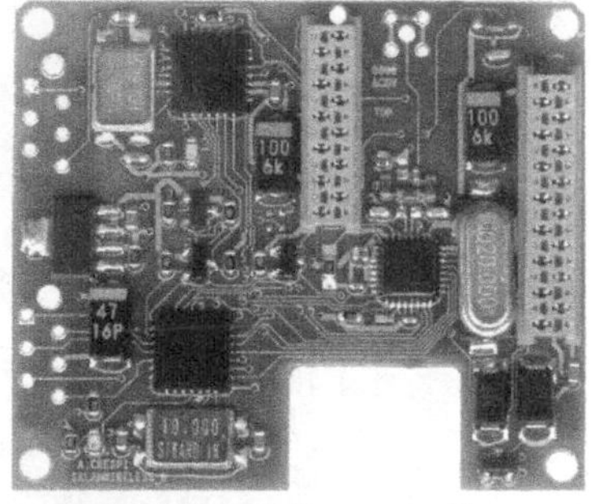

(d) Mocomotion controller

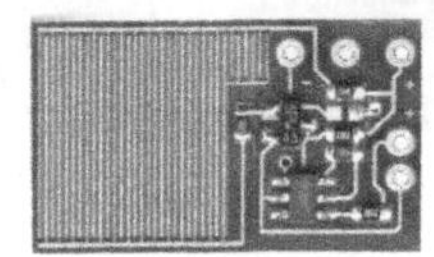

(e) Water sensor

Fig. 3.12 Pictures of the printed circuit boards of the robot (real size)

elements to ease the detection of water leakages. It has a sensitive surface of about 1 cm^2 consisting of several parallel tracks, half of which are connected to the power source through a resistor. When water (or a big amount of moisture) is on this surface, it acts like a resistor between the power source and the base of an NPN transistor, which begins to conduce. When water is detected, the circuit blinks an LED fixed through the top of the element, therefore, allowing the user to immediately detect the leakage and its position (i.e., the concerned element). The LED blinking is implemented using a PIC10F200 microcontroller, which in normal conditions accepts an incoming control signal for the LED and transparently replicates it on the output, hence permitting the LED to be used for other purposes. The introduction of this water detector dramatically simplified the handling of water leakages in the robot, which can now be localized without unmounting all the elements.

The 2.83 W DC motor (Faulhaber 1724 T 003 SR) has a maximum torque of 4.2 mN m and drives a gearbox with a reduction factor of 125. It is approximately four times faster and stronger than the motor used in the previous generation of elements, therefore, allowing higher amplitudes and oscillation frequencies to be reached. The output axis of the gears is fixed to the connection piece, which is inserted into the next element. Six wires are inserted into the axis and connected to the power boards of two adjacent elements: two are used for the external power, two for the I^{2}C bus, one for the power switch, and the last one is reserved for future usage and currently unconnected.

3.3.2 Limb Elements

The limb elements (Fig. 3.13) have been designed mainly as legs for the salamander robot, but can indeed be used for other purposes (e.g., the pectoral fins of the BoxyBot fish robot [12]). Each limb element includes a pair of identical circuits (one

Fig. 3.13 Internal view of a limb element (real size)

for the left limb and one for the right one). The design is unlike a real animal limb: this element has an axis capable of continuous rotation as output (and thus only one degree of freedom), similarly to robots using whegs (i.e., wheel-legs, see [28, 32]). This gives to the element both flexibility (it can be used for other purposes than legs) and simplicity (only one motor and gearbox per limb).

These elements are based on the same electronics of the body elements; however, as the printed circuits are also used as mechanical support for the gears and the motor, the components are differently distributed between the circuits. Additionally, an infrared LED/phototransistor couple allows the detection of the absolute position of the output axis (using a hole in the last wheel of the gearbox), to automatically align it when powering up the robot.

3.3.3 Locomotion Controller Circuit

The locomotion controller circuit has been designed to meet the following criteria:

- To provide a simple but flexible locomotion controller with low energy consumption. It should be possible to implement on it any control algorithm, following the needs of the user (in the work presented here, CPG-based controllers have been implemented on it).
- To have bidirectional radio communication capabilities (both on ground and under water) for remote control and measure.

The circuit is placed inside an empty body element (i.e., without the motor and the gearbox); a variation of the same circuit without the radio communication functions has been used for controlling the BoxyBot fish robot [12]. A block schema of the controller electronics can be seen in Fig. 3.14. The circuit is based on a PIC18F2580 microcontroller, which is master on the I^2C bus of the robot. It can implement a locomotion controller (e.g., a CPG) and sends out the setpoints to the motor controllers of each element in real time. The main microcontroller communicates, using a local serial line, with a PIC16LF876A microcontroller, which controls a nRF905 radio transceiver. The radio communication is handled by this separate microcontroller for simplicity, and because the PIC18F2580 cannot handle hardware SPI and I^2C at the same time. The antenna is internal to the element and consists of a simple $\lambda/4$ wire (where λ is the wavelength of the used frequency). The radio system uses the 868 MHz ISM band: preliminary experiments showed that a 10 mW signal (the power transmitted by the nRF905) on this frequency can penetrate in water up to at least 30 cm (the maximum tested depth). The more common 2.4 GHz band has not been used because it is heavily absorbed by the water. The maximal bandwidth is approximately 50 kbps, largely enough to send control commands and parameters to the online trajectory generator.

The software running on the locomotion controller can easily be reprogrammed with an external programming connector placed on the element.

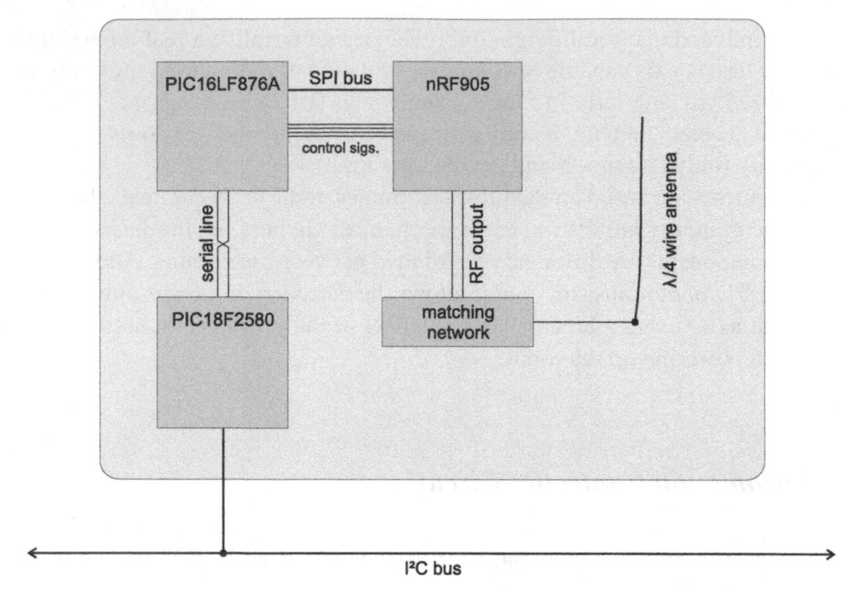

Fig. 3.14 Block schema of the electronics of the locomotion controller circuit

3.4 Experiments

Several experiments have been done to characterize the locomotion of the robot. We first present the measures of how the locomotion speed is related to the input drive of the CPG, then compare how the robot locomotion is similar to the one of the real salamanders in kinematic terms (focusing on lateral displacements of the body).

3.4.1 Speed as Function of Drive

The speed of the robot has been measured for 18 different drive values (between 1.0 and 3.0 for walking, and between 3.001 and 5.0 for swimming, with a step of 0.25). Each measure has been repeated five times, giving a total of 90 measures.

For walking, the speed has been measured by taking the time used to travel a given distance (i.e., 2 m). For swimming, the procedure was the same, but the distance was reduced to 1 m and the measure only started after an acceleration space of approximately 50 cm, to approach steady-state swimming as close as possible. The results are plotted in Fig. 3.15.

For walking, the speed almost linearly increases over the whole range of drives, from 3.54×10^{-2} m s^{-1} for $d = 1$ to 1.19×10^{-1} m s^{-1} for $d = 3$. For swimming, the speed increases linearly from 1.20×10^{-1} m s^{-1} for $d = 3.001$ to 1.40×10^{-1} m s^{-1} for $d = 3.5$, then stabilizes around this value up to $d = 4$. For $d > 4$, the speed

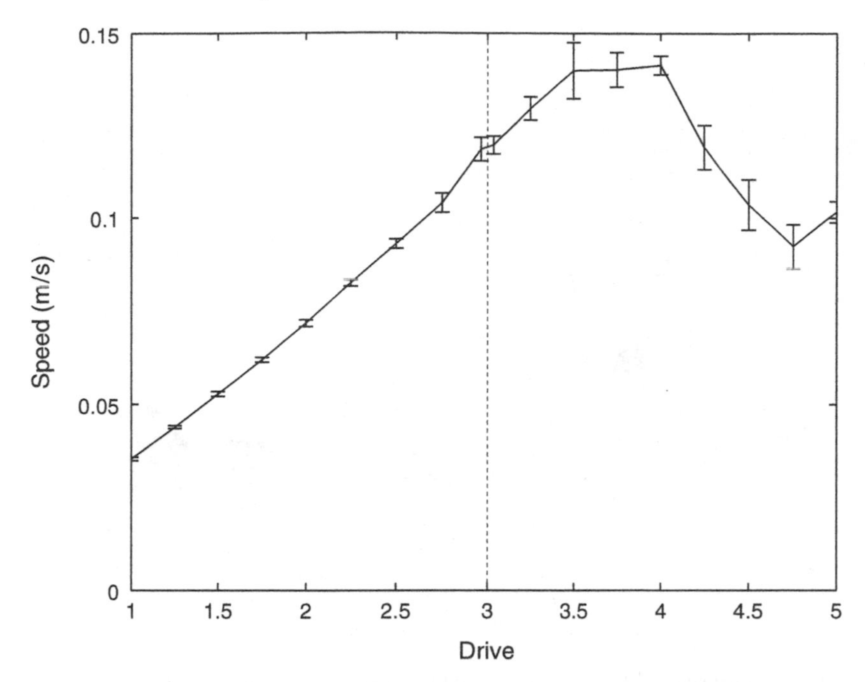

Fig. 3.15 Speed of the salamander robot for walking ($d \leq 3$) and swimming ($d > 3$)

decreases, with a minimal value of $9.24 \times 10^{-2}\,\mathrm{m\,s^{-1}}$ for $d = 4.5$. The decrease is mainly owing to the torque limits of the motors, which are not high enough to follow the desired trajectories in water at high frequency, with the consequence that the speed saturates, and then for higher drives, the traveling wave begins to deform, with the resulting decrease of speed.

3.4.2 Kinematic Measurements

To compare the movements of the robot with those of the real animal, kinematic measurements have been done on the robot using a custom video tracking system. These data have been compared with kinematic recordings of *Pleurodeles waltlii* salamanders, which have been provided by Isabelle Delvolvé (INSERM, Bordeaux).

The robot was filmed from above at 15 frames $\mathrm{s^{-1}}$ with a Basler A602fc-2 camera using an 8 mm C-mount lens. The frame data acquired over an IEEE1394 link was processed in real time with a custom program using the ARTag library [15] to extract the (x, y) coordinates of the markers (sort of 2D barcodes, see Fig. 3.16) placed on the robot. For body elements, the markers were placed at the rotation center of the output axis; for head and limb elements, they were placed at the same distance from the element's border than on body elements. The coordinates have been exported

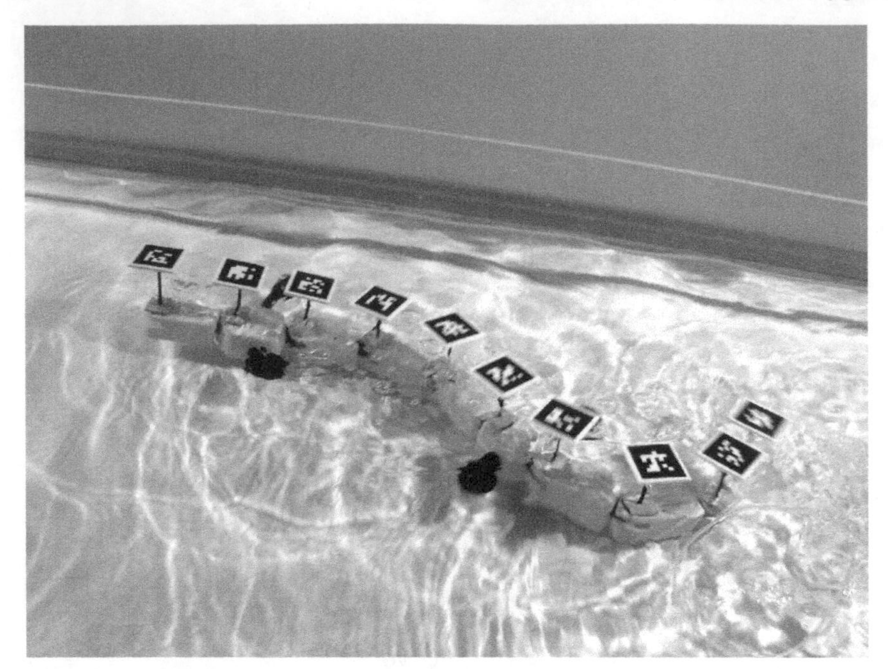

Fig. 3.16 The salamander robot swimming with the tracking markers fixed on it

in CSV files and then imported in MATLAB for processing and analysis, as for the salamander. The tracking markers had a size of $55 \times 55\,\text{mm}^2$. To minimize motion blur, the exposure time of the camera was set around 2 ms, using two 500 W halogen projectors for lighting. For walking, they were fixed on the top of the robot with double-sided adhesive tape. For swimming, they were fixed on a PVC support having the same size as the marker and placed 75 mm above the robot (using a rigid PVC cylinder of diameter 4 mm) to ensure that the markers were always out of the water during tracking (Fig. 3.16). The measures were repeated five times for each drive level. For walking, the camera field of view was always containing two complete cycles; for swimming, this varied between two and five cycles.

For illustration, snapshots from videos (without the tracking markers) for one locomotion cycle for walking and swimming can be seen in Figs. 3.17 and 3.18, respectively.

A detailed comparison of the lateral displacements during a complete locomotion cycle can be found in Figs. 3.19 and 3.20. The two gaits are easily distinguished: walking uses standing waves, whereas swimming uses a traveling wave along the body. As it can be seen in the figures, the generated waves are similar for the salamander and the robot, although there are some small differences.

The envelopes of lateral displacement of each marker (relative to the direction of motion) measured with the video tracking are plotted in Figs. 3.21 (walking) and 3.22 (swimming), with the corresponding data of the animal. For walking, the motion is qualitatively similar for the robot and the animal; both have minimal lateral

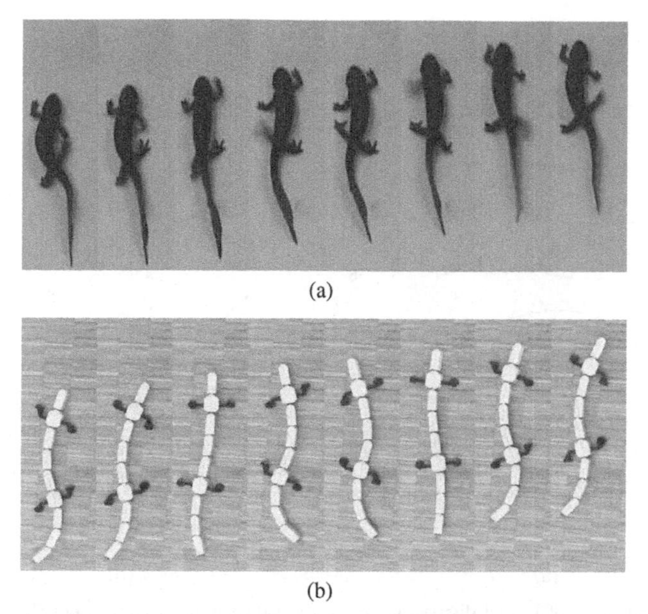

(a)

(b)

Fig. 3.17 Snapshots from videos for salamander (**a**) and robot (**b**) walking. The time step between the snapshots is 0.12 s for the salamander and 0.20 s for the robot

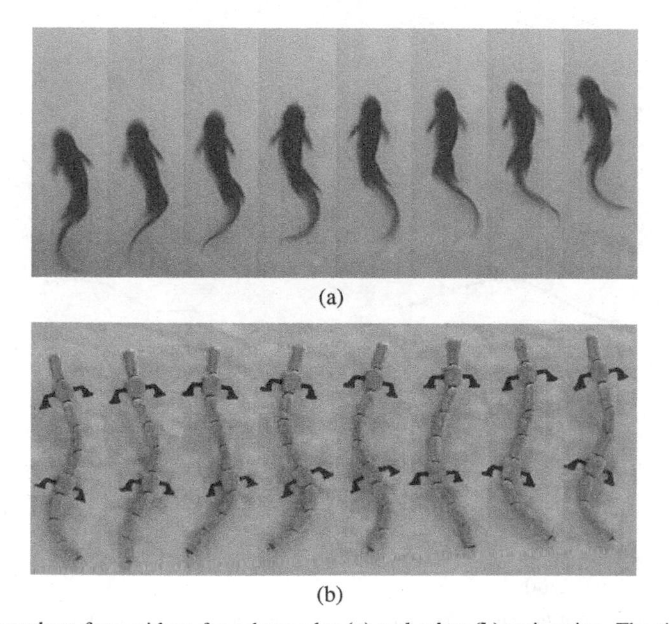

(a)

(b)

Fig. 3.18 Snapshots from videos for salamander (**a**) and robot (**b**) swimming. The time step between the snapshots is 0.04 s for the salamander and 0.12 s for the robot

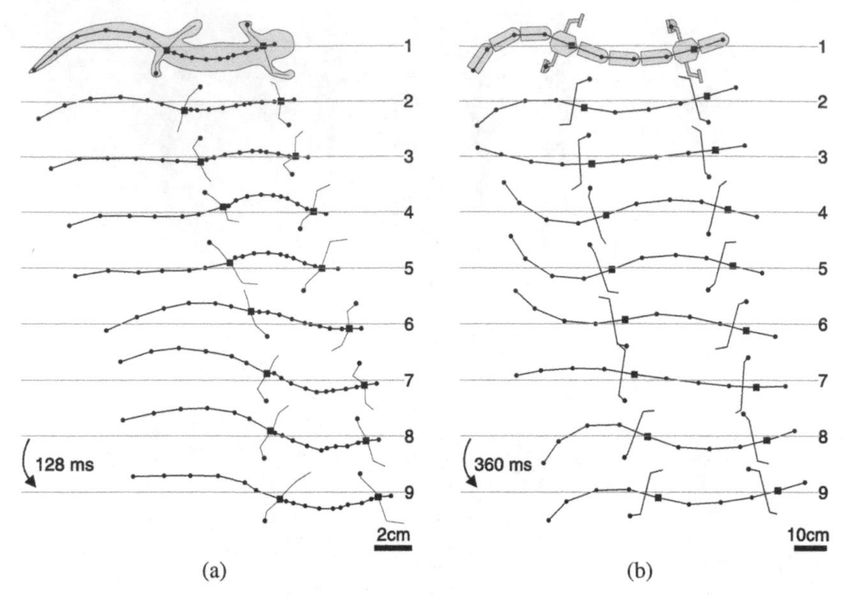

Fig. 3.19 Comparison of lateral displacements of the salamander (**a**) and robot (**b**) during walking. Velocities were $0.06\,\mathrm{m\,s^{-1}}$ (0.34 body lengths $\mathrm{s^{-1}}$) for the animal and $0.06\,\mathrm{m\,s^{-1}}$ (0.07 body lengths $\mathrm{s^{-1}}$) for the robot ($d = 2.0$)

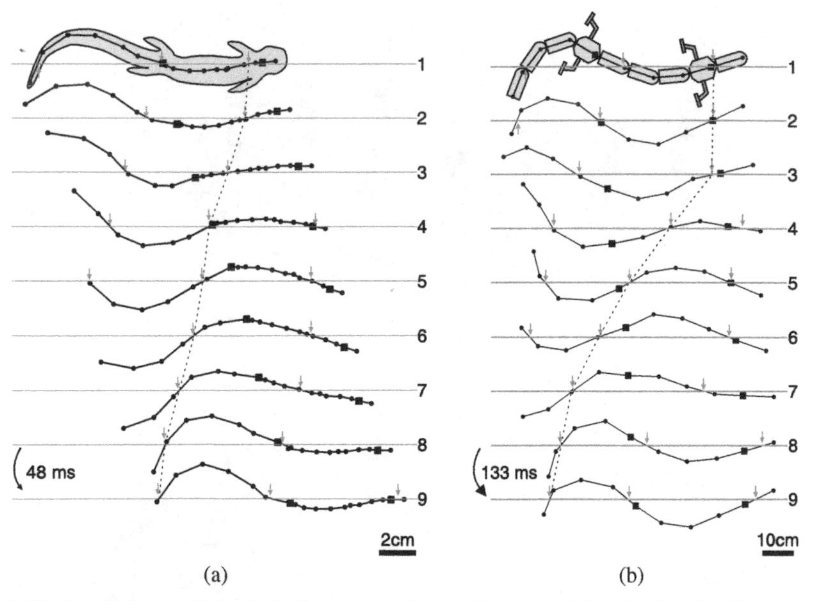

Fig. 3.20 Comparison of lateral displacements of the salamander (**a**) and robot (**b**) during swimming. Velocities were $0.17\,\mathrm{m\,s^{-1}}$ (0.89 body lengths $\mathrm{s^{-1}}$) for the animal and $0.11\,\mathrm{m\,s^{-1}}$ (0.13 body lengths $\mathrm{s^{-1}}$) for the robot ($d = 4.0$)

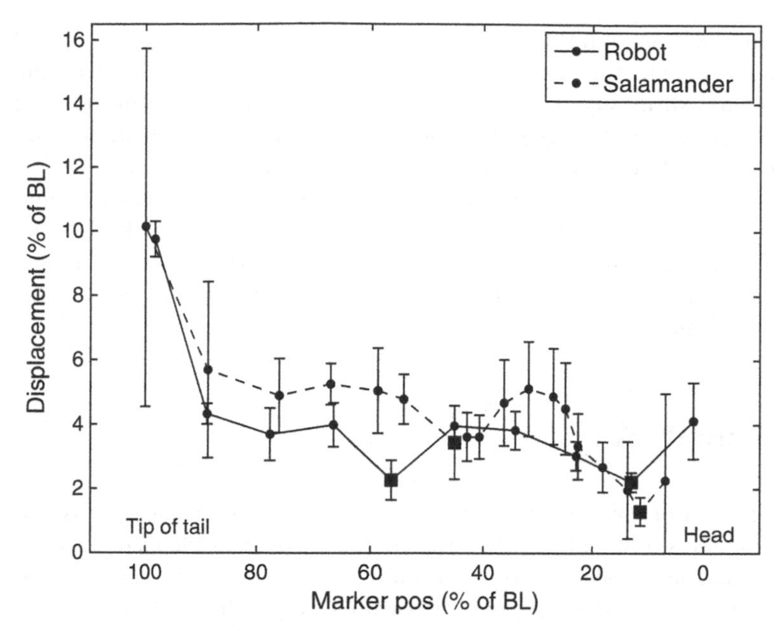

Fig. 3.21 Lateral displacements of the robot and real salamander during walking

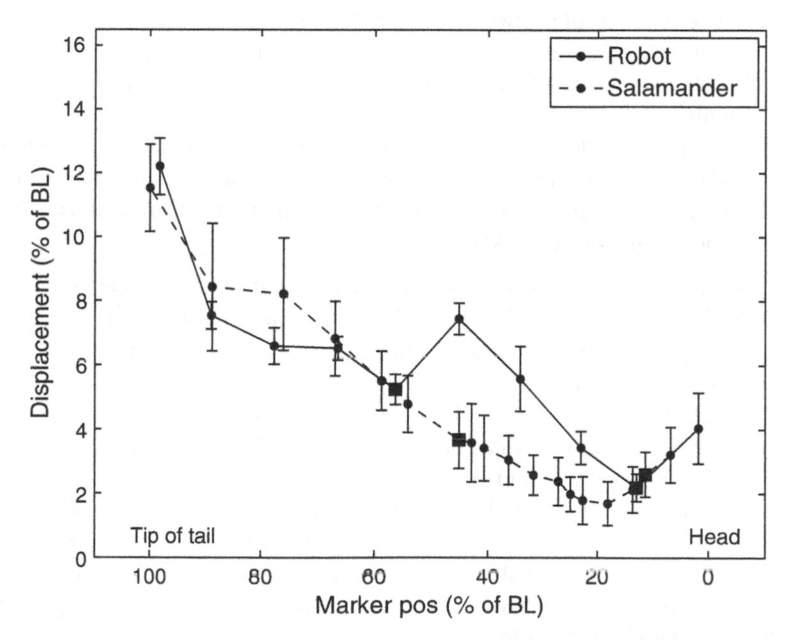

Fig. 3.22 Lateral displacements of the robot and real salamander during swimming

displacements at the girdles (which are not exactly at the same relative position). The main differences are the displacements of the tail (the salamander maintains the tip of its tail mostly straight, while the robot moves it) and of the head (the robot lacks a joint in the neck, therefore producing a greater lateral displacement of the head compared to the real salamander that makes compensatory neck movements). For swimming, the motion is also similar for the robot and the salamander, but with more differences than for walking. Particularly, the lateral displacements of the robot are higher than those of the salamander between the girdles. This can be explained by the lack of hinge joints in the limb elements of the robot and by their increased weight.

3.5 Future Work

Although most of the limitations of the first generation of salamander robot were addressed and solved with the current generation, some problems remain partially unsolved, and new weaknesses (related to previously unavailable features) appeared.

Despite the efforts to correct problems with waterproofing, small water leakages still happen periodically. They mainly concern body elements and owe to two main sources:

- The absence of a pair of screws (due to lack of space) at the horizontal center of the elements causes an insufficient compression of the O-ring. This problem is partially solved by using a silicone-based sealant around the O-ring when closing the elements.
- The forces applied to the output axis during vertical movements (e.g., when lifting the robot or when it has to overcome a small obstacle) can detach the brass axis from the flexible polyurethane rubber into which it is inserted, thus opening the way for possible water leakages.

These waterproofing problems have been addressed in a new third generation prototype (which is currently being tested) using a new closing system based on permanent magnets (instead of screws), thus also reducing the damages to the elements when unmounting them.

The limited computational capabilities of the PIC18F2580 used for the locomotion controller (a 8-bit microcontroller running at 40 MHz, obtaining a speed of 10 MIPS) required strong optimization of the code implementing the CPG locomotion controller. For instance, the CPG had to be implemented with fixed-point arithmetics and a large use of lookup tables to obtain an acceptable execution speed (almost entirely using the available RAM and program memory). We are currently actively testing a new locomotion controller based on a faster microcontroller (a 60 MHz 32-bit ARM microcontroller).

The following improvements to the current generation of salamander robot have been applied in the third generation elements now being tested:

- The replacement of the current global I^2C bus with a faster and more reliable CAN bus. The I^2C is now used only inside elements for local communications between the different components.
- Adding the possibility to retrieve status information from the battery monitoring/protection circuit, therefore, allowing the user to know the charging state of all the batteries and to estimate the energy consumption of the robot and of each individual element.
- Integrating a new microcontroller (a PIC18F2580) in the elements, for interfacing to the CAN bus, and providing local computational capabilities.
- Having the possibility to remotely reprogram the trajectory generator program using the radio link. The user has now the possibility to replace the program without using any special cable or programmer. This is especially important in the new robot that does not feature an external programming connector on the head.
- Having more status information and diagnostic possibilities (both locally on the elements, using RGB LEDs, and remotely).
- Adding infrared distance/obstacle sensors on the head to locally close the control loop and allow the robot to be autonomous in simple environments.
- Adding a degree of freedom to the limb elements, which are currently rigid and interrupt the wave along the chain of elements; this will result in a better locomotion.

Further improvements to be implemented are the following:

- Implement a really distributed CPG, with each pair of oscillators placed in their own element, with couplings through the CAN bus. The new local microcontrollers and the use of a CAN bus will ease the implementation of this kind of controller.
- Potentially adding a supplementary degree of freedom to have the possibility to lift the body (i.e., to remove the current limitation to planar locomotion).
- Adding other sensors (e.g., accelerometers or light sensors) and vision capabilities.

3.6 Conclusion

3.6.1 Realization of an Amphibious Salamander Robot

Amphibious robotics is a rather challenging topic: the complete waterproofing requirements of amphibious robots, for instance, completely influence all aspects in the design process, rendering it more complex comparing to dry-land robots. However, the advantages of amphibious robots are evident: an amphibious robot like a salamander can deal with difficult environments, which include water in any form (e.g., rain, partially flooded terrains, mud, etc.). This is a clear advantage, for instance, for outdoor robots used for inspection or exploration tasks. In our salamander robot, we used a modular approach that uses only two different types of elements (i.e., body and leg elements). The same modules can be used to construct other types of robots, some of which have been realized [11, 12].

3.6.2 Central Pattern Generators in Robots

Central pattern generators are more and more used for the control of robots. This biologically inspired control technique is well suited for the control of complex robots having multiple degrees of freedom, as CPGs can generate coordinated control signals for all the joints, receiving only simple low-dimensional inputs. This means that they are a good control method for implementing interfaces to be used by human operators for the interactive control of such robots. They also provide a sort of "abstraction layer" that can hide the complexity of the robot to the end user, also rendering it possible to control different types of robots with the same set of control signals. Finally, CPGs can also easily deal with continuously changing input signals possibly featuring discontinuities, like those that can be provided by a human operator or learning algorithm. In this work, a CPG model for controlling a salamander robot, inspired from the spinal cord structure of real salamanders, has been developed and successfully implemented to run on the onboard locomotion controller. Two drive signals are sufficient to modulate speed, direction, and type of gait; this also simplifies the control by a human user or higher level controller. One of the advantages of the use of CPGs is the possibility to easily include sensory feedback in the control loop, but this has still to be implemented in the salamander robot (as the current generation does not include any sensor, except for the motor incremental encoder).

3.6.3 Contributions to Biology

Finally, this robot proved to be a useful tool for verifying biological hypotheses. As explained in Sect. 3.1.1, robots could provide, for example, a "body" with which models of locomotion controllers can be tested to verify whether they actually generate locomotion in a real environment. We presented a model of central pattern generator that explains the locomotion control in salamanders and the transition between walking and swimming using a single control parameter. This model has been successfully implemented and tested on the salamander robot, demonstrating that it can actually generate forward motion of a body, using real actuators, in a real environment.

References

1. Ayers, J., Crisman, J.: The lobster as a model for an omnidirectional robotic ambulation control architecture. In: R. Beer, R. Ritzmann, T. McKenna (eds.) Biological neural networks in invertebrate neuroethology and robotics, pp. 287–316. Academic Press (1993)
2. Breithaupt, R., Dahnke, J., Zahedi, K., Hertzberg, J., Pasemann, F.: Robo-salamander – an approach for the benefit of both robotics and biology. In: Proceedings of the 5th International Conference on Climbing and Walking Robots (CLAWAR 2002) (2002)
3. Cabelguen, J., Bourcier-Lucas, C., Dubuc, R.: Bimodal locomotion elicited by electrical stimulation of the midbrain in the salamander Notophthalmus viridesecens. The Journal of Neuroscience **23**(6), 2434–2439 (2003)
4. Chirikjian, G., Burdick, J.: Design, implementation, and experiments with a thirty-degree-of-freedom 'hyper-redundant' robot. In: ISRAM 1992 (1992)
5. Choi, H., Ryew, S.: Robotic system with active steering capability for internal inspection of urban gas pipelines. Mechatronics **12**, 713–736 (2002)
6. Cohen, A.: Evolution of the vertebrate central pattern generator for locomotion. In: A.H. Cohen, S. Rossignol, S. Grillner (eds.) Neural Control of Rhythmic Movements in Vertebrates. Jon Wiley & Sons (1988)
7. Conradt, J., Varshavskaya, P.: Distributed central pattern generator control for a serpentine robot. In: ICANN 2003 (2003)
8. Crespi, A., Badertscher, A., Guignard, A., Ijspeert, A.: An amphibious robot capable of snake and lamprey-like locomotion. In: Proceedings of the 35th international symposium on robotics (ISR 2004) (2004)
9. Crespi, A., Badertscher, A., Guignard, A., Ijspeert, A.: AmphiBot I: An amphibious snake-like robot. Robotics and Autonomous Systems **50**(4), 163–175 (2005)
10. Crespi, A., Badertscher, A., Guignard, A., Ijspeert, A.: Swimming and crawling with an amphibious snake robot. In: Proceedings of the 2005 IEEE International Conference on Robotics and Automation (ICRA 2005), pp. 3035–3039 (2005)
11. Crespi, A., Ijspeert, A.: AmphiBot II: An amphibious snake robot that crawls and swims using a central pattern generator. In: Proceedings of the 9th International Conference on Climbing and Walking Robots (CLAWAR 2006) (2006)
12. Crespi, A., Lachat, D., Pasquier, A., Ijspeert, A.: Controlling swimming and crawling in a fish robot using a central pattern generator. Autonomous Robots **25**, 3–13 (2008)
13. Delvolvé, I., Bem, T., Cabelguen, J.M.: Epaxial and limb muscle activity during swimming and terrestrial stepping in the adult newt, *Pleurodeles waltl*. Journal of Neurophysiology **78**, 638–650 (1997)
14. Dowling, K.: Limbless locomotion: Learning to crawl with a snake robot. Ph.D. thesis, Robotics Institute, Carnegie Mellon University, Pittsburgh, PA (1997)
15. Fiala, M.: ARTag revision 1. A fiducial marker system using digital techniques. Tech. Rep. NRC 47419, National research council Canada, institute for information technology (2004)
16. Frolich, L., Biewener, A.: Kinematic and electromyographic analysis of the functional role of the body axis during terrestrial and aquatic locomotion in the salamander *Ambystoma tigrinum*. Journal of Experimental Biology **62**, 107–130 (1992)
17. Fukuoka, Y., Kimura, H., Cohen, A.: Adaptive dynamic walking of a quadruped robot on irregular terrain based on biological concepts. The International Journal of Robotics Research **22**(3–4), 187–202 (2003)
18. Gao, K.Q., Shubin, N.: Late jurassic salamanders from northern China. Nature **410**, 574–577 (2001)
19. Hiraoka, A., Kimura, H.: A development of a salamander robot - design of a coupled neuro-musculoskeletal system. In: Proceedings of the Annual Conference of the Robotics Society of Japan, Osaka (2002). (Paper in Japanese)
20. Hirose, S.: Biologically Inspired Robots (Snake-like Locomotors and Manipulators). Oxford University Press (1993)

21. Hirose, S., Fukushima, E.: Snakes and strings: New robotic components for rescue operations. In: B. Siciliano, D. Paolo (eds.) Experimental Robotics VIII: Proceedings of the 8th International Symposium ISER02, pp. 48–63. Springer-Verlag (2002)
22. Ijspeert, A., Crespi, A., Ryczko, D., Cabelguen, J.M.: From swimming to walking with a salamander robot driven by a spinal cord model. Science **315**, 1416–1420 (2007)
23. Klaassen, B., Paap, K.: GMD-SNAKE2: A snake-like robot driven by wheels and a method for motion control. In: Proceedings of 1999 IEEE International Conference on Robotics and Automation (ICRA 1999), pp. 3014–3019. IEEE (1999)
24. Lee, T., Ohm, T., Hayati, S.: A highly redundant robot system for inspection. In: Proceedings of the conference on intelligent robotics in the field, factory, service, and space (CIRFFSS '94), pp. 142–149. Houston, Texas (1994)
25. McIsaac, K., Ostrowski, J.: A geometric approach to anguilliform locomotion: Simulation and experiments with an underwater eel-robot. In: Proceedings of 1999 IEEE International Conference on Robotics and Automation (ICRA 1999), pp. 2843–2848. IEEE (1999)
26. Miller, G.: Neurotechnology for biomimetic robots, chap. Snake robots for search and rescue. Bradford/MIT Press, Cambridge London (2002)
27. Paap, K., Dehlwisch, M., Klaassen, B.: GMD-snake: a semi-autonomous snake-like robot. In: Distributed Autonomous Robotic Systems 2. Springer-Verlag (1996)
28. Quinn, R.D., Nelson, G., Bachmann, R., Kingsley, D., Offi, J., Ritzmann, R.: Insect designs for improved robot mobility. In: Proceedings of the 4th International Conference on Climbing and Walking Robots (CLAWAR 2001) (2001)
29. Roth, G., Nishikawa, K., Naujoks-Manteuffel, C., Schmidt, A., Wake, D.B.: Paedomorphosis and simplification in the nervous system of salamanders. Brain, Behavior and Evolution **42**, 137–170 (1993)
30. Roth, G., Nishikawa, K., Wake, D.B.: Genome size, secondary simplification, and the evolution of the brain in salamanders. Brain, Behavior and Evolution **50**, 50–59 (1997)
31. Saito, M., Fukaya, M., Iwasaki, T.: Serpentine locomotion with robotic snakes. IEEE Control Systems Magazine **22**, 64–81 (2002)
32. Saranli, U., Buehler, M., Koditschek, D.: RHex – a simple and highly mobile hexapod robot. The International Journal of Robotics Research **20**(7), 616–631 (2001)
33. Stefanini, C., Orlandi, G., Menciassi, A., Ravier, Y., La Spina, G., Grillner, S., Dario, P.: A mechanism for biomimetic actuation in lamprey-like robots. In: Proceedings of the First IEEE/RAS-EMBS International Conference on Biomedical Robotics and Biomechatronics (BioRob 2006), pp. 579–584 (2006)
34. Takayama, T., Hirose, S.: Development of HELIX: a hermetic 3D active cord with novel spiral swimming motion. In: Proceedings of TITech COE/Super Mechano-Systems Symposium 2001, pp. D-3 (2001)
35. Takayama, T., Hirose, S.: Amphibious 3D active cord mechanism "HELIX" with helical swimming motion. In: Proceedings of the 2002 IEEE/RSJ International Conference on Intelligent Robots and Systems (IROS 2002), pp. 775–780. IEEE (2002)
36. Umetani, Y., Hirose, S.: Biomechanical study of active cord mechanism with tactile sensors. In: Proceedings of the 6th international symposium on industrial robots, pp. c1-1–c1-10. Nottingham (1976)
37. Webb, B.: What does robotics offer animal behaviour? Animal Behaviour **60**, 545–558 (2000)
38. Webb, B.: Can robots make good models of biological behaviour? Behavioral and brain sciences **24**, 1033–1050 (2001)
39. Webb, B.: Robots in invertebrate neuroscience. Nature **417**, 359–363 (2002)
40. Webb, B., Reeve, R.: Reafferent or redundant: Integration of phonotaxis and optomotor behavior in crickets and robots. Adaptive Behavior **11**(3), 137–158 (2003)
41. Wilbur, C., Vorus, W., Cao, Y., Currie, S.: Neurotechnology for biomimetic robots, chap. A Lamprey-Based Undulatory Vehicle. Bradford/MIT Press, Cambridge London (2002)
42. Worst, R.: Robotic snakes. In: Third German Workshop on Artifical Life, pp. 113–126. Verlag Harri Deutsch (1998)
43. Yamada, H., Chigisaki, S., Mori, M., Takita, K., Ogami, K., Hirose, S.: Development of amphibious snake-like robot ACM-R5. In: Proceedings of the 36th International Symposium on Robotics (2005)

Chapter 4
Multilocomotion Robot: Novel Concept, Mechanism, and Control of Bio-inspired Robot

Toshio Fukuda, Tadayoshi Aoyama, Yasuhisa Hasegawa, and Kosuke Sekiyama

4.1 Introduction

In the background of robotics development, there are two explicit expectancies toward robots as follows.

1. Executing repetitive tasks human get tired to do.
2. Executing dangerous or hard work human cannot do on behalf of humans.

From such background, industrial robots were developed to conduct tasks that were hard physically for humans, such as conveyancing heavy materials or implementing oppressive work such as marathon uncomplicated operation. Besides, it is dangerous for workers to put tasks in practice in construction site, nuclear power plant, or outer space, hence it is desired to develop the robot substitute for human workers.

On the other hand, in the age of an aging society, the prospective role of robots is turning gradually from just working machines doing monotonous work in factories to partners who support human life, and putting to practical use of such "partner robot" is expected. To realize this kind of function, it is necessary to accomplish the ability to recognize circumstances and also to achieve the ability to move toward the objective point autonomously in the various environments. "Bio-inspired robot" imitating motion of living creatures can be considered prepartner robot, because the working area of a partner robot is living environments.

There have been many successful researches related to bio-inspired robot, which realize dynamic and skillful motions performed by creatures. Kimura et al. developed the dog-type robot *TEKKEN*, and the motion of the virtual spring-damper system of each leg and the rolling motion of the body are mutually entrained through rolling motion feedback to the CPGs, and thus can generate adaptive three-dimensional walking [18]. Furthermore, to create a self-contained quadruped robot capable of walking on natural ground, several new reflexes and responses had to be developed in addition to those developed in previous studies [26]. Floyd and Sitti developed the lizard-type robot, and varied parameters with the focus being a maximization of the ratio of the lift to power through simulation and experimentation, and

A. Adamatzky and M. Komosinski (eds.), *Artificial Life Models in Hardware*,

realized running on the water [12]. Mori and Hirose developed the snake-type robot *ACM-R3* and showed by experiment that it has the possibility of various movement methods [27, 28]. Especially, brachiation is focused as an interesting locomotion form performed by long-armed apes by using their arms to swing from branch to branch. Brachiation is a dynamic and skillful action mundanely performed by some kinds of apes [34]. A pioneering research analyzed dynamics of brachiation and proposed on using a six-link brachiation robot [15]. Fukuda et al. [17, 32, 33] developed a two-link brachiation robot, "Brachiator II," and proposed a heuristic method to find feasible motions. A self-scaling reinforcement learning algorithm was also proposed to achieve robustness against some disturbances [19]. As an analytical approach, Nakanishi et al. proposed the "Target Dynamics Method," in which the target dynamics are used as inputs to the controller; they investigated the "ladder," "swing up," "rope," and "leap" problems [30].

However, many researches related to bio-inspired robot were specific to just one type of locomotion mode. In addition, robots developed in these work were designed to be specialized to perform such locomotion.

Here, focusing on animals in the nature and considering the intrinsic differences between the previous bio-inspired robot and them, it is seen that diversity of locomotion is important. They are capable of performing several kinds of locomotion by stand-alone and accommodating the alteration of environment by choosing the adequate locomotion from multiple locomotion modes. Such diversity of locomotion, which was not pursued in the previous work regarding bio-inspired robotics, is a significant challenge to achieve the above-mentioned goal.

This chapter continues as follows. In Sect. 4.2, we describe the diversity of locomotion in animals at first, and then propose the novel concept of bio-inspired robotics, multilocomotion robot. multilocomotion robot is designed to perform several kinds of locomotion such as biped walking, quadruped walking, and brachiation. In Sect. 4.3, we introduce three robots (Gorilla Robot I, II, III) built as the prototype of multilocomotion robot; the link structure and control system including sensors and actuators are described in detail. In Sect. 4.4, we determine an optimal structure for a quadruped robot, which will allow the robot's joint torque sum to be minimized. And we discuss about the hardware of Gorilla Robot as a quadruped robot on slopes. In Sect. 4.5, we introduce previous works that used the hardware proposed in this chapter. Finally, we summarize this chapter in Sect. 4.6.

4.2 Multilocomotion Robot

4.2.1 Diversity of Locomotion in Animals

There are great many types of living creatures on the earth (Fig. 4.1). The number of species that have been identified so far is about 1.5 million. It is said that it becomes no fewer than 30 million if unidentified living creatures in the places that has not

Fig. 4.1 Biological diversity

been explored yet such as the rain forest are included. They have adapted to the environment and built a variety of unique society in the 40 billion-year evolutionary history from birth of the earth.

Here, let the environment on the earth be classified into three factors (air, ground, and hydrospace) and let locomotion pattern of animals in each circumstances be described.

At first, to move in the air, it is necessary to generate the higher lift force than its own weight. As for the living creatures, flying animals like birds and a part of insects (snake doctor or butterfly, cicada, bee) acquire the lifting force by flapping their own wings. Fast free movement irrelevant to the environmental condition of the ground is possible by moving in the air, hence it has possibility to collect nourishments effectively. On the other hand, ambulatory movement on the ground is performed by actuating joints to move limbs and a body to take advantage of the friction force and gravity. As for ambulation on the ground, differently from movement in the air or water, suitable locomotion pattern is dependent on the environmental condition as the ground includes desert, grassland, forest, and mountain terrain and the ground condition is of infinite variety. A part of insects (grasshopper or calicoback, ant, mantis) and quadruped animals (horse or camel, dog) and primates (human or monkey) walk to ambulate by using their own limbs. The living creatures without limbs such as serpent or earthworm can move by creeping. As just described, these living creatures acquire the locomotion suitable for the environmental condition and live to adapt to the circumstances. Lastly, in the water space such as sea or river, natant animals (e.g., fish, calamary, octopus, medusoid) move freely in water with their fin. Additionally, there are some animals that ambulate by walking on the bottom of the ocean, such as crustacea (shrimp or crab).

As mentioned earlier, the living creatures acquire a variety of locomotion modes according to the environment. These locomotion pattern and physical structure are related significantly. The dominant dynamics of the system during movement is dependent on both the locomotion pattern and on the physical structure [1]. It is conceivable that the locomotion modes and physical structure are coevolved to accommodate to the circumstances in evolutionary history.

In addition, it should be noticed that the locomotion pattern that many of the living creatures acquired is not single. For instance, birds, which basically possess not only wings but also legs, cannot obtain the nourishment freely. Snakes moves by creeping; however, some of them are able to swim in water. Many of the insects, which make up 75% of animals, have not only six legs but also wings and can move freely in air. By selecting multiple locomotion patterns, these animals have the high ability to ambulate.

As described earlier, it can be said that animals possess the diversity of locomotion and that such diversity is one of the important characteristics to live in nature.

4.2.2 Multilocomotion Robot

Diversity of locomotion, which has not been pursued in previous work of bio-inspired robotics, is one of the significant points to realize so as to put life-supporting robots like a human partner into practical use.

Considering both the mechanism necessary to realize tasks that human conduct on a daily life and locomotion mode to accomplish, in evolutionary term, the physical structure that is situated in between primates such as chimpanzee and orangutan, human seem to be most appropriate [10, 34].

Thus, as the new concept of bio-inspired robot, we propose multilocomotion robot shown in Fig. 4.2. Multilocomotion robot is the novel bio-inspired robot that can perform in stand-alone several kinds of locomotion such as biped walking and quadruped walking, brachiation.

4.2.2.1 Biped Walking

In all living creature with limbs in nature, the animals that acquire the biped walking are part of mammal such as the anthropoid closely related to human (human, chimpanzee, orangutan, gibbon, and gorilla). These creatures acquire biped locomotion by accommodating to the circumstances in the evolutionary process. Acquirement of biped walking made their hands free, hence it became possible to conduct some tasks by hands during locomotion. Consequently, it is thought that their intelligence is superior to other animals and they accomplish quite high ability to adapt to the dynamic environment.

Biped walking is normally modeled as an inverted pendulum [9]. The dynamics of an inverted pendulum is intrinsically unstable; hence, to achieve biped walking,

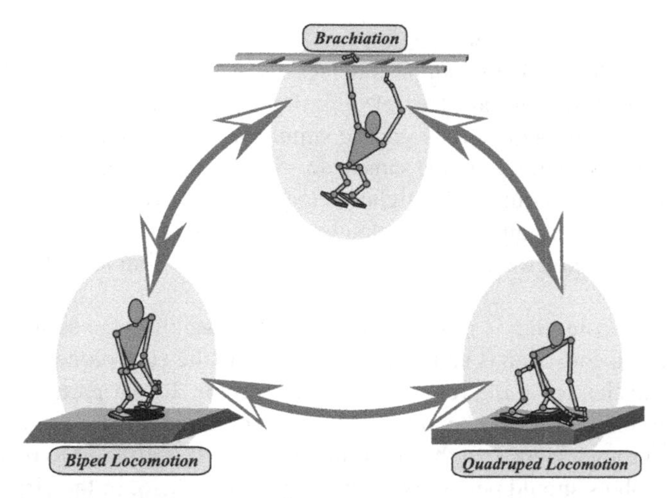

Fig. 4.2 Concept of the multilocomotion robot

it is necessary to control the body dynamics skillfully so as to compensate for the walking stability. However, the height of COM (Center of Mass) is kept at a higher position than that for quadruped locomotion or creeping, thus it has merit that the energy efficiency to ambulate can be quite high by taking advantage of the gravity effects. Also, as, for biped walking, the view point is kept in high position, it is possible to observe larger range of environment than does quadruped walking.

From the fact that most of human routine tasks are conducted in bipedal posture, the necessity to realize biped walking is explicit. In addition, the residential environment is built based on human body structure, hence biped locomotion is suited for life-supporting robots. Recently, lots of researches aiming to realize biped walking with humanoid robots are conducted actively. It can be said that it is one of the most developed area in recent years. Besides, accomplishment of faster locomotion by realizing running is anticipated [29].

4.2.2.2 Quadruped Walking

Animals performing quadruped locomotion are familiar to us [29]. Quadruped walking is locomotion mode that many of the animals living on the ground employ.

It is known that there are several gait pattern in quadruped walking, such as crawl gait, pace gait, trot gait, bounce gait, and gallop gait [13]. Crawl gait is the walking pattern employed by mainly most of the animals ambulating at low velocity, such as reptile and amphibian. In crawl gait, more than three feet contact the ground constantly and there is the period of four-feet-contact. Thus, the walking stability is quite high; however, it does not suit for fast locomotion in terms of energy efficiency.

Most of mammals (cat or dog, horse) at low velocity perform gait that body is supported by two legs constantly and each leg moves forward in the following order: right rear-leg, right front-leg, left rear-leg, left front-leg. As the walking velocity increases, the number of supporting legs decreases and trot gait [21] and pace gait are generated. In trot gait, legs of opposing corner – for example, right rear-leg and left front-leg – contact the ground simultaneously by turns at foot-contact. Meanwhile, in pace gait, legs in the same side – for example, right rear-leg and right front-leg – do at foot-contact. At higher velocity, the period of no-feet-contact is shown, namely gallop gait [20]. It is known that animals realize the high energy efficient locomotion by switching these gait pattern according to walking velocity or state [22].

Quadruped locomotion is more stable than biped walking, hence it can be said that quadruped locomotion is suited for ambulation in the environment where there are many obstacles or on the rough and rugged terrain. To put partner robots into practical use, robots need to be able to ambulate on not only flat ground but also the irregular ground. Hence, in the environment where it is difficult to realize biped locomotion, robots should perform quadruped walking. Also, in the circumstances where the degree of spatial freedom is limited such as a narrow cave or tunnel, quadruped walking is effective.

4.2.2.3 Brachiation

Brachiation, which is one of the motions in tree, is locomotion mode employed by arboreal animals such as gibbon and orangutan. It is the locomotion pattern that animals dangle on the branch and swing from branch to branch like pendulum as shown in Fig. 4.3.

It is know that dynamics of brachiation is modeled as a pendulum [3,4,11]. As for locomotion on the ground, COM is basically in upper position, whereas for brachiation, motion is performed dangling on the branch and COM is in lower position. Chang et al. built the system to observe the gibbon brachiation, and estimated the COM trajectory and alteration of mechanical energy [5].

Brachiation enhances the ambulating ability of multilocomotion robot. It is difficult to perform biped or quadruped locomotion without the rigid ground. However, if there is something to hold and dangle such as branch or pipe, it is possible to ambulate by using it.

Fig. 4.3 Brachiation of a gibbon

4.3 Gorilla Robot

To build a robot performing various locomotion mode mentioned earlier, various devices of design is needed, such as link structure, mass distribution, actuators, gear ratio, sensors selection, and range of joint motion. For instance, to make biped and quadruped walking (locomotion on the ground with legs) and brachiation (locomotion with arms) compatible, both arms and legs that can generate high enough torque to be able to support its own weight. Besides, as swing leg and swing arm is needed to move forward in short period, knee joints and elbow joints need high angular velocity. Therefore, to realize multi-locomotion robot as shown in Fig. 4.2, it is necessary to develop a balanced robot capable of satisfying high torque and high angular velocity. However, in previous work, design guideline of such type of robots has not been clarified yet.

Hence, as an animal close to the concept of multilocomotion robot, we select male gorilla (order: Primates, family: Pongidae, genus: *Gorilla*; Fig. 4.4). Gorillas normally ambulate in particular type of quadruped locomotion named knuckle walk. It often employs biped walking to carry something by hand and sometimes performs brachiation. In addition, as body size of mature male gorilla is about 1.7–1.8 m, it is conceivable that gorilla is suited for the model of multilocomotion robot in the human residential environment.

We built and developed three gorilla robots as the prototype of multilocomotion robot. Controllers of biped walking, quadruped walking, and brachiation have been composed and whole control system of multilocomotion robot has been developed from both sides of software and hardware.

In the following, hardware architecture of three gorilla robots and control system is explained.

Fig. 4.4 Western lowland gorilla. *Left*: Shape. *Right*: Skeletal preparations

4.3.1 Gorilla Robot I

The overview of first prototype of multilocomotion robot, Gorilla Robot I, is shown in Fig. 4.5. This robot was designed based on skeletal preparations of mature male gorilla by taking into consideration degree of freedom necessary to perform biped, quadruped walking, and brachiation. Robot link structure and joint placement are shown in Fig. 4.6. Shoulder joints and hip joints are primarily classified into ball joint. In recent years, realization of ball joint has begun to be investigated [31]; however, its development is on the way and there are many problems to solve in order to realize endurance and performance. Hence, in this research, these joints are composed of two joints – pitch and roll axis, and the mechanical structure was designed as follows: 5 DOF leg (2 DOF hip, 1 DOF knee, 2 DOF ankle), 5 DOF arm (2 DOF shoulder, 2 DOF elbow, 1 DOF wrist).

Motor output and reduction gear ratio was decided according to the numerical simulation and motion analysis of each locomotion. Each joint is actuated by DC servo motor (DC servo actuator, RH series, Harmonic Drive Systems Inc.) and as driver amplifier, TITech Driver Version 2 (Okazaki Sangyo Co.) is employed. The operating system of controller is real-time operating system, VxWorks (Wind River Systems Inc.). Figure 4.7 shows the system architecture of Gorilla Robot I. Computer to control, AD/DA board, counter board, and power is set outside the robot and connected with wires.

As the counter board (Interface PCI-6201) counts the pulse from encoder embedded in each actuator, joint angle is obtained at all time. Force sensors (wrists: IFS-67M25A50, ankles:IFS-100M40A100, NITTA CORPORATION) are installed on the wrists and ankles of the robot. This makes it possible to measure floor reaction force and ZMP (Zero Moment Point) during biped or quadruped locomotion and to observe the reaction force from bar and impact force at hand-contact in brachiation.

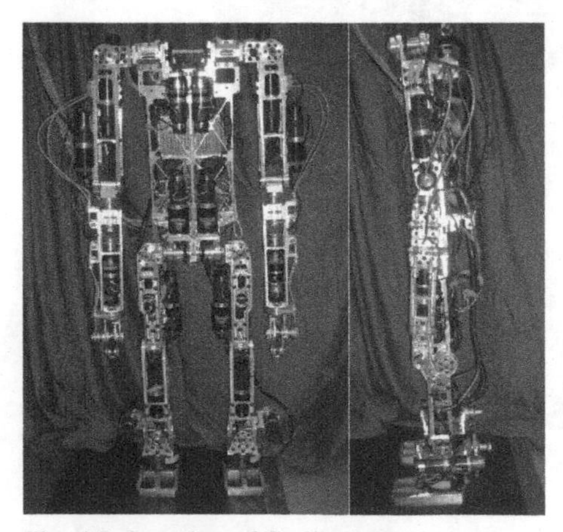

Fig. 4.5 Overview of Gorilla Robot I

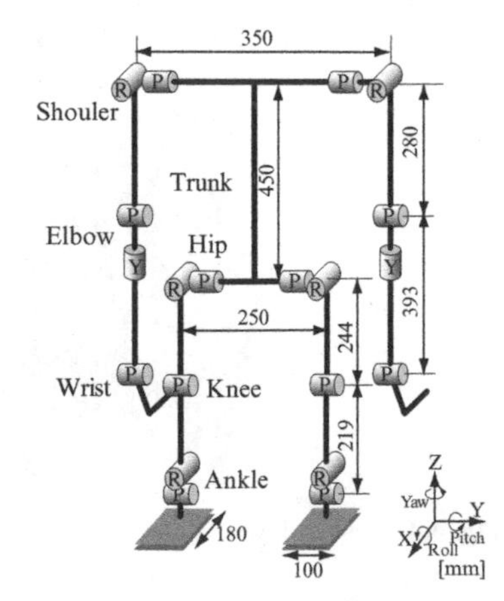

Fig. 4.6 Link structure of Gorilla Robot I

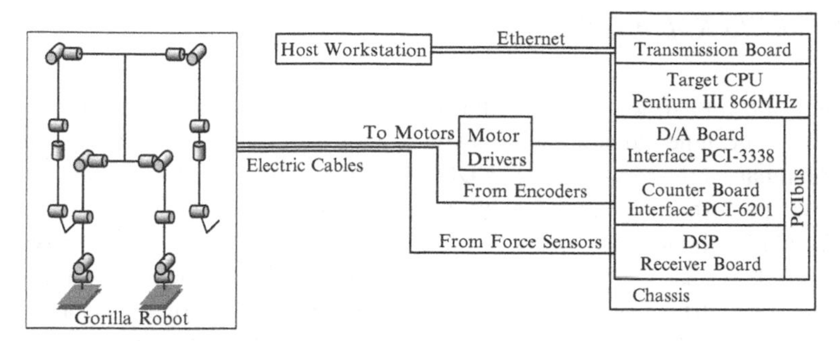

Fig. 4.7 Control system of Gorilla Robot I and II

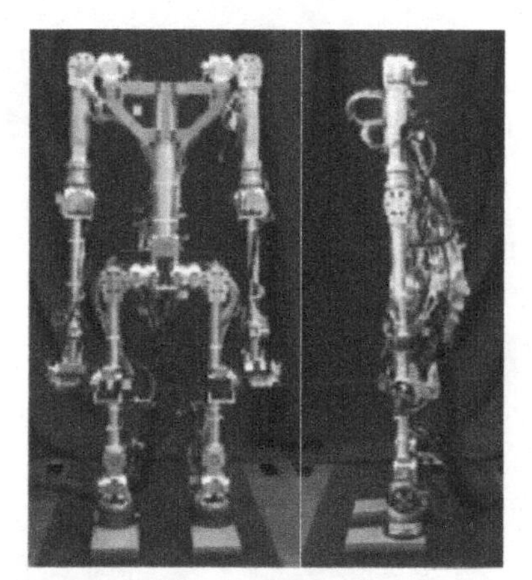

Fig. 4.8 Overview of Gorilla Robot II

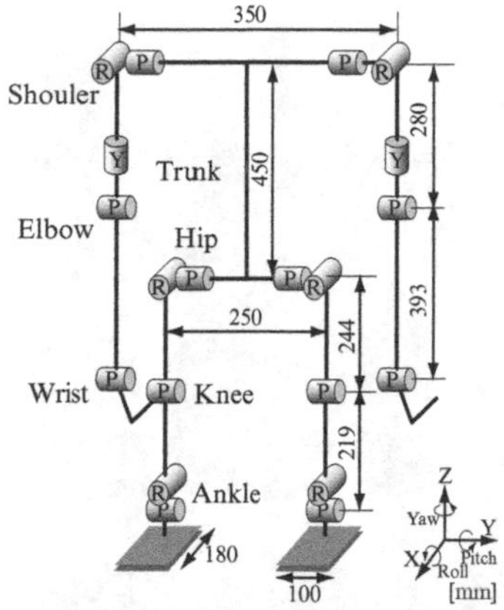

Fig. 4.9 Link structure of Gorilla Robot II

Consequently, this robot could not perform brachiation as it weighted too heavy, over 30.0 kg. To solve this problem, we built Gorilla Robot II, which is lighter than Gorilla Robot I.

4.3.2 Gorilla Robot II

Overview and link structure of Gorilla Robot II is shown in Figs. 4.8 and 4.9. The control system of this robot is same as that of Gorilla Robot I shown in Fig. 4.7.

Improved point in design is to exchange the pitch axis and yaw axis of elbow joint. Besides, the weight of robot is decreased to 20.0 kg with almost the same link structure as Gorilla Robot I.

AC servo drive (Σ-mini series, Yasukawa Electric Co.) force sensors are installed at wrists and ankles of the robot as motors and drivers to actuate each joint.

4.3.3 Gorilla Robot III

Overview and link structure of Gorilla Robot III is shown in Figs. 4.10 and 4.11. This robot is designed to add the following 4 DOF to 20 DOF of Gorilla Robot II: 2 DOF lumbar (roll and yaw axis), 1 DOF hip joints (yaw axis). Gorilla Robot III was able to perform more smooth motion by using the added DOF.

Its height is about 1.0 m and weight is about 22.0 kg. This robot is driven by 24 AC motors of 20–30 W with 100–200 times of speed reduction by harmonic gears. AC servo drive (Σ-mini series, Yasukawa Electric Co.) force sensors are installed at wrists and ankles of the robot as motors and drivers to actuate each joint (Fig. 4.12).

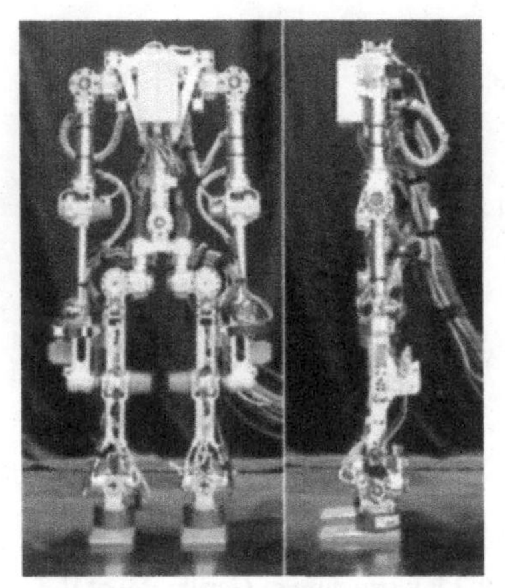

Fig. 4.10 Overview of Gorilla Robot III

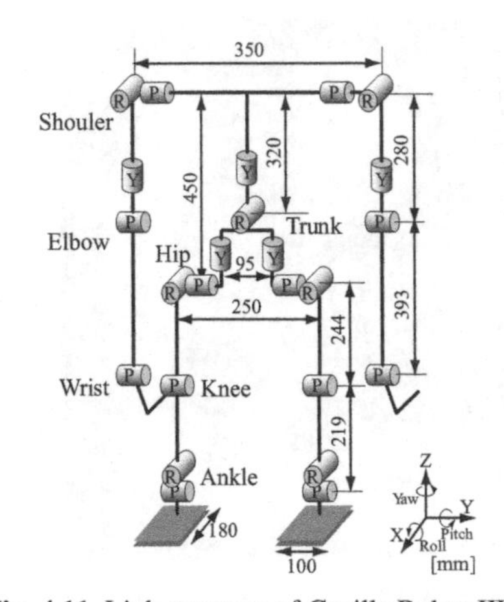

Fig. 4.11 Link structure of Gorilla Robot III

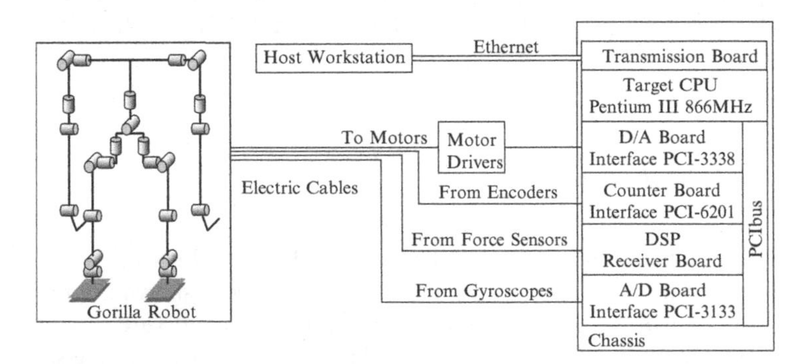

Fig. 4.12 Control system of Gorilla Robot III

4.4 Evaluation of the Gorilla Robot on Slopes as Quadruped Hardware

4.4.1 Evaluation of Joint Torque in Quadruped Walk on a Slope

4.4.1.1 Derivation Method of Joint Torque

The intermittent crawl gait is adopted as the basic gait pattern in this work, because the intermittent crawl gait has been successfully used by robots walking on steep slopes and it is a widely used control method [36]. The intermittent crawl gait is composed of two phases: three-point ground contact and four-point ground contact. In the four-point ground contact phase, joint torque has to be calculated using the dynamics, because the motion of the COG has to be considered. We calculated joint torque using "open dynamics engine (ODE)" [35]. ODE is a library used for simulating articulated rigid body dynamics without the calculation of complex equations. In the three-point ground contact phase, the joint torque can be calculated by statics because the COG motion can be ignored. Calculations of static torque are expressed later. In general, to control the position and posture, six or more degrees of freedom are needed. Thus, link structure consists of six degrees of freedom in each leg as shown in Fig. 4.13. The origin of the coordinate axes is placed at the COG position. Additionally, we express a position vector on the top of the leg by $X = (x,y,z)^T$ and a joint angle vector by, $q = (q_1, q_2, q_3, q_4, q_5, q_6)^T$. Then, forward kinematics is expressed as follows:

$$X = f(q). \tag{4.1}$$

Small displacement of the joint angle and position of the end effector are expressed as follows:

$$\delta X = J(q)\,\delta q, \tag{4.2}$$

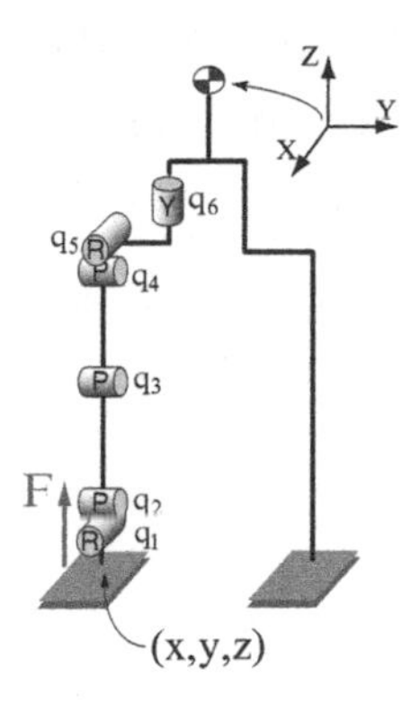

Fig. 4.13 Link model

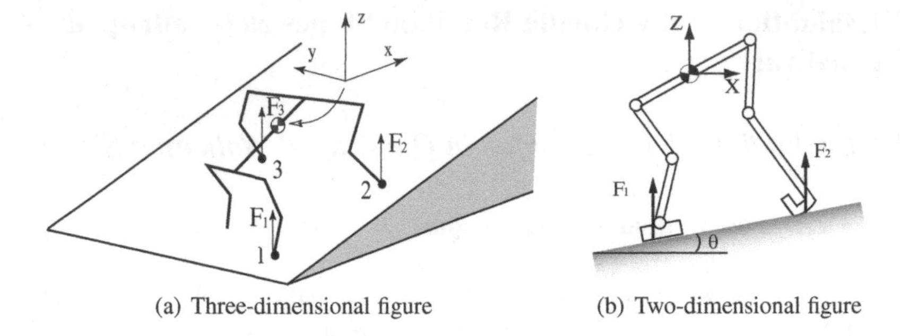

(a) Three-dimensional figure (b) Two-dimensional figure

Fig. 4.14 Definition of coordinates and grounding points

where $J(q)$ is a Jacobean matrix, δX is a small displacement vector of position, δq is a small displacement vector of angle. The principle of virtual work provides the relation between the force from an end effector and the joint torque as follows:

$$\tau = J(q)^T F, \tag{4.3}$$

where $\tau = (\tau_1, \tau_2, \tau_3, \tau_4, \tau_5, \tau_6)^T$ is a vector of joint torque, $F = (f_x, f_y, f_z)^T$ is a vector of ground-reaction-force. Equation (4.3) gives a joint torque vector from a vector of ground-reaction-force.

Next, we will explain the calculation method of the ground-reaction-force vector to derive the torque vector. As shown in Fig. 4.14, the right-handed standard coordinates is set at the position of the center of gravity and the X axis parallels the horizontal surface. We define the three grounding points as follows: 1, 2, and 3 position vectors for each grounding points) as $r_1 = (x_1, y_1, z_1)^T$, $r_2 = (x_2, y_2, z_2)^T$, and $r_3 = (x_3, y_3, z_3)^T$, and ground-reaction-forces of each grounding point as F_1, F_2, F_3.

As the ground-reaction-forces vector parallels the gravity vector, components of F_1, F_2, F_3 are expressed as follows:

$$F_1 = (\|F_1\| \sin\theta, 0, \|F_1\| \cos\theta)^T \tag{4.4}$$

$$F_2 = (\|F_2\| \sin\theta, 0, \|F_2\| \cos\theta)^T \tag{4.5}$$

$$F_3 = (\|F_3\| \sin\theta, 0, \|F_3\| \cos\theta)^T, \tag{4.6}$$

where θ is a slope angle, $\|F_1\|, \|F_2\|, \|F_3\|$ are normed vector of F_1, F_2, F_3. From equilibrium of force and momentum, (4.7) and (4.8) are derived as follows:

$$\sum_{i=1}^{3} F_i = Mg \tag{4.7}$$

$$\sum_{i=1}^{3} F_i \times r_i = 0, \tag{4.8}$$

where $g = (0, 0, g)$ is a gravitational acceleration vector.

Simultaneous equations with respect to $||F_1||$, $||F_2||$, $||F_3||$ are given by substituting the components of the respective vector to (4.7) and (4.8) as follows:

$$||F_1|| + ||F_2|| + ||F_3|| = Mg, \tag{4.9}$$

$$y_1||F_1|| + y_2||F_2|| + y_3||F_3|| = 0, \tag{4.10}$$

$$\sum_{i=1}^{3}(x_i \cos\theta - z_i \sin\theta))||F_i|| = 0. \tag{4.11}$$

From (4.9)–(4.11), $||F_1||$, $||F_2||$, and $||F_3||$ are obtained as follows:

$$||F_1|| = \frac{Mg\left[y_3\left(x_2 \cos\theta + z_g \sin\theta\right) - y_2\left(x_3 \cos\theta + z_g \sin\theta\right)\right]}{A}, \tag{4.12}$$

$$||F_2|| = \frac{Mg\left[-y_3\left(x_1 \cos\theta + z_g \sin\theta\right) - y_1\left(x_3 \cos\theta + z_g \sin\theta\right)\right]}{B}, \tag{4.13}$$

$$||F_3|| = \frac{1}{y_3 - y_2}\left(y_2 - ||F_1||\right), \tag{4.14}$$

where

$$A = (y_2 - y_3)\,(x_1 \cos\theta + (y_3 - y_1)\,z_g \sin\theta) + (y_1 - y_2)\,(x_2 \cos\theta + z_g \sin\theta) + (x_3 \cos\theta + z_g \sin\theta) \tag{4.15}$$

$$B = (y_2 - y_3)\,(x_1 \cos\theta + (y_3 - y_1)\,z_g \sin\theta) + (y_1 - y_2)\,(x_2 \cos\theta + z_g \sin\theta) + (x_3 \cos\theta + z_g \sin\theta)\,. \tag{4.16}$$

Ground reaction force vectors can be calculated by substituting $||F_1||$, $||F_2||$, and $||F_3||$ from (4.12), (4.13), and (4.14) into (4.4), (4.5), and (4.6), respectively. In addition, the toque vector of each leg can be calculated by substituting F_1, F_2, and F_3 into (4.3).

4.4.1.2 Evaluated Value

We define a torque cost function as squared torque divided by the stride length S as follows:

$$C_s = \frac{1}{S}\int_0^T \tau^t \tau\, dt, \tag{4.17}$$

where T is the cycle time of walking, $\tau = (\tau_i)^t$ $(i = 1, \cdots, 24)$ is the 24 degrees of joint torque vector that is the sum of all joint torque in the robot. Squared torques are used by Channnon et al. as a cost function to derive the optimal biped gait that enables smooth walking and low energy consumption [6]. In addition, Kiguchi et al. make use of squared torque as an evaluated function of walking energy under the assumption that most of the energy is consumed in acceleration and deceleration motions [25]. In this chapter, we also use C_s as a cost function for the criteria of energy reduction and smooth walking motion.

Table 4.1 Parameters of Gorilla Robot III

Parameter	Units	Value
l_1	m	0.68
l_2	m	0.52
l_3	m	0.45
α	deg	22.8

Table 4.2 Parameter setup

Parameter	Units	Value
Total link length, L	m	1.2
Velocity, V	m/s	0.05
Stability margin, S_{M}	m	0.05
Height of COG, H	m	0.45
Mass of robot, M	kg	24
Time of cycle, T	sec	Variable
Stride length, S	m	Variable
Angle of torso, α	deg	Variable

4.4.2 Simulation Analysis

4.4.2.1 Simulation Setting

We first defined several parameters related to walking motion. When walking, we assume that the robot will take largest stride, S, possible as long as its physical constraints allow it to do so. To provide the uniform simulation conditions for a respective parameter set, the moving velocity V (m s^{-1}) is set to constant. Thus, the time T (s) of a cycle is expressed as $T = \frac{S}{V}$. In other words, the walking cycle time will decrease as the walking stride length increases. Moreover, we decided the stability margin to design the COG trajectory. Let stability margin S_M be 0.05 m. In what follows, the employed physical parameters of the simulations are adopted from the structure of Gorilla Robot III (Fig. 4.10). The physical parameters of Gorilla Robot III in Fig. 4.15 are shown in Table 4.1. Limb length is variable under $l_1 + l_2 = L(const)$ (Fig. 4.15). Let L be 1.2 m, l_3 be 0.45 m, and mass of robot be 24 kg. The angle of torso α is determined from limb leg length as follows:

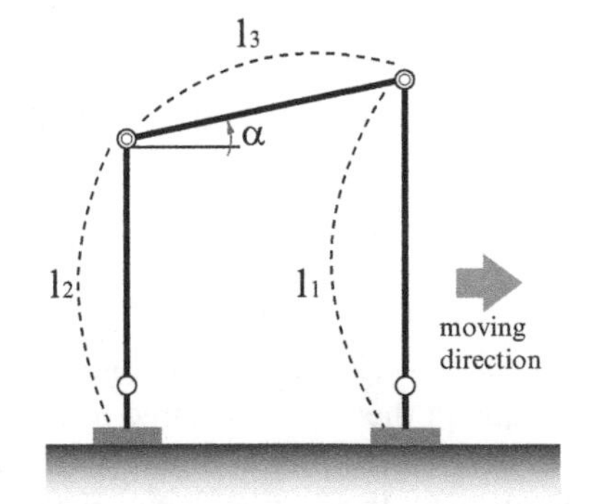

(a) Case1: the length of the rear leg is shorter than the length of front leg ($l_1 > l_2$)

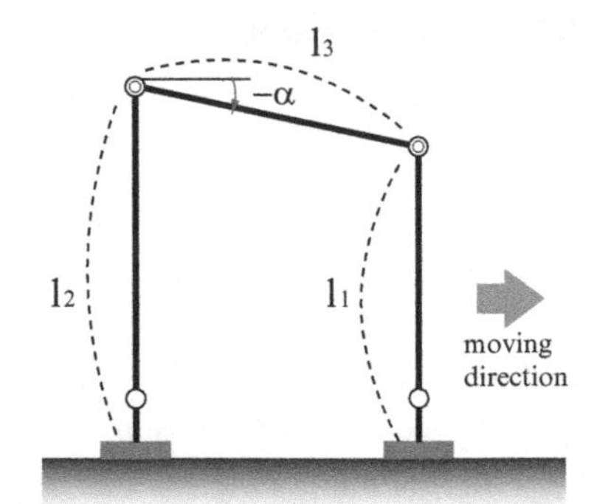

(b) Case2: the length of the rear leg is longer than the length of fore leg ($l_1 < l_2$)

Fig. 4.15 Definition of link parameter

$$\alpha_1 = \arcsin\left(\frac{l_1 - l_2}{l_3}\right). \tag{4.18}$$

Here, let the front range of motion be S_f and the rear range of motion be S_r. When angle of torso is α_1, the relation of range is $S_\mathrm{f} > S_\mathrm{r}$ or $S_\mathrm{f} < S_\mathrm{r}$ except the condition of $l_1 = l_2$. Therefore, stride length is fitted to the shorter range of motion. Alternatively, if we set the angle of torso α_2 as the front and rear range equal such as $S_\mathrm{f} = S_\mathrm{r}$, stride length is longest. In the basic gait pattern, joint torque is influenced by static torque supporting the COG in the three-point contact phase while dynamic torque moves the COG during the four-point contact phases. In this work, as the moving velocity is constant, a walking cycle becomes longer as stride length increases. As the effects of static torque increase as the walking cycle lengthens, the required torque that the angle of torso is α_1 is smaller than the torque required that the angle of torso is α_2. The angle of torso is set as α_1, because the purpose of this research is the reduction of joint torque. Table 4.2 shows these parameters.

In the simulation, C_s is calculated based on various conditions where the "ratio of rear leg length (RRL)" and slope angle are changed. The trajectory of a joint angle is calculated by inverse kinematics obtained from the COG position and a torso posture. We assume that the COG of a robot is fixed in the robot body. The joint angle is controlled by PI controller. Simulation results are evaluated in the following two cases: *Case1* is situation where the length of the rear leg is longer than that of the front leg (Fig. 4.15a), *Case2* is the situation where the length of the rear leg is shorter (Fig. 4.15b).

4.4.2.2 Simulation Result

Figure 4.16 shows the cost function C_s of each RRL and the slope angle based on the simulation. If we focus on a two-dimensional plane with constant RRL, it can be confirmed that RRL will be small because the slope angle minimizing C_s is larger. The simulation result is qualitatively evaluated in two cases in Sect. 4.4.2.1. Traveling upslope, the C_s of *Case1* is larger than *Case2*. Alternatively, the C_s of *Case2* is larger than *Case1* when traveling downslope.

Here, we consider the simulation result. As shown in a Fig. 4.17a, both legs are set to torque τ of the sum total output by each leg. As the force used for kicking the ground can be expressed as $F_1 = \frac{\tau}{l_1}, F_2 = \frac{\tau}{l_2}$, force can take out with the torque with the same one where a leg is shorter. As the ratio of the projecting point of the COG is $R_1 : R_2$ on the slope, as shown in Fig. 4.17b, the ratio of the force concerning fore- and hind-legs is set to $R_2 : R_1$. Therefore, a robot with short legs can take out force with the same output. On sloping ground, as the load on hind legs increases as the slope angle increases, a robot that has a higher ratio of hind leg force has advantages in high load conditions.

Furthermore, in this section, as the simulation is analyzed under constant movement velocity conditions, the length of the walking cycle and the stride

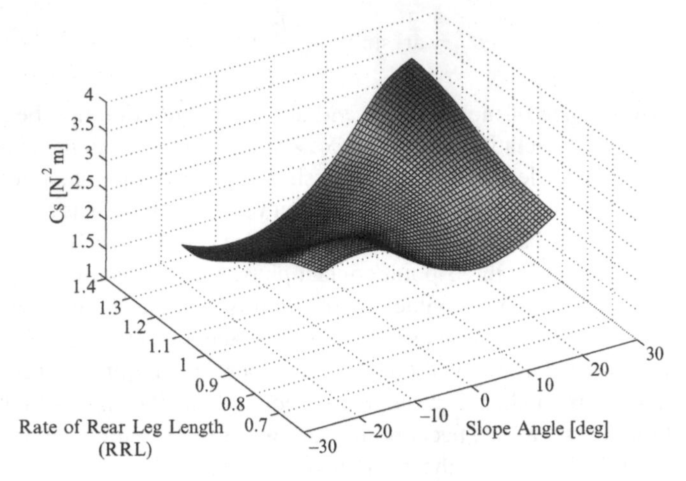

Fig. 4.16 Simulation result

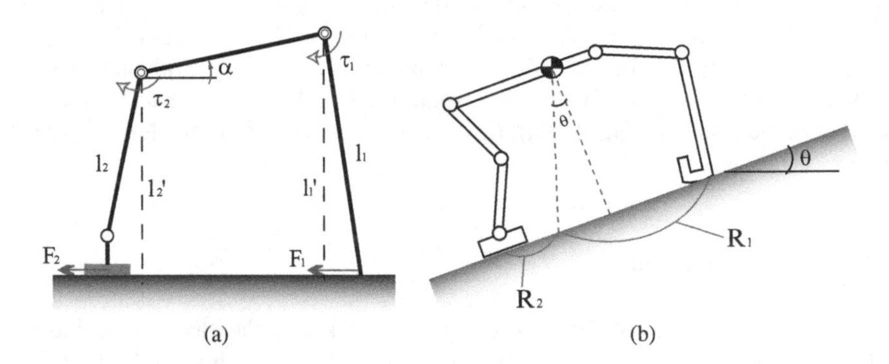

Fig. 4.17 Relationship between torque and length of legs

length increase in tandem. Thus, an increase of stride length leads to increasing time in the three-point contact phase. Owing to the absence of COG motion during the three-point contact phase, the increasing time of this phase implies further influence of static torque for C_S. On the contrary, a decrease of stride length can speed up walking motion, which extends the influence of dynamic torque for C_S. The RRL can change the influence of static and dynamic torque, because stride length geometry is decided by RRL. In other words, optimal RRL can be expressed as the physical structure that minimizes the sum of static and dynamic torque.

4.4.3 Experiment

4.4.3.1 Evaluated Value of Experiment

An experimental joint torque τ_i^{ex} is given as follows:

$$\tau_i^{\mathrm{ex}} = \frac{\tau_{\mathrm{r}}}{3} v_c m, \quad (4.19)$$

where τ_r is the rated torque, v_c is the order voltage, and m is the gear ratio.

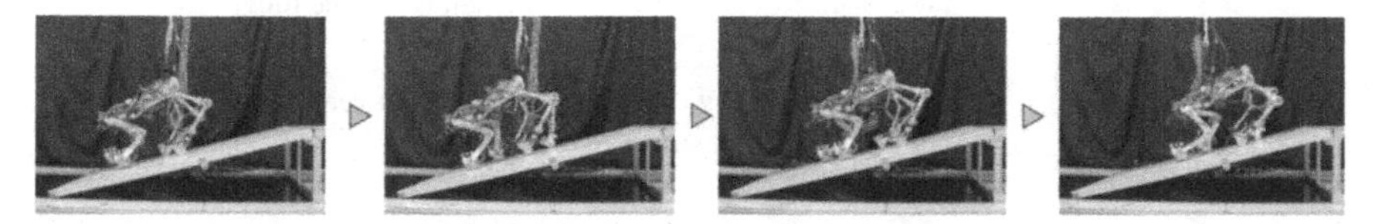

Fig. 4.18 Snapshots of experiment (slope Angle: 15°)

τ_i^{ex} is calculated from (4.19) for each sampling time during one cycle of quadruped walking. Experimental torque vector τ^{ex} can be calculated from τ_i^{ex}, and experimental cost C_s^{ex} is calculated using (4.17). The experiments were carried out based on the understanding that all parameters would be equal to those used in the simulation. However, the link length is fixed at $l_1 = 0.52, l_2 = 0.68$ based on "Gorilla Robot III," because link length of this robot cannot be modified, and the C_s^{ex} is measured experimentally when the slope angle of a slope is set at every 2.5° in $-15°$ to 15°.

4.4.3.2 Experimental Result

Experimental results are shown in Table 4.3. A snapshot of one experiment is shown in Fig. 4.18.

From Fig. 4.19, the experimental results and simulation results show that the RRL is 0.76 (Fig. 4.16), the same as for "gorilla robot III." Experimental values were not close to the numerical simulation results when the slope is steep. One of the reasons for errors between the simulation and the experiment is that we did not consider slippage in the simulation. In the real experiment, stick-slip phenomena were observed at the contact point of the robot's foot at large slope angles. In this experimental environment, the robot does not slip without pitch but with pitch. Thus, the robot needs to output large amounts of torque. Alternatively, C_s^{ex} is close to C_s when the slope angle is $-12.5°$ to 12.5°. In particular, the experimental result that C_s in 5° slope is less than that in 0° slope agrees with the simulation result to be most characteristic. These experimental results confirm that the simulation result agrees with the results obtained from an actual robot on a small slope angle and that the RRL can be indexed to reduce C_s of each slope angle.

Table 4.3 Experimental results

Slope angle (°)	C_s^{ex} ($\times 10^4$N^{2}m)
−15	4.00
−12.5	3.73
−10	3.50
−7.5	3.34
−5	3.28
−2.5	3.11
0	3.04
2.5	2.88
5	2.77
7.5	2.87
10	3.20
12.5	3.34
15	3.90

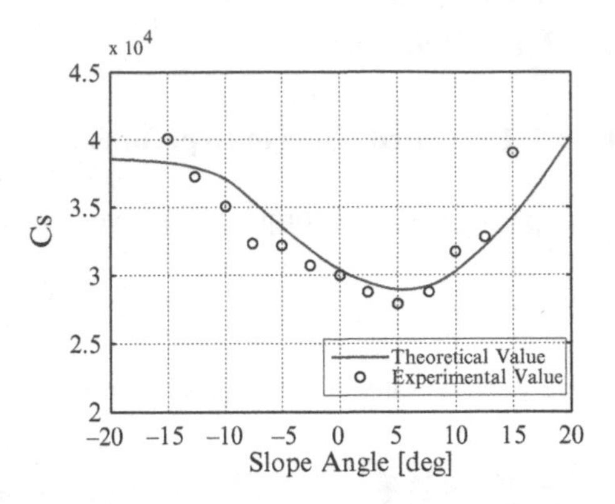

Fig. 4.19 Comparison between simulation and experiment

4.4.4 Discussion

In this section, the optimal ratio of limb length of a quadruped robot designed to reduce joint torque on a slope was derived by analysis. Numerical simulations C_s analyzed in each limb length and in each slope angle for a robot walking on a slope and the optimal RRL was derived. The results indicate that the C_s increases as the slope angle increases if the length of the rear leg is longer than that of the front leg. Alternatively, the C_s decreases as the slope angle increases if the length of the rear leg is shorter. In other words, the robot that has rear leg lengths shorter than its front leg lengths, such as Gorilla Robot III, requires reduced C_s when going forward upslope and going backwards downslope. Additionally, experimental results conducted with an actual robot showed that results of numerical simulation were sufficiently precise to be of future use.

4.5 Previous Works Using the Gorilla Robot

Some types of locomotion related to the Gorilla Robot have been designed until now. Kajima et al. realized smooth continuous brachiation by applying elbow-bending action to decrease the impact forces and by using the excess mechanical energy after the end of the locomotion phase [23]. Moreover, Kajima et al. proposed an energy-based control method that enables the stable continuous brachiation for irregular bar intervals [24]. Yonode et al. generated appropriate COM trajectory to avoid rolling motion in a ladder and realized three types of vertical ladder climbing locomotion: static gait, pace gait, and trot gait [37]. The snapshots of a ladder climbing in pace gait are shown in Fig. 4.20. These control architectures are developed independently in each locomotion. When we consider a transition motion connecting one locomotion to another, two designed controllers corresponding to each locomotion type are not enough. A new control algorithm that covers control properties of two locomotion controllers should be developed because the intermediate motion cannot be realized by fusing control outputs from two controllers. Based on this notion, we have proposed a novel control method named "Passive Dynamic Autonomous Control (PDAC)" [7]. The approach of PDAC is to describe the robot dynamics as a one degree-of-freedom autonomous system around a contact point, using an interlocking so that the robot could keep the robot as inherent dynamics. Three-dimensional natural dynamic walking is achieved by applying PDAC as shown in Fig. 4.21 [8, 14]. We also designed a controller for a quadrupedal walk [2] and brachiation [16] using the same PDAC. The snapshots of the quadruped walk are shown in Fig. 4.22 and the brachiation are shown in Fig. 4.23.

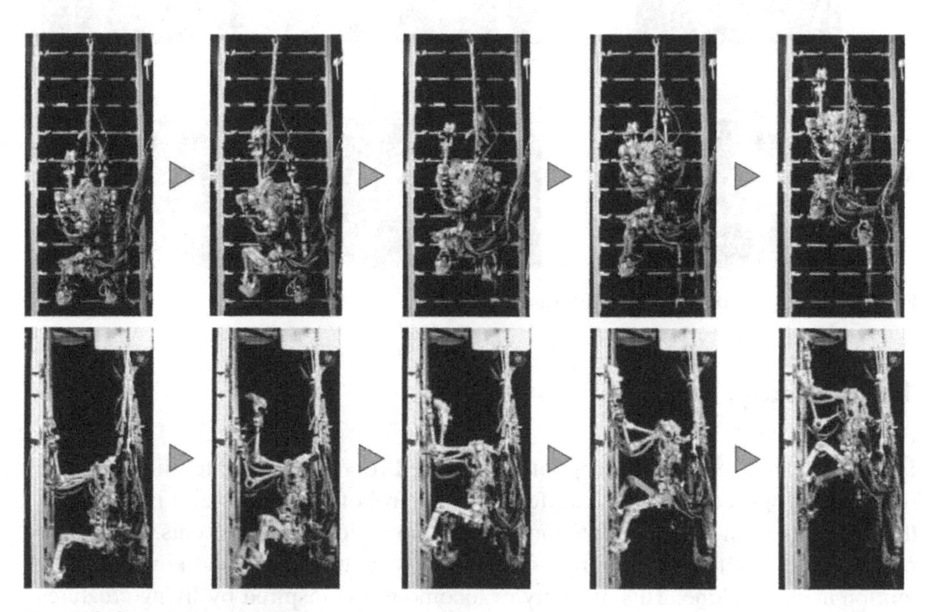

Fig. 4.20 Snapshots of climbing ladder motion

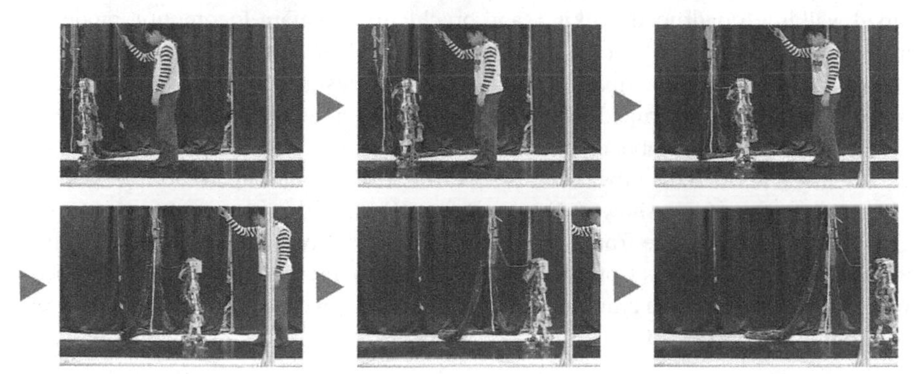

Fig. 4.21 Snapshots of the bipedal walking using PDAC

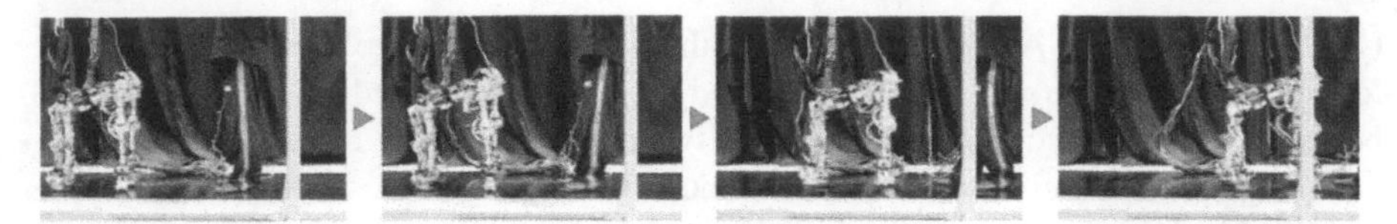

Fig. 4.22 Snapshots of the quadrupedal walking using PDAC

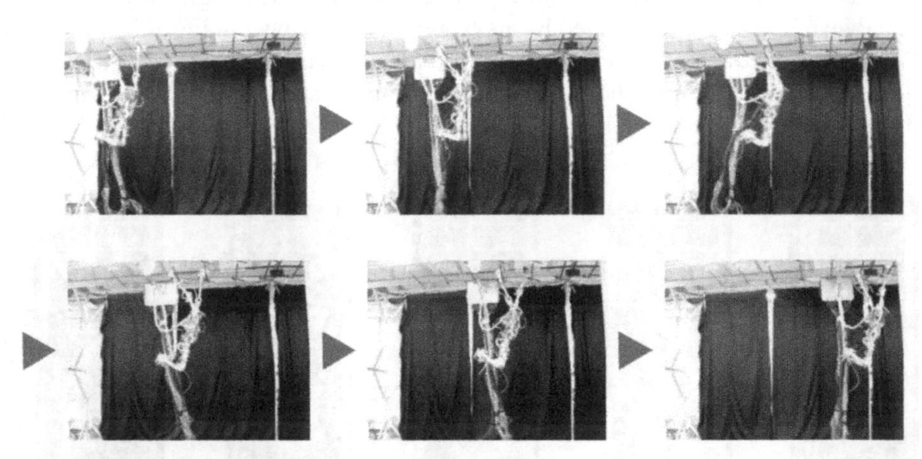

Fig. 4.23 Snapshots of the brachiation using PDAC

4.6 Summary

In this chapter, the novel concept of bio-inspired robotics named multilocomotion robot was proposed. To realize a life-supporting robot in the future, it is necessary that the robots can ambulate autonomously in the various environments. Multilocomotion robot has the high ability to ambulate by achieving several kinds of locomotion in stand-alone. This diversity of locomotion is inspired by living creatures on earth. We mentioned about diversity of locomotion in animals and described biped walking, quadruped walking, and brachiation as main locomotion mode of multilocomotion robots. Second, three types of Gorilla Robot developed as multilocomotion robot were described. Gorilla Robot was designed to be able to perform biped locomotion, quadruped locomotion, brachiation, etc. The hardware configuration of each Gorilla Robot and control system architecture were presented. Third, we discussed about the hardware of Gorilla Robot III as quadruped robot on slopes. The results of analysis showed Gorilla Robot III should go forward upslope and go backwards downslope from viewpoint of reduction joint torque. Finally, we introduced several previous works using the Gorilla Robot: biped walking, quadruped walking, brachiation, and climbing ladder motion. And we also introduced the novel controlling method PDAC.

References

1. Alexander, R.M.: Exploring Biomechanics: Animals in Motion. W. H. Freeman, San Francisco, CA (1992)
2. Asano, Y., Doi, M., Hasegawa, Y., Matsuno, T., Fukuda, T.: Quadruped walking by joint-interlocking control based on the assumption of point-contact. Transactions of the Japan Society of Mechanical Engineers, Series C **73**(727), (in Japanese) (2007)
3. Bertram, J.E.A., Chang, Y.H.: Mechanical energy oscillations of two brachiation gaits: Measurement and simulation. American Journal of Physical Anthropology **115**, 319–326 (2001)
4. Bertram, J.E.A., Ruina, A., Cannon, C.E., Chang, Y.H., Coleman, M.J.: A point-mass model of gibbon locomotion. Journal of Experimental Biology **202**, 2609–2617 (1999)
5. Chang, Y.H., Bertram, J.E.A., Lee, E.V.: External forces and torques generated by the brachiating white-handed gibbon (hylobates lar). American Journal of Physical Anthropology **113**, 201–216 (2000)
6. Channon, P.H., Hopkins, S.H., Pham, D.T.: A variational aproach to the optimization of gait for a bipedal robot. Journal of Mechanical Engineering Science(Part C) **210**, 177–186 (1996)
7. Doi, M., Hasegawa, Y., Fukuda, T.: Passive dynamic autonomous control of bipedal walking. In: Proceedings of IEEE/RSJ International Conference on Humanoid Robots, pp. 811–829 (2004)
8. Doi, M., Hasegawa, Y., Fukuda, T.: 3D dynamic walking based on the inverted pendulum model with two degree of underactuation. In: Proceedings of IEEE/RSJ International Conference on Intelligent Robots and Systems, pp. 2788–2793 (2005)
9. Donelan, J.M., Kram, R., Kuo, A.D.: Simultaneous positive and negative external mechanical work in human walking. Journal of Biomechanics **35**, 117–124 (2002)
10. Eimerl, S., DeVore, I.: The Primates. Time-Life Books (1966)
11. Fleagle, J.: Dynamics of a brachiating siamang (hylobates (symphalangus) syndactylus). Nature **248**(445), 259–260 (1974)
12. Floyd, S., Sitti, M.: Design and development of the lifting and propulsion mechanism for a biologically inspired water runner robot. IEEE Transactions on Robotics **24**(3), 698–709 (2008)
13. Formal'sky, A., Chevallereau, C., Perrin, B.: On ballistic walking locomotion of a quadruped. International Journal of Robotics Research **19**(8), 743–761 (2000)
14. Fukuda, T., Doi, M., Hasegawa, Y., Kajima, H.: Multi-locomotion control of biped locomotion and brachiation robot, In: Fast Motions in Biomechanics And Robotics: Optimization And Feedback Control. Springer-Verlag, pp. 121–145 (2006)
15. Fukuda, T., Hosokai, H., Kondo, Y.: Brachiation type of mobile robot. In: Proceedings of IEEE International Conference on Advanced Robotics, pp. 915–920 (1991)
16. Fukuda, T., Kojima, S., Sekiyama, K., Hasegawa, Y.: Design method of brachiation controller based on virtual holonomic constraint. In: Proceedings of IEEE/RSJ International Conference on Intelligent Robots and Systems, pp. 450–455 (2007)
17. Fukuda, T., Saito, F., Arai, F.: A study on the brachiation type of mobile robot (heuristic creation of driving input and control using cmac). In: Proceedings of IEEE/RSJ International Conference on Intelligent Robots and Systems, pp. 478–483 (1991)
18. Fukuoka, Y., Kimura, H., Cohen, A.H.: Adaptive dynamic walking of a quadruped robot on irregular terrain based on biological concepts. International Journal of Robotics Research **22**(3-4), 187–202 (2003)
19. Hasegawa, Y., Fukuda, T., Shimojima, K.: Self-scaling reinforcement learning for fuzzy logic controller – Applications to motion control of two-link brachiation robot. IEEE Transactions on Industrial Electronics **46**, 1123–1131 (1999)
20. Herr, H.M., McMahon, T.A.: A galloping horse model. International Journal of Robotics Research **20**(1), 26–37 (2001)
21. Herr, H.M., McMahon, T.A.: A trotting horse model. International Journal of Robotics Research **19**(6), 566–581 (2001)

22. Hoyt, D.F., Taylor, C.R.: Gait and the energetics of locomotion in horses. Nature **292**(16), 239–240 (1981)
23. Kajima, H., Doi, M., Hasegawa, Y., Fukuda, T.: A study on brachiation controller for a multi-locomotion robot – Realization of smooth, continuous brachiation. Advanced Robotics **18**(10), 1025–1038 (2004)
24. Kajima, H., Hasegawa, Y., Doi, M., Fukuda, T.: Energy-based swing-back control for continuous brachiation of a multilocomotion robot. International Journal of Intelligent Systems **21**(9), 1025–1038 (2006)
25. Kiguchi, K., Kusumoto, Y., Watanabe, K., Izumi, K., Fukuda, T.: Enegy-optimal gait analysis of quadruped robots. Artificial Life and Robotics **6**, 120–125 (2002)
26. Kimura, H., Fukuoka, Y., Cohen, A.H.: Adaptive dynamic walking of a quadruped robot on natural ground based on biological concepts. International Journal of Robotics Research **26**(5), 475–490 (2007)
27. Mori, M., Hirose, S.: Development of active cord mechanism acm-r3 with agile 3D mobility. In: Proceedings of IEEE/RSJ International Conference on Intelligent Robots and Systems, pp. 1552–1557 (2001)
28. Mori, M., Hirose, S.: Three-dimensional serpentine motion and lateral rolling by active cord mechanism acm-r3. In: Proceedings of IEEE/RSJ International Conference on Intelligent Robots and Systems, pp. 829–834 (2002)
29. Muybridge, E.: Animals in Motion. Dover, NY(1957)
30. Nakanishi, J., Fukuda, T., Shimojima, K.: A brachiating robot controller. IEEE Transactions on Robotics and Automation **16**(2), 109–123 (2000)
31. Okada, M., Nakamura, Y.: Development of cybernetic shoulder – A three dof mechanism that imitates human shoulder-motions. Journal of the Robotics Society of Japan **18**, 690–698 (2000) (in Japanese)
32. Saito, F., Fukuda, T., Arai, F.: Swing and locomotion control for two-link brachiation robot. In: Proceedings of IEEE International Conference on Robotics and Automation, pp. 719–724 (1993)
33. Saito, F., Fukuda, T., Arai, F.: Swing and locomotion control for a two-link brachiation robot. IEEE Control System Magazine **14**(1), 5–12 (1994)
34. Simons, E.L.: Primate Evolution: An Introduction to Man's Place in Nature. Macmillan, NY (1972)
35. Smith, R.: Open dynamics engine(ode) (2008). URL `http://www.ode.org/`
36. Tsukagoshi, H., Hirose, S., Yoneda, K.: Maneuvering operations of the quadruped walking robot on the slope. In: Proceedings of IEEE/RSJ International Conference on Intelligent Robots and Systems, pp. 863–869 (1996)
37. Yoneda, H., Sekiyama, K., Hasegawa, Y., Fukuda, T.: Vertical ladder climbing motion with posture control for multi-locomotion robot. In: Proceedings of IEEE/RSJ International Conference on Intelligent Robots and Systems, pp. 3579–3584 (2008)

Chapter 5
Self-regulatory Hardware: Evolutionary Design for Mechanical Passivity on a Pseudo Passive Dynamic Walker

Hiroshi Yokoi and Kojiro Matsushita

5.1 Introduction

The functioning of our body is the resultant of the lifelong survival competition. The body possesses the special physical characteristic shape, size, and material, and has been living throughout using this characteristic by achieving the function of the movement and the sense. Lot of functions have been added to the simple reflective system that emerged in the first generation. After the competition, the basic function of the reflective system of sense and action has been changed to be integrated and over wrapped by the new reflective system. The basic function necessary for survival has been achieved. The performance of this simple reflective system gives the chance for surviving, and we have the simple way to design an individual with good performance who can adapt to the environment.

We focused on the process to acquire the structure and the material characteristics of the leg as an example of a simple reflective system. Moreover, the result of examining how the function of walking develops is described. We discuss evolution design system by using simple genetic algorithm (GA) that allow to achieve the function of the leg. We also demonstrate how to optimize a set of parameters in the competition between various designs of hardware.

Various experiments were conducted on the relation of the change in the biped structure and the walking functions. The experimental result shows that the phenomenon of absorption of the kinetic energy with active rhythm generation is the main mechanism of a passive pendulum movement, and the viscoelasticity of the body and the discharges confirmed to stabilized walking generations are the central characteristics. Particularly, the self regulatory function as an autonomous adjustment of motion by the passive element was developed by using Pseudo Passive Dynamic Walker of various types by arranging the viscoelasticity of the body in the main joint of the leg (waist, knee, and ankle).

A. Adamatzky and M. Komosinski (eds.), *Artificial Life Models in Hardware*,

5.2 Background

Conventional robots are designed for multiple purposes. So they implicitly tend to have high specifications (i.e., multiple degrees of freedom, fast information processing, and high energy consumption), and their movements are precisely manipulated by their control for the purpose of achieving all required tasks. Meanwhile, biomimic robots are mimics of biological system. So, their control and physical structure implicitly indicate their high adaptability to their tasks and environments and, therefore, tend to require smaller energy consumption and lower control complexity than the conventional robots. The representative instance is the passive dynamic walker [3, 6]. The robot does not have any motors and sensors, but has a well-designed structure based on the human characteristics during walking: passive hip joints, latch knee joints, and semi-circular feet. Then, the robot maintains balance and walks down a slope by exploiting gravity as the driving force. The physical structure itself self-regulates its locomotion.

Thus, we believe that it is necessary to implement self-regulatory physical structures in a design process, and such concept is referred to as embodiment [2,4] in the field of embodied artificial intelligence and robotics [7].

Embodiment is basically defined as special features in the body, causing high adaptability to its tasks and environments in robotics. There is increasing evidence that the exploitation of a robot's embodiment can increase its energy efficiency and reduce the complexity of its control architecture. However, embodiment has only been investigated using conventional robots and its implications have not been systematically quantified. Then, the current agreement in embodied cognitive science indicates that embodiment can emerge in robots by the biologically inspired reproductive process. One of the most successful of these applications was the work of Sims [8, 9], in which artificial creatures were automatically designed within a three-dimensional physics simulation. The simulation generated a variety of locomotive creatures with unique morphologies and gaits, some of which have little analogy in the biological world. This suggested that the interdependence between morphology and control plays an important role in the evolution of locomotion. However, the Sims' work did not achieve "legged" locomotion by the design system.

Thus, evolutionary design has not yet overcome these technical difficulties and the use of simulation. Therefore, as the best solutions at present, biologically inspired knowledge should be appropriately applied to unknown parts of the system. For example, legged locomotion has not yet been emerged spontaneously in coupled evolution – only crawling. It seems that more detail of the morphological composition is required in its design process and, then, the evolution leads to more locomotion than crawling. So to make further progress, we need to gain a better understanding of how to synthesize complex yet functional morphologies, and biologically inspired knowledge will be useful for improving our designs: Alexander [1] has shown a variety of locomotion types using different structures such as curved surfaces, universal joints, and parallel links (structure); Vogel [11] stressed the importance of material properties such as muscle, ligament, tendon, and bone (material properties). We believe that such biologically inspired knowledge contributes to generating more concrete design of locomotive robots and provide variety of self-regulatory mechanisms.

5.3 Evolutionary Design System of Legged Robots

In Fig. 5.1, a conceptual figure of the proposed evolutionary design system is shown. The system has three main features: coupled evolution, interactive interface, and evaluation. The coupled evolution architecture is an automatic design system of the morphology and controller of a legged robot. The design process is based on a biologically similar reproductive process, so that we assume that the design system provides a developmental environment to emerge embodiment of legged robots. The reproductive process is basically computed with a GA so that its meta-heuristic characteristic helps designers to interact with evolutionary processes. Thus, the designers conduct trial-and-error activities by observing acquired behavior and physical structures through the interactive interface architecture. Then, evaluation criteria (i.e., fitness function and energy efficiency) are used to specify their physical features. Through the repetition of the two architectures, the evolutionary design system facilitates concrete design solutions involving legged locomotion and the emergence of embodiment (i.e., self-regulatory hardware).

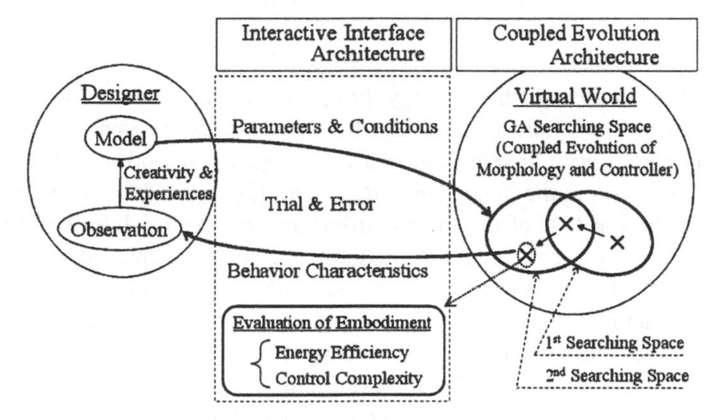

Fig. 5.1 Conceptual figure of evolutionary design system

5.3.1 Three-dimensional Physics World

The design system is implemented using Open Dynamics Engine (ODE) [10], which is an open-source physics engine library for the three-dimensional simulation of rigid body dynamics. The environmental configuration of the design system is given as sampling time 0.01 s, gravity 9.8 m s^{-2}, friction 1.0, ground spring coefficient 5,000 N m^{-1}, ground damper coefficient 3,000 N s m^{-1}.

5.3.2 Coupled Evolution Part

The coupled evolution part is characterized by a general GA process, a gene structure representing morphology and controller, and evaluation methods. The general GA process starts with random genes and conducts 100–300 generations using a population size of 100 for each run. After all generations, an evolutionary process ends, and the next evolutionary process starts with new random genes. Table 5.1 lists setting values for the GA.

Table 5.1 Values for GA parameters

Parameter	Value
Seed	30–100
Generation	300
Population	100–200
Num. of gene locus	50–100%
Crossover	5–10%
Mutation	5–10%

The GA process uses an elite strategy that preserves constant numbers of higher fitness in the selection/elimination process due to its local convergence. Here, fitness is defined as the distance traveled forward for a constant period (i.e., 10 s). At the end of each generation, the genes are sorted from highest to lowest fitness value. The genes in the top half of the fitness order are preserved, while the others are deleted. The preserved genes are duplicated, and the copies are placed in the slots of the deleted genes. The copied genes are crossed at 5–10% and mutated at 5–10%.

The gene structure is a fixed-length gene and presents morphological and control parameters. Each locus contains a value ranging from −1.00 to +1.00 at an interval of 0.01. Figure 5.2 shows locus IDs corresponding to design parameters: L, W, H, M0, M1, M2, M3, M4, k, c, amp, and cycle. Morphology of a legged robot in the design system consists of five basic kinds of design components in Table 5.2: joint type (compliant/actuated), joint axis vector, link size, link angle, and link mass. These physical components are viewed as basic components of a biological system.

Based on the physiological knowledge of gait control, the controller of a legged robot consists of simple rhythmic oscillators. The characteristics are mainly determined with two types of parameters: amplitude and frequency (Table 5.3). Pre-conditional, all the oscillators have the same wavelength, and the contra-lateral oscillators are in anti-phase. Here, we assume that a simple controller leads to forming special physical features on evolved legged robots (embodiment).

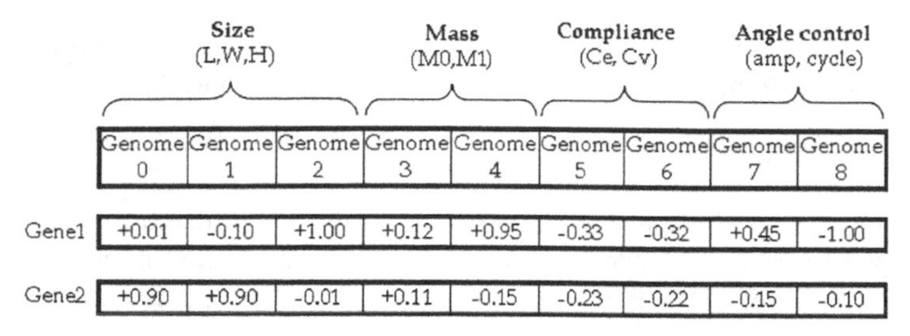

Fig. 5.2 Concept figure of gene structure

Table 5.2 Basic morphological configuration

		Link 1	Link 2
Size	Length [m]	0.1	0.1
	Width [m]	0.1	0.1
	Height [m]	0.1 to 0.5	0.1 to 0.5
Absolute Angle at Pitch (y axis) [rad]		$-\pi / 3$ to $\pi / 3$	$-\pi / 3$ to $\pi / 3$
Mass [kg] (Total Mass X [kg])		X * 10-90%	X * 10-90%

Table 5.3 Basic joint configuration

			Joint 1	Joint 2
Type	Compliance	Elasticity [N/m]	10^{-2} to 10^{+4}	10^{-2} to 10^{+4}
		Viscosity [Ns/m]	10^{-2} to 10^{+4}	10^{-2} to 10^{+4}
	Angle Control	Amplitude [rad]	0 to $\pi/2$	0 to $\pi / 2$
		Cycle [sec]	0.5 to 1.5	
Axis Vector		X	-1.0 to +1.0	-1.0 to +1.0
		Y	-1.0 to +1.0	-1.0 to +1.0
		Z	-1.0 to +1.0	-1.0 to +1.0

5.3.3 Evaluation Methods

Energy consumption and energy efficiency are applied as evaluation methods. It contributes to qualifying evolved legged robots. In physics, mechanical work is equivalent to the amount of energy transferred by a force, and is calculated by multiplying the force by the distance or by multiplying the power (in Watt) by the time (in seconds). Then in the case of a motor, time and rotational distance are related with its angular speed and the torque, which causes angular speed to increase, is regarded as mechanical work. So, power W in rotational actuation is calculated with the following equation:

$$W = \text{torque (N m)} \times 2\pi \times \text{angular velocity (rad s}^{-1}\text{)}. \tag{5.1}$$

Energy consumption for a walking cycle is represented as follows:

$$J = T \sum_{i=0}^{N} \int_0^T |2\pi T r_i(t) \dot{\theta}_i| \, dt, \tag{5.2}$$

and energy efficiency is computed as energy consumption per meter (in this equation, total mass is ignored because it is set as a common characteristic among all robots):

$$J/m = J/Dis. \tag{5.3}$$

In (5.1)–(5.3), t is time in seconds, N is a number of actuated points, Tr_i is a torque in point i, Dis is a distanced traveled for a walking cycle (in meters), dt is a sampling time (in seconds), $\dot{\theta}_i$ is an angular velocity (rad s^{-1}), T is a walking cycle (in seconds).

5.4 Evolutionary Design of Biped Robots

Evolutionary design of biped robots is conducted in this section: the first evolutionary design aims at verifying emergence of embodiment; the second evolutionary design aims at specifying self regulatory mechanisms.

5.4.1 Morphological and Control Configuration for Biped Robots

Biped robots are constructed using nine rigid links (Fig. 5.3): an upper torso, a lower torso, a hip, two upper legs, two lower legs, and two feet. These body parts are, respectively, connected at torso, upper hip, lower hip, knee, and ankle joints, and the robots have eight degrees of freedom.

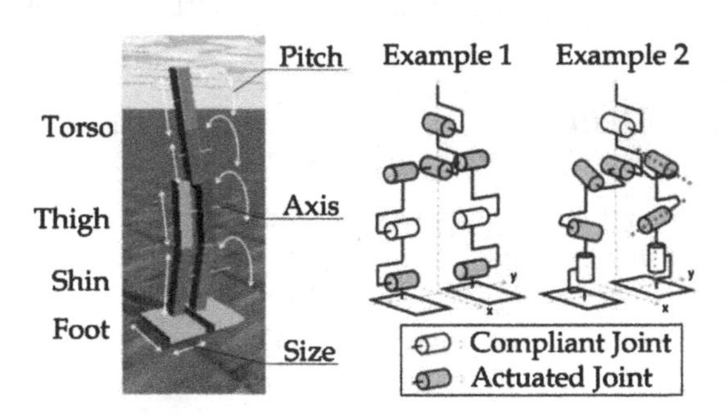

Fig. 5.3 Morphological configuration

Table 5.4 lists control parameters (i.e., amplitude and frequency) and Table 5.5 lists morphological parameters (i.e., size, weight, absolute angle of each link and selection of whether it is oscillatory or compliant, as well as its elasticity coefficient and viscosity coefficient if the joint is compliance or amplitude and frequency if the joint is a oscillator, and axis vector of each joint). In addition to this setting, joint settings are constrained to be contra-laterally symmetric around the xz plane (Fig. 5.4).

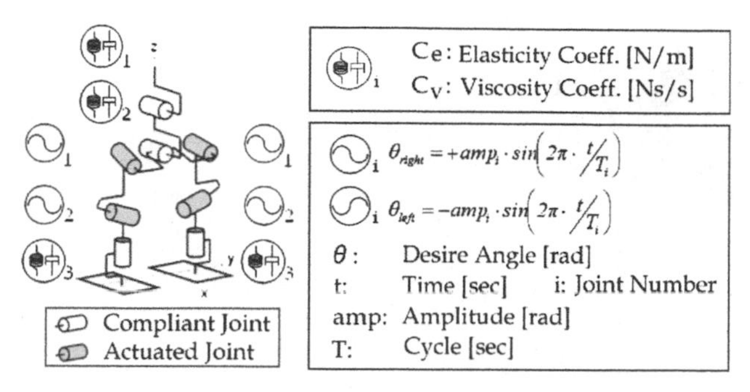

Fig. 5.4 Control configuration

Table 5.4 Characteristic of joints (searching parameters colored in blue)

	Torso	Hip	Knee	Ankle
Type				
Compliance				
Elasticity coeff. [N m^{-1}]	$10^{-2}-10^{+4}$	$10^{-2}-10^{+4}$	$10^{-2}-10^{+4}$	$10^{-2}-10^{+4}$
Viscosity coeff. [N s m^{-1}]	$10^{-2}-10^{+4}$	$10^{-2}-10^{+4}$	$10^{-2}-10^{+4}$	$10^{-2}-10^{+4}$
Actuation (angle control)				
Amplitude [rad]	0 to $\pi/2$	0 to $\pi/2$	0 to $\pi/2$	0 to $\pi/2$
Cycle [s]	0.5 to 1.5	0.5 to 1.5	0.5 to 1.5	0.5 to 1.5
Axis vector				
X	1	−1.0 to +1.0	−1.0 to +1.0	−1.0 to +1.0
Y	0	−1.0 to +1.0	−1.0 to +1.0	−1.0 to +1.0
Z	0	−1.0 to +1.0	−1.0 to +1.0	−1.0 to +1.0

Table 5.5 Characteristic of links (searching parameters colored in gray)

		Upper/ lower Torso	Hip	Thigh	Shin	Foot
Size	Length [m] (X axis)	0.1	-	0.1	0.1	0.1 to 0.5
	Width [m] (Y axis)	0.1	-	0.1	0.1	0.1 to 0.5
	Height [m] (Z axis)	0.1 to 0.5	-	0.1 to 0.5	0.1 to 0.5	0.05
	Radium [m]	-	0.05	-	-	-
Absolute angle at pitch (y) axis [rad]		-π/3 to π/3	-	-π/3 to π/3	-π/3 to π/3	-π/3 to π/3
Parallel displacement on y axis[m]		-	-	-	-	0 to 0.2
Total Mass 20 [kg] (a+2b+2c+2d+e=100%)		e	a	b	c	d

5.4.2 Results of First Evolutionary Design

We focus on the most successful nine seeds: that is, those are the biped robots that traveled forward more than 7 m for 10 s (Fig. 5.5). Table 5.6 lists the performance of

Fig. 5.5 A walking scene

the best nine biped robots: the second column reports their distance traveled forward for 10 s; the third column, their walking cycle; the fourth column, their angular velocity of oscillators; the fifth column, their energy efficiency; the six column, their numbers of contralateral set of actuated joints (i.e., four types – torso hip, knee, ankle joints).

The biped robots with less number of actuated joints indicate high energy efficiency. As shown in Table 5.7, the further analysis indicates that hip joints tend to become actuated joints, and knee joints tend to be compliant joints. Especially, the characteristics of the compliant joints are categorized into three types: free joint, suspension joint, and fixed joint corresponding to the degree of elasticity and viscosity.

Table 5.6 Performance of best nine biped robots in order of energy efficiency. Energy efficiency is computed with average torque 25 N m, and lower values indicate better performance

Seed	Distance [m]	Cycle [s]	Angular velocity [rad/s]	Energy efficiency [J/m]	Number of actuated DOFs
09	13.0	1.02	0.50	6.0	2
11	7.3	1.05	0.32	7.0	1
02	7.7	1.17	0.38	7.7	2
00	10.6	1.05	0.55	8.2	3
14	7.0	1.01	0.45	10.0	2
22	11.9	1.06	0.99	13.1	3
08	9.2	1.04	0.81	13.8	3
21	7.1	1.08	0.69	15.2	3
17	7.2	1.04	0.88	19.1	3

According to the effects of the compliance, the best nine biped robots can be categorized into two types of walks: the biped robot walks without using any compliance (i.e., actively controlled walker) and the biped robot walks with using some effects of compliance (i.e., compliant walker). The differences are shown in Figs. 5.6 and 5.7. Figure 5.6 shows joint angle trajectories of both walkers, and Fig. 5.7 presents results of frequency analysis on the transitions. It reveals that the knee oscillation of the compliant walker is induced by oscillators at other joints, 1 Hz frequency. This can be considered as self-regulation. The amplitude of 2 Hz

Table 5.7 Characteristics of compliant joints among best nine biped robots, where C_e is an elasticity coefficient, in $\mathrm{N\,m^{-1}}$, C_v is a viscosity coefficient, in $\mathrm{N\,s\,m^{-1}}$

Types	Condition	Num. of compliant joints				
		Torso	Hip	Knee	Ankle	Total
Free joint	0<=Ce<10 && 0<=Cv<10	1	0	1	1	3
Suspension joint	10<Ce<=100 && 0<=Cv<10	0	0	5	0	5
Fixed joint	Ce>100 || Cv>=10	3	1	1	1	6
Total num.		4	1	7	2	14

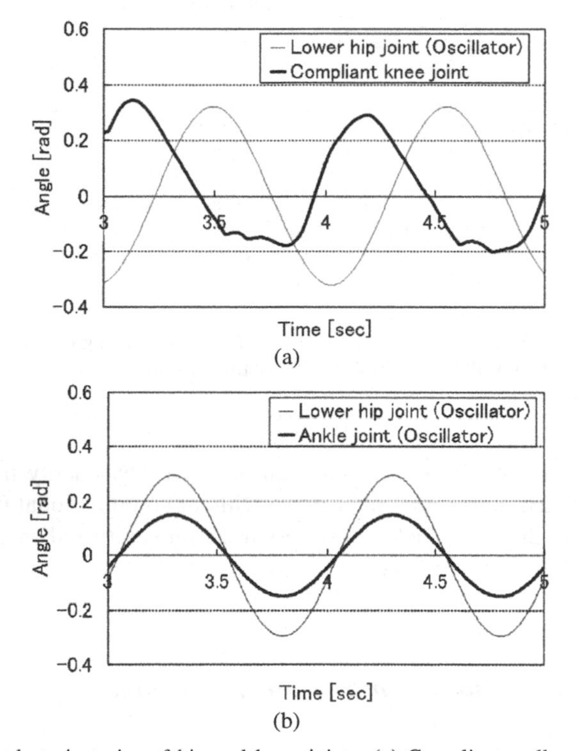

Fig. 5.6 Joint angle trajectories of hip and knee joints: (**a**) Compliant walker and (**b**) Actively controlled walker

oscillations in Fig. 5.7a realizes ground impact absorption (self-stabilization) with high compliance. That is, these two appropriate states of compliant joints indicate passive and dynamical functions during locomotion. Therefore, such type of the robots can be called pseudo-passive dynamic walkers. Moreover, these two functions serve as examples of the computational trade-off possible between morphology

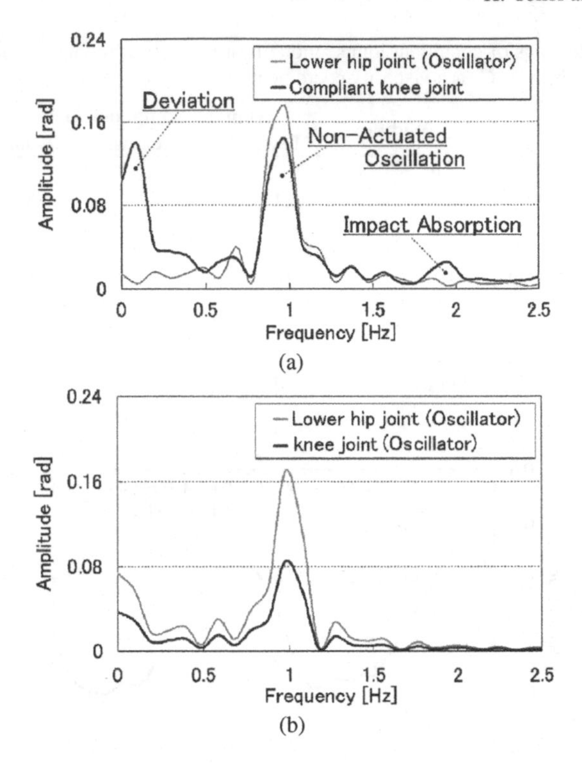

Fig. 5.7 Frequency analysis (i.e., discrete Fourier transform) of joint angle trajectories of hip and knee joints: (**a**) Compliant walker and (**b**) Actively controlled walker

and controller, because compliant joints can be moved by energy input channels other than controlled motors and filter noise without computational power. It suggests that the embodiment, which reduces control complexity and energy consumption, is emerged in the design system.

5.4.3 Additional Setup Condition for the Second Evolutionary Design

The second evolutionary design is conducted for clarifying the embodiment: compliance. Basically, for the purpose of narrowing its solution space to specify physical structures exploiting compliance, an additional condition (i.e., restriction of the numbers of actuated joints) is added to the design system. Table 5.8 indicates joint configurations for the second evolutionary design, and a scheme for joint-type selection is as follows: one of four types of joint structures (i.e., either set of torso, hip, knee, and ankle becomes an actuated joint and other sets of the joints are compliant)

Table 5.8 Joint configuration (searching parameters colored in gray)

	Torso	Hip	Knee	Ankle
Type				
Compliance				
Elasticity Coeff. [N/m]	$10^{-2}-10^{+4}$	$10^{-2}-10^{+4}$	$10^{-2}-10^{+4}$	$10^{-2}-10^{+4}$
Viscosity Coeff. [Ns/m]	$10^{-2}-10^{+4}$	$10^{-2}-10^{+4}$	$10^{-2}-10^{+4}$	$10^{-2}-10^{+4}$
Actuation (angle control)				
Amplitude [rad]	0 to $\pi/2$	0 to $\pi/2$	0 to $\pi/2$	0 to $\pi/2$
Cycle [sec]	0.5–1.5	0.5–1.5	0.5–1.5	0.5–1.5
Selection of Joint Type				
0–3				
0	Act.	Comp.	Comp.	Comp.
1	Comp.	Act.	Comp.	Comp.
2	Comp.	Comp.	Act.	Comp.
3	Comp.	Comp.	Comp.	Act.

is selected for a walker. The evolutionary design is conducted using 100 different random seeds, is run for 100 generations, and the population is comprised of 100 individuals.

5.4.4 Results of the Second Evolutionary Design

The evolutionary design generated six notable walks. In this section, their walking characteristics are described according to their joint structures.

Hip actuated walkers arose the most (i.e., 55 out of 100 seeds). The gaits can be characterized into three notable types: statically stable, dynamically unstable, and dynamically stable walks. The statically stable walk (Fig. 5.8a) is achieved most often among the hip actuated walkers: they increase their mechanical stability and keep a narrow amplitude range in their oscillation so that their COG-X always remains within their supporting polygon while walking. Figure 5.8b shows a dynamically unstable walk. It walks with a tottering gait because the edges of its feet randomly contact the ground. So, its performance is unstable even on the flat plane. Meanwhile, Fig. 5.8c shows a dynamically stable walk. The main feature is the axis vector of its hip joint: the oscillations at the hip axis synchronously move not only the legs in the sagittal plane but also the torso in the lateral plane. So, the hip actuation mechanism keeps its balance and forward locomotion (i.e., self-regulatory mechanism). Overall, the hip actuated walkers tend to exploit compliance only a little, and achieve their walks mainly by their actuation.

Knee actuated walkers are rarely generated (i.e., three out of 100 seeds). Basically, the walker rotates the knee at the x axis, and does not fall over for 6 s.

Ankle actuated walkers are also hardly generated (i.e., 12 out of 100 seeds). Its walk exploits compliance: the ankle is actuated and, then, the compliance in the walker synchronously moves the hip, knee, and torso joints by the actuation.

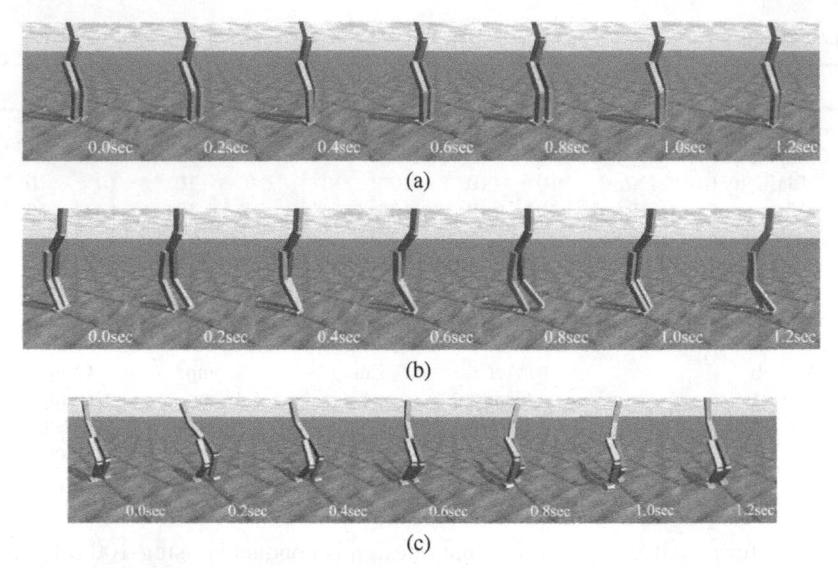

Fig. 5.8 Representative hip actuated walkers: (**a**) statically stable walk, (**b**) dynamically unstable walk, (**c**) dynamically stable walk

Fig. 5.9 Representative ankle actuated walker

Figure 5.9 shows the walking scene. It is observed that the physical structure of the walker is regarded as a laterally oscillating spring while walking. Thus, it indicates that compliance contributes to its stable walk.

Torso actuated walkers are produced as the second most common solutions (i.e., 29 out of 100 seeds). Figure 5.10 shows the representative walking scene. The walker transfers a torso oscillation (actuation) in the lateral plane to hip oscillations (compliance) in the sagittal plane by exploiting its joint structure and material properties and achieves stable walking.

Overall, we illustrated the relations between physical structures, distances traveled, and energy consumption; the best fitness from 100 independent evolutionary runs is plotted on a two-dimensional graph, see Fig. 5.11, where energy consumption is vertical axis, distance traveled is horizontal axis, and markers represent joint structures. It is characteristic that each type of walkers is distributed around a certain

Fig. 5.10 Representative torso actuated walker

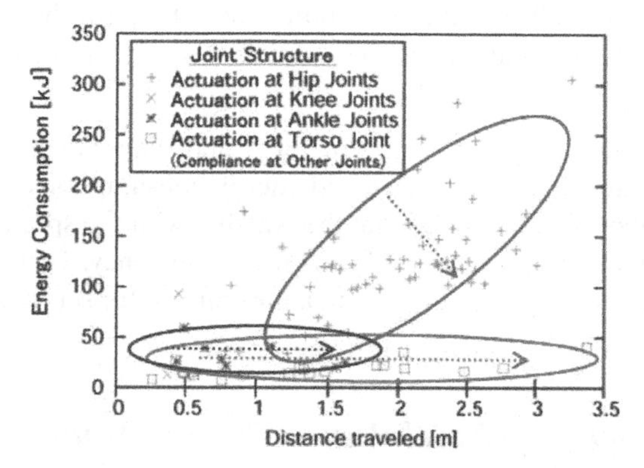

Fig. 5.11 A physical representation of embodiment. It illustrates the relations between joint structure, energy consumption, and distance traveled. *Circles* indicate distribution of four types of walkers, and *arrows* indicate the tendency of specific physical characteristics

area on the graph: the hip actuated walkers around the center; the knee actuated walkers around zero; the ankle actuated walkers around the left bottom; the torso actuated walkers around the bottom.

Table 5.9 Best performance of four types of walkers

Actuated Joint	Hip	Knee	Ankle	Torso
Number of best fitness (Total 100 seeds)	56	3	12	29
Max distance traveled [m] (For 6 s)	3.04	0.48 (fall)	1.62	3.42
Energy consumption [kJ] (For 6 s)	120.	–	25.	45.
Energy efficiency [kJ/m^{-1}]	39.4	–	15.4	13.1

Table 5.9 shows parameters of the best performance, that is, their distances traveled, energy consumption, and energy efficiencies, in each type. In terms of the rate of solutions generated, the evolutionary design generated hip actuated walkers the most. However, the torso and ankle actuated walkers achieved higher energy efficiencies (energy consumption divided by distance traveled), so the generating rate does not relate to the emergence of embodiment.

For physical features, higher fitness (i.e., distance traveled) of each walker tends to have a more specific physical feature:

1. The ankle actuated walkers have high compliance at hip for the sagittal rotation, knee and torso for the lateral rotation;
2. The hip actuated walkers have low compliance at knee, ankle, and torso (i.e., only the hip is joints with mobility);
3. The torso actuated walkers have high compliance at hip for the sagittal rotation, low compliance at knee and ankle (i.e., the thigh and shin are regarded as one link).

Figure 5.11 demonstrate the joint structures and material properties (special physical features) and distance traveled and energy consumption energy efficiency (evaluation of embodiment). We see that the walkers with the special physical features perform better in distance traveled and energy efficiency. That is, it indicates a physical representation of the embodiment, a mechanism for self-regulation.

5.4.5 Development of a Novel Pseudo Passive Dynamic Walker

The best embodiment in the representation (Fig. 5.11) illustrates a structure, which is high compliance at the hip, low compliance at the knee and ankle, and actuation at the torso, and which achieves stable locomotion by transferring actuation power from lateral oscillation at the actuated torso to sagittal oscillations at the compliant hip. Then, such biped robots are simplified, and a novel PPDW is designed as shown in Fig. 5.12 (left). Figure 5.12 (right) shows walking mechanism of the PPDW: it exploits its own physical features (i.e., actuation, gravitational, and inertial forces) for taking steps.

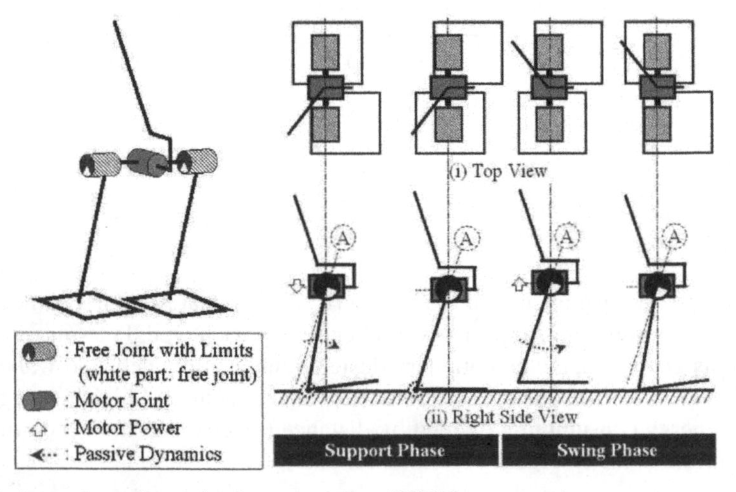

Fig. 5.12 A novel pseudo passive dynamic walker (PPDW)

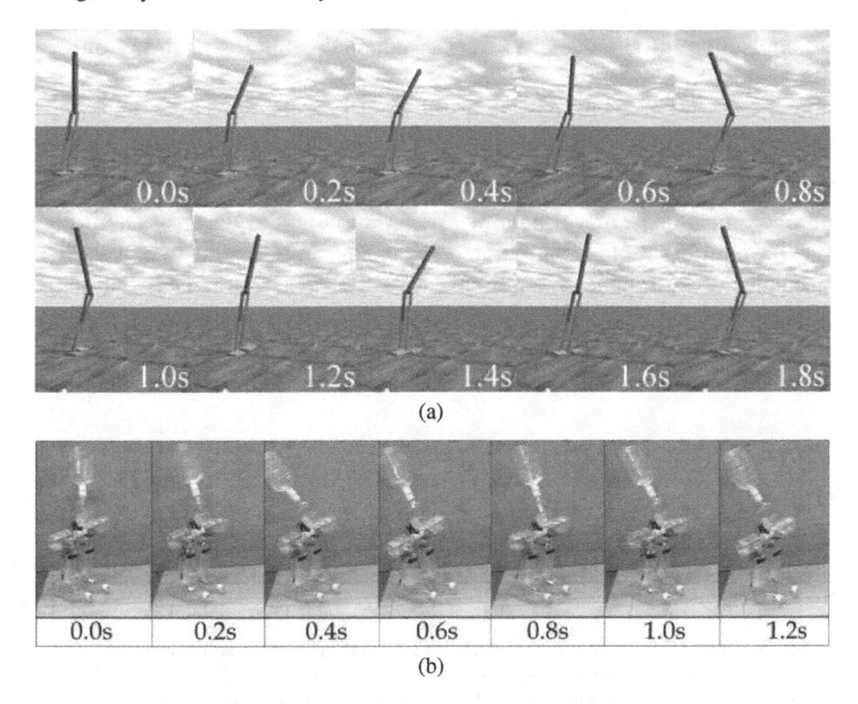

Fig. 5.13 Walk scenes of the PPDW: (**a**) virtual world, (**b**) real world

The PPDW is developed for the verification of the embodiment in both virtual and real world. The ODE is used for the implementation of virtual PPDW. A robotic development kit is applied for the development of real PPDW. The robotic development kit characterizes using plastic bottles as frames of robot structure, RC servomotors as actuated joints, and hot glues for connecting them [5]. This unique approach has advantages of shorting machining and building time and enabling easy assembly and modification for everyone. It is not for developing precisely operated robots but the kit is durable enough to realize the desired behavior.

As a result, the PPDW was developed in both virtual and real world, the PPDW achieved the desired stable locomotion (Fig. 5.13). It is significant that the real PPDW developed with the developmental kit that characterizes rough design, but it performed the desired locomotion. Therefore, the embodiment indicates high robustness, and low-cost stable locomotion on flat plane.

5.5 Conclusion

In this paper, we focused on self-regulatory functions of legged robots. We proposed a design process for emerging embodiment – an evolutionary design system – in the three-dimensional virtual world and it characterized the following:

1. A biologically reproductive process (i.e., coupled evolution of morphology and controller),
2. An interactive interface for designers to tune parameters, and
3. Evaluation methods of their embodiments for specifying their functions.

Then, we explored self-regulatory functions in various solution spaces representing morphological and control parameters. Eventually, the design system clearly illustrated a physical representation, and automatic motion adjustments (i.e., passive mechanism) were specified among the evolved robots: a torso oscillation in the lateral plane actuated passive hip joints. Such biped robots were regarded as novel pseudo passive dynamic walkers, and their self regulatory locomotion was investigated in both virtual and real world.

References

1. Alexander, R.: Principles of Animal Locomotion. Princeton University Press, Princeton (2002)
2. Brooks, R.: Cambrian Intelligence: The Early History of the New AI. MIT Press, Cambridge, MA (1999)
3. Collins, S., Wisse, M., Ruina, A.: A three-dimensional passive-dynamic walking robot with two legs and knees. The International Journal of Robotics Research 20, 607–615 (2001)
4. Gibson, J.: The Ecological Approach to Visual Perception, Houghton-Mifflin, Boston (1979)
5. Matsushita, K., Yokoi, H., Arai, T.: Plastic-bottle-based tobots in educational robotics courses – understanding embodied artificial intelligence, Journal of Robotics and Mechatronics 19, 212–222 (2007)
6. McGeer, T.: Passive dynamic walking. International Journal of Robotics Research 9, 62–82 (1990)
7. Pfeifer, R., Scheier, C.: Understanding Intelligence. MIT Press, Cambridge, MA (1999)
8. Sims, K.: Evolving Virtual Creatures. Computer Graphics Annual Conference Proceedings, pp 43–50 (1994)
9. Sims, K.: Evolving 3D morphology and behavior by competition. In: R. Brooks and P. Maes, Editors, Artificial Life IV Proceedings, MIT Press, Cambridge, MA, pp 28–39 (1994)
10. Smith, R.: Open Dynamic Engine. URL: http://www.ode.org/, (2000)
11. Vogel, S.: Cats's Paws and Catapults. W.W. Norton, NY (1998)

Chapter 6
Perception for Action in Roving Robots: A Dynamical System Approach

Paolo Arena, Sebastiano De Fiore, and Luca Patané

6.1 Introduction

For a mobile robot working in the real world, the ability to interpret information coming from the environment is crucial, both for its survival and for its accomplishing tasks. The cognitive capabilities of a robotic system are defined by the way in which the information gathered from sensors are processed to produce a specific action or behavior. Two broad classes can be distinguished: the *cognitivist approach* based on symbolic information processing and the *emergent systems approach* that is directed to the application of dynamical systems connected to the principles of self-organization [39]. Among the numerous solutions proposed by researchers, a great part is located in between the two main approaches.

In the recent past, research was directed towards endowing a robot with capabilities of self-creating an internal representation of the environment. Experiments in real world differ from applications in structured environments because they are mostly dynamically changing, so that it is impossible to program robot behaviors only on the basis of a priori knowledge. Moreover, the control loop must be able to process the different stimuli coming from the environment in a time that must be compatible with the real time applications. To solve these open problems, research activities have been focused on suitable solutions obtained taking inspiration from nature and applying biological principles to develop new control systems for perception–action purposes.

Recently inside the research activity of the EU Project SPARK [37], a bio-inspired framework for action-oriented perception has been proposed [5, 13].

Perception is no longer considered as a stand-alone process, but as an holistic and synergetic one tightly connected to the motor and cognitive system [14]. Perception processes are indivisible from action: behavioral needs provide the context for the perceptual process, which, in turn, works out the information required for motion control.

A. Adamatzky and M. Komosinski (eds.), *Artificial Life Models in Hardware*,

In this work, the perceptual system, based on principles borrowed by insect neurobiology, nonlinear dynamics, and complex system theory, has been realized by means of chaotic dynamical systems controlled through a new technique called *Weak Chaos Control* (WCC).

WCC technique allows to create perceptual states that can be managed by the control system because it directly is related to the concept of embodiment and situatedness [38]. The biological principles that inspired the proposed approach are based on Freeman's theories. His approach recognizes the existence of internal (mental) motivationally driven pattern formation [16, 19]. Cerebral cortex processes information coming from objects identified in the environment from receptors by enrolling dedicated neural assemblies. These are nonlinear dynamical coupled systems whose collective dynamics constitutes the mental representation of the stimulus. Freeman and co-workers, in their extensive experimental studies on the dynamics of sensory processing in animals [16–18, 24], conceive a dynamical theory of perception. Through electroencephalogram (EEG), Freeman evaluated the action potentials in the olfactory bulb and he noticed that the potential waves showed a typical chaotic behavior. So he came to the conclusion that the emergence of a mental pattern, in correspondence to an incoming stimulus, is the result of a chaotic dynamics in the sensory cortex in cooperation with the limbic system that implements the supporting processes of intention and attention [16].

The application of chaotic models as basic blocks to reproduce adaptive behaviors [4, 11, 36] is an interesting aspect of the current research activity on this field. The idea is that the chaos provides the right properties in terms of stability and flexibility, needed by systems that evolve among different cognitive states. In particular, perceptual systems dynamically migrate among different attractors that represent the meaning of the sensory stimuli coming from the environment.

Moreover, other studies consider the role of noise as an added value to reactive systems that otherwise could not be able to escape from the deadlock situations produced by local minima [28]. Chaotic dynamics can further improve reactive system capabilities, in fact chaotic systems generate a wide variety of attractors that can be controlled guiding the transit from one to another. Freeman's studies lead to a model, called K-sets, of the chaotic dynamics observed in the cortical olfactory system. This model has been used as dynamic memory for robust classification and navigation control of roving robots [20–22, 24].

The architecture proposed here, taking into consideration the relevant principles previously underlined, is based on the control of chaotic dynamics to learn adaptive behaviors in roving robots. The main aim is to formalize a new method of chaos control applied to solve problems of perceptual state formation.

The WCC approach following these guidelines uses a chaos control technique applied to a multiscroll chaotic system [27]. Therefore, the WCC is a general technique that can be applied to several chaotic systems [7, 41]. All sensory signals are mapped as different potential reference dynamics used to control the chaotic system. This creates association between sensory information and a particular area located within the multiscroll phase plane, in a way that reflects the topological position of the robot within the sensed environment. Moreover, thanks to the addition of a learning stage (called in the following *Action Selection layer*), the technique allows to implicitly include within the robot control system the real robotic structure and dimensions, including the sensor position, thus situating the robot within the environment. The improvements provided by the Action Selection layer are shown in several works [5, 10, 25]. Here a bio-inspired adaptive structure, based on an unsupervised, reward-based method derived by the *Motor Map* paradigm, has been used. This network, inspired by the paradigm of Kohonen Nets [23], is able to plastically react to localized excitation by triggering a movement (like the motor cortex or the superior colliculus in the brain) [30].

The aim of this chapter is to present in detail the WCC navigation technique using the Motor Map paradigm to select the robot action. Moreover, the proposed navigation control technique, based on the action-oriented perception paradigm, has been implemented in the FPGA-based board and tested on a roving robot in a real environment.

In the next section we will describe the control architecture that reproduce the sensing–perception–action loop. Section 6.3 is devoted to illustrate the hardware used to implement the proposed architecture. In Sect. 6.4, the implementation is explained in more details, while the results are reported in Sect. 6.5.2. Finally, in Sect. 6.7, we will draw the conclusions and present the possible developments that we are currently investigating.

6.2 Control Architecture

The sensing–perception–action loop is at the basis of research in the cognitive system field. In this chapter we propose an implementation of the action-oriented perception paradigm applied to the navigation control of an autonomous roving robot.

A basic scheme of the architecture is shown in Fig. 6.1, where two main blocks can be distinguished:

Perceptual system creates an internal representation of the environment from the sensory inputs.

Action selection network learns the suitable action associated to each environmental situation on the basis of the task to be accomplished.

When a robot is placed in an unknown environment, it is subject to a huge amount of external stimuli. To explore the area avoiding obstacles, the robot, sensing the environment, can create an internal representation of the stimuli in relation to its body. Therefore, when no stimuli are perceived (i.e., there are no active sensors), the system evolves in a chaotic behavior and the robot continues to explore the environment performing a random action determined by the chaotic evolution of a dynamical system representing the core of the perceptual formation mechanism. When external stimuli are perceived, the dynamical system evolution is enslaved into low order dynamics that depends on the contribution of each active sensor. The behavior that

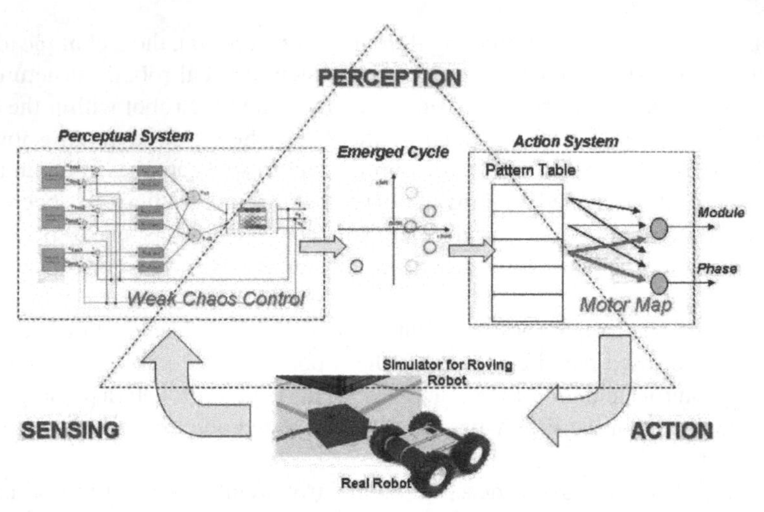

Fig. 6.1 Control architecture used to implement the sensing-perception-action loop on a roving robot

correspond to this perception will plastically depend on the characteristics emerging dynamics, as a result of a learning algorithm. The learning mechanism is here driven by a Reward Function (RF), designed on the basis of the mission to be accomplished (in our case navigation tasks). In the following sections, the main blocks are described in details.

6.2.1 Perceptual System

The core of this layer is the WCC method [6]. The crucial advantage of this approach is the possibility to create a compact representation of the environment that can be used for real-time navigation purposes in mobile robots. Moreover, this method permits to handle with a great number of sensors. It emulates the perceptual processes of the brain in which specific cerebral patterns emerge depending on the perceived sensorial stimuli. To model this behavior, a chaotic system, proposed by Chen [27], has been used as a plastic layer in which perceptual states can emerge. The chaotic behavior of the Chen's multiscroll system can be enslaved to regular periodic patterns (i.e., emergence of perceptual states) by using the sensory stimuli as reference control signals. The control mechanism has been realized with a feedback on the state variables x and y controlled in order to follow the reference cycles. This multiscroll system can be viewed as a generalization of the Chua's double scroll attractor represented through saturated piecewise linear functions and of other circuits able to generate n-scrolls [29]. It is able to generate one-dimensional (1D) n-scrolls, two-dimensional (2D) $n \times m$-grid scrolls, or three-dimensional (3D) $n \times m \times l$-grid scroll chaotic attractors by using saturated function series. In this paper, a 2D multiscroll system has been chosen. It is described by the following differential equations [27]:

$$\begin{cases} \dot{x} = y - \frac{d_2}{b} f_1(y;k_2;h_2;p_2,q_2), \\ \dot{y} = z, \\ \dot{z} = -ax - by - cz + d_1 f_1(x;k_1;h_1;p_1,q_1) \\ \qquad + d_2 f_1(y;k_2;h_2;p_2,q_2), \end{cases} \tag{6.1}$$

where the following so-called saturated function series (PWL) $f_1(x;k_j;h_j;p_j,q_j)$ has been used:

$$f_1(x;k_j;h_j;p_j;q_j) = \sum_{i=-p_j}^{q_j} g_i(x;k_j;h_j), \tag{6.2}$$

where $k_j > 0$ is the slope of the saturated function, $h_j > 2$ is called *saturated delay time*, p_j and q_j are positive integers, and

$$g_i(x;k_j;h_j) = \begin{cases} 2k_j & \text{if } x > ih_j + 1, \\ k_j(x - ih_j) + k_j & \text{if } |x - ih_j| \le 1, \\ 0 & \text{if } x < ih_j - 1, \end{cases}$$

$$g_{-i}(x;k_j;h_j) = \begin{cases} 0 & \text{if } x > -ih_j + 1, \\ k_j(x + ih_j) - k_j & \text{if } |x + ih_j| \le 1, \\ -2k_j & \text{if } x < -ih_j - 1. \end{cases}$$

System (6.1) can generate a grid of $(p_1 + q_1 + 2) * (p_2 + q_2 + 2)$ *scroll attractors*. Parameters p_1 (p_2) and q_1 (q_2) control the number of *scroll attractors* in the positive and negative direction of the variable x (y), respectively. The parameters used in the following ($a = b = c = d_1 = d_2 = 0.7$, $k_1 = k_2 = 50$, $h_1 = h_2 = 100$, $p_1 = p_2 = 1$, $q_1 = q_2 = 2$) have been chosen according to the guidelines introduced in [27] to generate a 2D 5×5 grid of scroll attractors. An example of the chaotic dynamics of system (6.1) is given in Fig. 6.2, also where a 3D grid of scrolls is shown.

In our approach, the perceptual system is represented by the multiscroll attractor of (6.1), while sensorial stimuli interact with the system through periodic inputs that can modify the internal chaotic behavior. As one of the main characteristics of perceptive systems is that sensorial stimuli strongly influence the spatial-temporal dynamics of the internal state, a suitable scheme to control the chaotic behavior of the multiscroll system on the basis of sensorial stimuli should be adopted.

Chaos control refers to a process wherein a tiny perturbation is applied to a chaotic system to realize a desirable behavior (e.g., chaotic, periodic, and others). Several techniques have been developed for the control of chaos [15].

A commonly used chaos control strategy is to select the desirable behavior among the unstable periodic orbits embedded in the system. The stabilization of one of the unstable periodic orbits can be obtained by applying a small perturbation when the chaotic orbit passes close to the desired behavior.

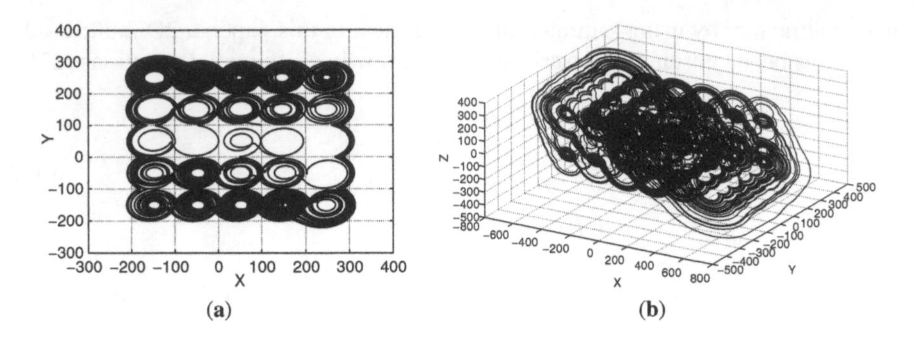

Fig. 6.2 (**a**) Projection of the 5×5 grid of *scroll attractors* in the plane *x*-*y*. (**b**) 3D-scroll attractor, 5×5×5 grid of scrolls

In general, the strategies for the chaos control can be divided in two classes: closed loop and open loop. The first class includes those methods that select the perturbation based on a knowledge of the state variables and oriented to control a prescribed dynamics. The two most important feedback methods are the Ott–Grebogi–York (OGY) [15,31] approach and the Pyragas technique [32]. The second class includes those strategies that consider the effect of external perturbations on the evolution of the system. In the OGY's method, the changes of the parameter are discrete in time as this method is based on the use of Poincar maps.

In view of our application, a continuous-time technique like the Pyragas's method is a suitable choice [32]. In this method [32, 33], the following model is taken into account:

$$\frac{\mathrm{d}y}{\mathrm{d}t} = P(y,x) + F(t), \qquad \frac{\mathrm{d}x}{\mathrm{d}t} = Q(y,x), \tag{6.3}$$

where y is the output of the system (i.e., a subset of the state variables) and the vector x describes the remaining state variables of the system. $F(t)$ is the additive feedback perturbation that forces the chaotic system to follow the desired dynamics. Pyragas [32, 33] introduced two different methods of permanent control in the form of feedback. In the first method, that is used here, $F(t)$ assumes the following form:

$$F(t) = K[\widehat{y}(t) - y(t)], \tag{6.4}$$

where $\widehat{y}$ represents the external input (i.e. the desired dynamics), and K represents a vector of experimental adjustable weights (adaptive control). The method can be employed to stabilize the unstable orbits endowed in the chaotic attractor reducing the high order dynamics of the chaotic system.

In the second method, the idea consists in substituting the external signal $\widehat{y}$ in (6.4) with the delayed output signal $y(t-\tau)$

$$F(t) = K[y(t-\tau) - y(t)], \tag{6.5}$$

where τ is a delay in time. This feedback performs the function of self-control.

In our case, a strategy based on (6.4) has been applied. The desired dynamics is provided by the periodic behavior associated with the sensorial stimuli. As more than one stimulus can be presented at the same time, the Pyragas method has been generalized to account for more than one external forcing.

Hence, the equations of the controlled multiscroll system can be written as follows:

$$\begin{cases} \dot{x} = y - \frac{d_2}{b} f_1(y;k_2;h_2;p_2,q_2) + \sum_i K_{x_i}(x_{r_i} - x) \\ \dot{y} = z + \sum_i K_{y_i}(y_{r_i} - y) \\ \dot{z} = -ax - by - cz + d_1 f_1(x;k_1;h_1;p_1,q_1) \\ \qquad + d_2 f_1(y;k_2;h_2;p_2,q_2), \end{cases} \tag{6.6}$$

where i is the number of external references acting on the system, x_{r_i}, y_{r_i} are the state variables of the reference circuits, which will be described in detail, and K_{x_i}, K_{y_i} represent the control gains. It can be noticed that the control acts only on the state variables x and y. The complete control scheme is shown in Fig. 6.3.

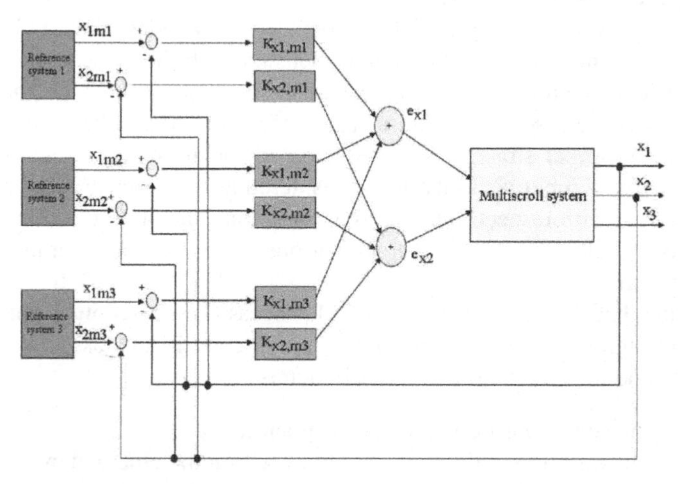

Fig. 6.3 The control scheme when three distinct reference signals (i.e., sensorial stimuli) are perceived by the multiscroll system

Each reference signal (x_{r_i}, y_{r_i}) is a periodic trajectory that represents a native cycle. It can be generated using a multiscroll system (6.1) with particular parameters ($a = b = c = 1$). The amplitude, pulse, and center position of the cycle depend on the initial conditions. The number of the multiscroll systems needed to generate the reference cycles is the same as the number of reference trajectories required. These reference signals can be built using a sinusoidal oscillator:

$$x_r(t) = A_{x_r} \sin(\omega_{xr} - \varphi_{x_r}) + x_{\mathrm{off}}$$

$$y_r(t) = A_{y_r} \sin(\omega_{yr} - \varphi_{y_r}) + y_{\mathrm{off}} \tag{6.7}$$

where $(x_{\text{off}}, y_{\text{off}})$ is the center of the reference cycle, ω is a frequency (in this work $\omega_{xr} = \omega_{yr} = 1$), φ_{x_r} and φ_{y_r} are the phases, A_{x_r} and A_{y_r} define the amplitude of the reference.

This solution allows to improve the performance of the control system: only the differential equations of the controlled system must be integrated, the reference trajectory are obtained using (6.7).

If the sensor stimuli are associated with the reference cycles, the WCC technique can be used to control robot navigation. A key point of this approach is that the reference cycle distribution in the phase plane *x*-*y* reflects the topological distribution of robot sensors; this contributes to include the geometrical embodiment of the robot itself within the environment. Moreover, the sensor range depicts the current robot operating space, which is dynamically encoded within the phase space of the multiscroll system. The link between reference signals and sensors are obtained through control gains K_{x_i} and K_{y_i}. The value of these control parameters are related to the amplitude of the sensory stimuli, and so a regular periodic pattern emerges as a function of the sensor readings.

Figure 6.4 shows an example of the association between sensors and reference signals used to control the chaotic system. Sensors with the same position on the robot are associated with reference cycles with the same position in the phase plane. Distance sensors are associated with the reference cycles that reflect their topological position. A similar strategy has been adopted for the target. When a target is within the range of robot visibility, it is considered as an obstacle located in a position symmetric with respect to the motion direction. The aiming action is guided by a reference cycle with a low gain, so that obstacle avoidance is a priority over reaching a target.

Each cycle that emerges from the control process (i.e., perceptual state) can be identified through its center position and shape. A code is then associated to each cycle and it is defined by the following parameters:

- x_q and y_q: the center position in the phase plane $x - y$
- $\bar{x}_q$: maximum variation of the state variable x within the emerged cycle
- $\bar{y}_q$: maximum variation of the state variable y within the emerged cycle

where q indicates the emerged cycle.

In this way, a few parameters provide an abstract and concise representation of the environment. To solve the robot navigation task, an action is performed by the robot according to the characteristics of the emerged pattern.

6.2.2 Action Selection Layer

The perceptual pattern obtained through the WCC technique is then processed by the action selection system (see Fig. 6.1).

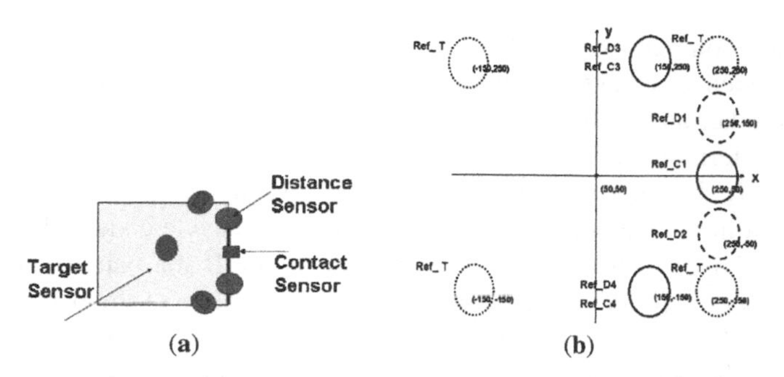

Fig. 6.4 (**a**) Map of the sensors equipped on the robot. (**b**) Reference cycles Ref_i linked to the sensors. Contact sensors are associated to the same cycles as the corresponding distance sensors, if they are placed in the same position on the robot. The T_i reference cycles are associated to the target sensor. The target is considered as an obstacle located in a symmetrical position with respect to the motion direction

The action block establishes the association between the emerged cycle and the consequent robot action. An action consists of two elements:

$$action = (module, phase). \tag{6.8}$$

The module and phase of an action determine, respectively, the motion step and the rotation angle to be performed by the robot. To perform this task, the Motor Map (MM) paradigm was applied. MMs are suitable to control the behaviors of the robot in an unknown environment, because they are adaptive, unsupervised structures and are simple enough to allow for a real-time learning. The MM is composed by two layers: one (V) devoted to the storage of input weights and another (U) devoted to the output weights. V represents the actual state of the system which is to be controlled and U determines the required control action. This allows the map to perform tasks such as motor control. Formally, a MM can be defined as an array of neurons mapping the space of the input patterns into the space of the output actions:

$$\varphi : V \longrightarrow U. \tag{6.9}$$

The learning algorithm is the key point to obtain a spatial arrangement of both the input and output weight values of the map. This is achieved by considering an extension of the Kohonen algorithm. At each learning step, when a pattern is given as input, the winner neuron is identified: this is the neuron that best matches the input pattern. Then, a neighborhood of the winner neuron is considered and an update involving both the input and output weights for neurons belonging to this neighborhood is performed. The learning procedure is driven by a reward function that is defined on the basis of the final aim of the control process [35]. A MM although very efficient to be trained could be difficult to be implemented in hardware because of the high number of afferent and efferent weights. As this work is finalized to the application on real roving robot prototypes, a simplified version is adopted [5]. The first difference is that the relationship of proximity is not considered among the neurons. Another and more important difference is that the afferent layer is substituted by a pattern table. Each emerged cycle, identified by a code, is stored in the pattern table (if it is not yet present) when the pattern emerges for the first time. Each element in the pattern table contains the emerged cycle code and the number of iterations from its last occurrence (defined as *age*). If the pattern table is full, the new element will overwrite the one containing the code of the pattern least recently used (LRU), that is, the one with the highest *age* value. This table permits to synthesize the perception of the environment. Therefore, the winning neuron of the afferent layer is replaced by the element q of the pattern table that contains the last emerged cycle. Moreover, the efferent (output) layer is now constituted of two weights for each element of the pattern vector. The element q is connected to the weights w_{qm} and w_{qp}, which represent, respectively, module and phase of the action associated with the pattern q (A_q). At each step, the robot does not perform the exact action suggested by the weights of q (w_{qm} and w_{qp}), but the final action is

$$A_q = (A_q(\text{module}), A_q(\text{phase})) = (w_{qm} + a_{s_q}\lambda_1, w_{qp} + a_{s_q}\lambda_2), \tag{6.10}$$

where λ_1 and λ_2 are random variables uniformly distributed in the range $[-1;1]$. The parameter a_{s_q} limits the searching area. Every time the pattern q emerges, a_{s_q} is reduced to focus the action search in a smaller range so as to guarantee the convergence of the efferent weights. When there are no inputs, the perceptual core of the robot (the multiscroll system) behaves chaotically. This implies that there are no emerged cycles and no entries in the pattern table. In this case, the robot explores the environment and its action depends on the position of the centroid of the particular chaotic scroll shown by the system during the simulation step. Of course, the exploration phase can also be performed using a forward motion, that is, not considering the chaotic wandering.

The unsupervised learning mechanism that characterizes the MM algorithm is based on a reward function (RF). This is a fitness function and it is the unique information that permits to determine the effectiveness of an action. In a random foraging task, a suitable choice for the RF is

$$\text{RF} = -\sum_i \frac{k_i}{D_i^2} - h_D D_T - h_A|\phi_T|, \tag{6.11}$$

where D_i is the distance between the robot and the obstacle detected by the sensor i, D_T is the robot–target distance, ϕ_T is the angle between the direction of the longitudinal axis of the robot and the direction connecting robot and target, and k_i, h_D, and h_A are appropriate positive constants determined during the design phase [9, 12].

6.3 Hardware Devices

The proposed method for robot navigation was tested in a simulation environment showing good results [8]. The choices done during the design of the control architecture, oriented to a real-time hardware implementation, simplify the implementation of the strategy in a high performing hardware to be embedded on a roving robot.

The first step followed consisted in the definition of the hardware framework. The system should be able to acquire information coming from a distributed sensory system. Moreover, the computational power have to be sufficient to obtain control command compatible with a real-time application, allowing in the mean time the data logging for debugging and post-processing performance analysis.

The two main options for the controller are a digital signal processor (DSP) or a field programmable gate array (FPGA). A microcontroller is not taken into account due to the limited resource and performance that could compromise the scalability of the system.

DSPs are currently used in several fields of applications: communications, medical diagnostic equipment, military systems, audio and video equipment, and countless other products, becoming increasingly common in consumers' lives.

DSPs are a specialized form of microprocessor, while FPGAs are a form of highly configurable hardware. In the past, the use of digital signal processors was ubiquitous, but now, with the needs of many applications outstripping the processing capabilities of DSPs (measured in millions of instructions per second (MIPS)), the use of FPGAs is growing rapidly. Thus, the comparison between digital signal processors and FPGAs focuses on MIPS comparison, which, while certainly important, is not the only advantage of an FPGA. Equally important, and often overlooked, is the FPGAs inherent advantage in product reliability and maintainability. Therefore, the device chosen to implement the WCC approach is an FPGA.

The particular FPGA device adopted is a Stratix II EP2S60 produced by Altera, which integrates a soft embedded processor, the NiosII. This solution was adopted to combine in a single device both the hardware reconfigurability of FPGA devices and the programming simplicity of microcontrollers. NiosII is, in fact, a 32-bit RISC digital microprocessor with a clock frequency of 50 MHz. It is a configurable "soft-core" processor. Soft-core means that the CPU core is offered in "soft" design form (not fixed in silicon), so that its functionalities, like the peripheral number and type or the amount of memory, can be easily modified according to the specific task [2].

The first attempt was carried out by using a development board provide by Altera [2]. This preliminary architecture led to the design of a new FPGA-based board, called SPARK board. The SPARK board has been developed during the SPARK project [37] and has been designed to fulfill the requirements needed by the SPARK cognitive architecture [13].

The mobile robot used for the experiment is called Rover II (see Fig. 6.5). It is a classic four-wheel drive rover controlled through a differential drive system by Lynxmotion, which has been modified to host a distributed sensory system. The complete control system was embedded on board and a wireless communication module was introduced for debugging purposes.

Fig. 6.5 Rover II: a dual drive roving robot endowed with a distributed sensory system

In the next section, more details about the robot and FPGA-based control architecture are given.

6.3.1 Spark Main Board

The FPGA-based development board preliminary used to assess the feasibility of the hardware control scheme showed some problems due to the big size of the board, the unused number of components, and mainly its lack of room within the FPGA to implement the needed algorithms. For these reasons, the design of a new specific board has been required.

Different approaches were taken into consideration during the design phase.

- One single board including a large FPGA (Stratix II EP2S180)
- One single board comprising two medium-size FPGA (Stratix II EP2S60)
- Two boards, each one with a medium-size EP2S60 FPGA connected via cables

After studying all the possibilities, the third option was adopted, as it is more robust in terms of developing time, costs, and flexibility. This solution permits us the use of only one board when the resources are enough for the application, and also two or even more boards when the algorithm requires more computational power. Figure 6.6 shows the SPARK 1.0 Main Board features in details. It represents a hardware platform for developing embedded systems based on Altera Stratix II EP2S60 FPGA with the following features:

1. Stratix II EP2S60F672C3N FPGA with more than 13,500 adaptive logic modules (ALM) and 1.3 million bits of on-chip memory
2. Eight expansion headers with access to 249 FPGA user I/O pins (3.3 V tolerant)
3. One expansion header with access to 16 FPGA user I/O pins (5 V tolerant)

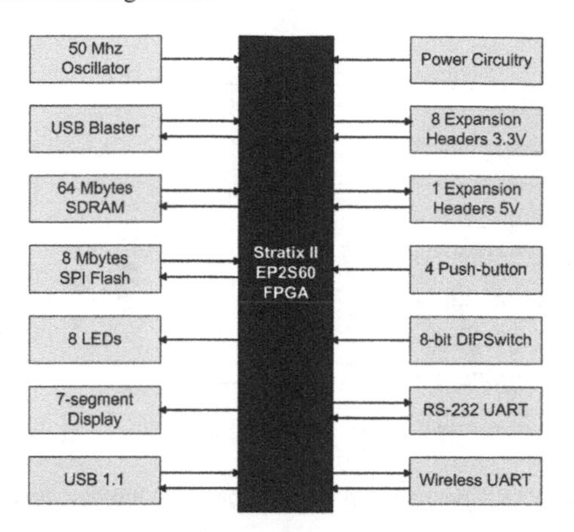

Fig. 6.6 SPARK board 1.0 block diagram

4. 1.2 V power supply
5. 5 V power supply
6. 12 V power supply connector
7. 3.3 V power supply
8. Eight LEDs connected to FPGA user I/O pins
9. Four push-button switches connected to FPGA user I/O pins
10. 32 Mbytes SDRAM
11. Dual seven-segment LED display
12. Integrated USB Blaster for connection with the ALTERA Software tools on a PC
13. 32 Mbytes of SDRAM
14. RS-232 serial connector with 5 V tolerant buffers
15. Wireless UART Communication Port
16. USB 1.1 Communication Port
17. 50 MHz clock signal for FPGA
18. 8 Mbytes SPI Flash Memory
19. Push-button switch to reboot the NIOS II processor configured in the FPGA
20. 8-bit DIP switch

A scheme with the component's locations on the SPARK board is shown in Fig. 6.7.

6.3.2 Rover II

The roving platform used for navigation experiments is a modified version of the dual drive Lynx Motion rover, called Rover II, whose dimensions are about $30 \times 30\,cm^2$. It is equipped with a bluetooth telemetry module, four infrared short

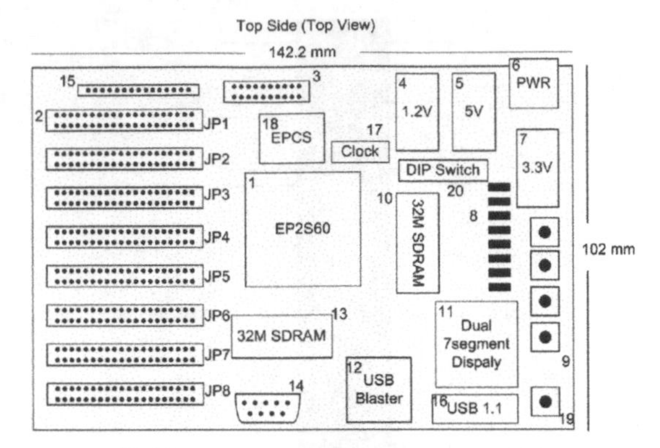

Fig. 6.7 Scheme of the SPARK board 1.0. A description of each block, numbered from 1 to 20, is reported in the text

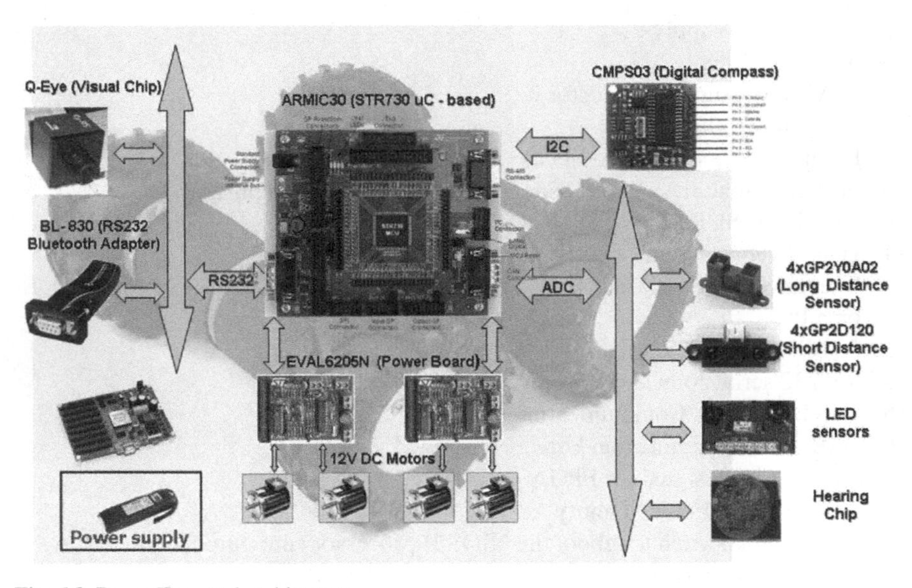

Fig. 6.8 Rover II control architecture

distance sensors Sharp GP2D120 (detection range 3–80 cm), four infrared long-distance sensors Sharp GP2Y0A02 (maximum detection distance about 150 cm), a digital compass, a low-level target detection system, a hearing board for cricket chirp recognition [34, 40] and with the Eye-RIS v1.2 visual system [3].

The complete control architecture, reported in Fig. 6.8, shows how the low-level control of the motors and the sensor handling is realized through a microcontroller STR730. This choice optimizes the motor performances of the robot, maintaining the high-level cognitive algorithms in the SPARK board and in the Eye-RIS visual system. Moreover, the Rover II can be easily interfaced with a PC through a bluetooth module, and this remote control configuration permits to perform some preliminary tests debugging the results directly on the PC.

In the following experiments, to fulfill a food retrieval task, distance sensors were used for obstacle avoidance, while a cricket-inspired hearing board was considered for the target detection issues [1].

Among the 5×5 possible reference cycles which can be used in the application reported here, only a subset of them has been effectively associated with a sensor. Their distribution in the phase plane x-y reflects the topological distribution of the robot sensors. Figure 6.9 shows a scheme of the robot equipped with four distance sensors and a target sensor. The corresponding reference cycles reported in the phase plane are related to the sensor positions. Distance sensors are directional, and each one is associated to a single reference cycle, while target sensor is characterized by an omnidirectional field of view, and for that reason it is associated to more than one (i.e., four) reference cycles [8].

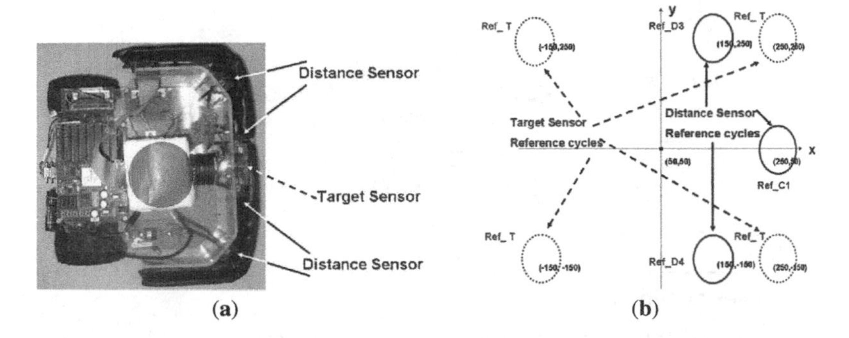

Fig. 6.9 Scheme of a Rover II robot (**a**) equipped with four distance sensors and a target sensor. In the phase plane $x-y$ (**b**), the reference cycles associated to each sensor are reported. Target sensor, due to its omnidirectional field of view, is associated to four reference cycles. Cycles can be generated by system (6.1) with following parameters: $a = b = c = d_1 = d_2 = 1$, $k_1 = k_2 = 50$, $h_1 = h_2 = 100$, $p_1 = p_2 = 1$, $q_1 = q_2 = 2$ and changing the offset. Equivalently, (6.7) can be used

The technique, based on placing reference cycles in the phase plane in accordance with the distribution of sensors on the robot, is important to directly connect the internal representation of the environment to the robot geometry. In our tests, only distance and hearing sensors have been used, although other sensors could be included, such us contact sensors, visual sensors, etc.

6.4 Hardware Implementation

In this section, the implementation of the framework for the control navigation algorithm presented in the previous section is discussed (see Fig. 6.10).

To realize the WCC approach for navigation control in hardware, the SPARK board was considered. Control algorithm was implemented on the Nios II soft-processor, and also customized VHDL (Very High Speed Integrated Circuits Hardware Description Language) blocks were used for the most time-consuming tasks.

In particular, the simulation and control of the multiscroll system are performed directly in a VHDL entity, named *sim_full*, while the NiosII microprocessor is devoted to handle the sensory system for the execution of the action selection layer and for the supervision of the activities of the VHDL entities implementing the WCC. When the simulation ends, the NiosII reads the parameters that identify the emerged cycle. Then it calculates the command to drive the roving robot. The simulation process implemented in the *sim_full* entity lasts for about 4.2 ms and the control algorithm running on NiosII for about 80 ms. Even if the RoverII robot could be driven via a bluetooth module, to reduce the communication time, the Spark board has to be mounted directly on the robot.

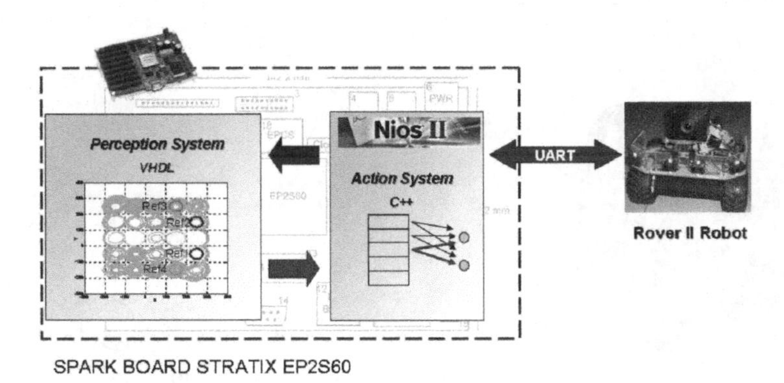

Fig. 6.10 Description of the framework used during the experiments

As discussed earlier, the control and simulation of the multiscroll system are made directly in VHDL entities. Other implementations were tested before testing the VHDL one. The first step was to simulate the chaotic system directly by using the NiosII in a C code algorithm. This solution works correctly, but the performances were not suitable for a real-time application. The high simulation time on NiosII (see Table 6.1) is probably due to the fact that its ALU (arithmetic logic unit) does not support natively floating point operations. This problem can be overcome by using custom instructions.

For these reasons, we went a step further; in fact, the "Soft-core" nature of the NiosII processor enables designers to integrate custom logic instructions (e.g., written in VHDL language) into the ALU. Similar to native NiosII instructions, custom instruction logic can take values from up to two source registers and optionally write back a result to a destination register. Also, in this case, the results were correct but the execution time, although decreased of one order of magnitude, remained too high for a real-time application. So, only the VHDL implementation is suitable as shown in Table 6.1.

Table 6.1 Performaces obtained with different implementations of the multiscroll control.

	Time[a]
Multiscroll control implemented on Nios II	10 s
Multiscroll control implemented on Nios II using custom instructions	1 s
Multiscroll control implemented in VHDL (*sim_full* entity)	4.2 ms

[a]The time is referred to the execution of 2,000 samples and the value of the integration step is 0.1

As the WCC is implemented directly in hardware, the role of the NiosII processors is limited to drive the activity of the block that implemented the *sim_full* entity to read the sensor values, to calculate the control gains, and to give motor commands to the robot on the basis of the actions generated by the action selection layer.

The main role of the *sim_full* entity consists in integrating the controlled third-order nonlinear multiscroll system for a given time. The integration algorithm implemented is a fourth order Runge–Kutta algorithm (RK4)[26]; the evolution time was fixed to 2,000 steps with an integration step of about 0.1. The evolution time length was experimentally chosen to guarantee the creation of "percepts" (i.e., low order dynamics like cycles) in the presence of sensorial stimuli. The output of the *sim_full* entity correspond to the behavior of the multiscroll system at the end of the integration process. These information are used by the Nios II to calculate robot actions.

For the implementation of the dynamical system, a decimal number arithmetic is needed. This algebra is not supported by the standard library of VHDL, and so the first step was to develop a dedicated library. As the fixed point operations are faster than the floating point ones, the fixed point arithmetic has been adopted. The total number of bits to codify the variable in the VHDL code is 24 bits. In particular, the most significant bit represents the sign: 14 bits are used to code the integer part and 10 bits code the fraction part.

In Fig. 6.11 the top level entity *sim_full* is shown. It is composed of two sub-entities:

- The *sim* block, which implements the simulation of the controlled multiscroll system adopting the RK4 algorithm with fixed integration step
- The *sim_machine*, which implements a finite state machine (FSM) that controls the activity of the *sim* entity.

The NiosII drives the *sim_full* entity through the input signals and provides the gains that are needed to control the simulated multiscroll system. Once the simulation is completed, it reads the results of the computation provided by the entity on its outputs.

Now we start with the description of the internal structure of the *sim* entity (Fig. 6.12). The *rk4*, *ram*, *structure_control_x*, and *structure_control_y* VHDL entities are the main components of this block. The function of the *rk4* entity is to implement the Runge–Kutta algorithm, and it needs the state vectors (x,y,z) and

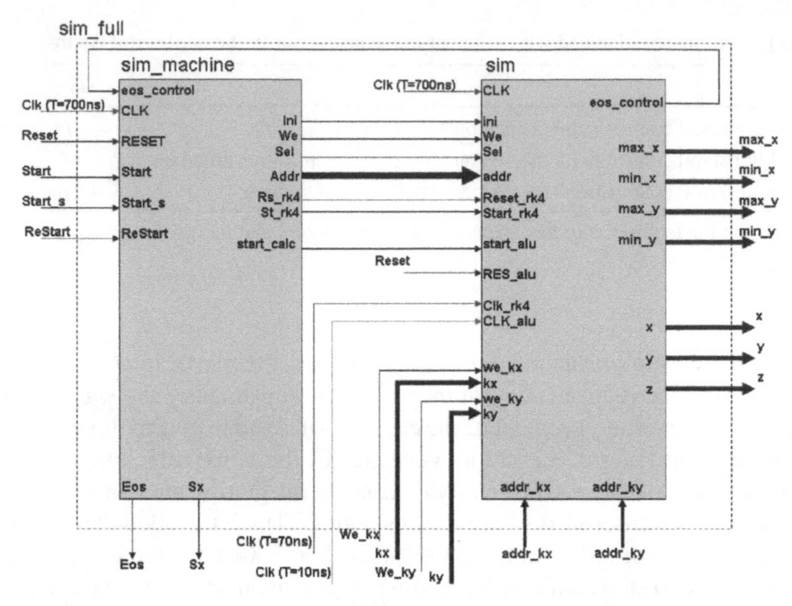

Fig. 6.11 VHDL Entity *sim_full*. The bold line represent a 24 bits length bus

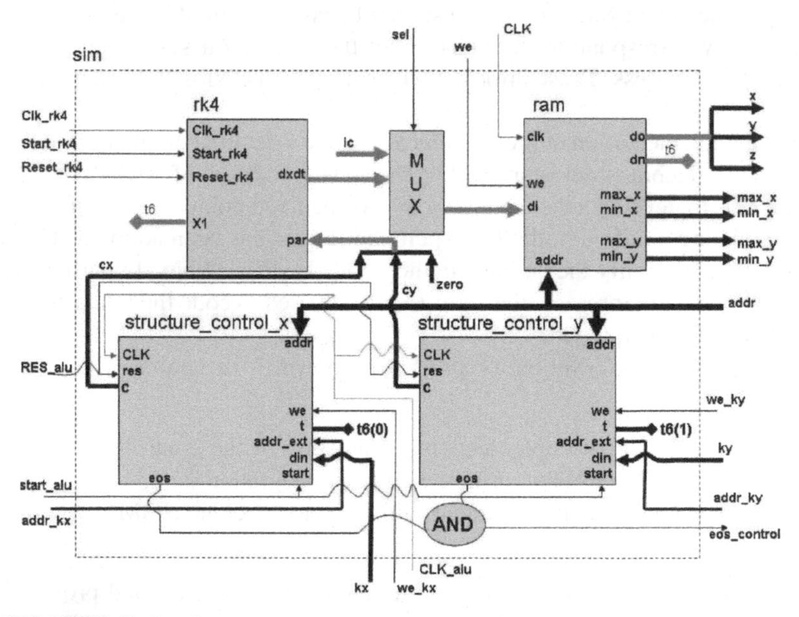

Fig. 6.12 VHDL Entity *sim*

the control signals to perform the integration. The function of the other inputs (i.e., *Clk_rk4*, *Start_rk4* and *Reset_rk4*) is to coordinate the activity of the entity.

The *ram* block is a memory that stores all the results of the integration: *di* is the data input, *do* is the data output, *dn* is the last sample produced and memorized, *we* is a write enable input, and *addr* is necessary to address the memory. The other outputs of this block represent the maximum and the minimum of the x and y state variables. This is an important feature that permits to determine the characteristics of the emergent cycle and to allow the use of the *sim_full* entity for the successive action selection steps.

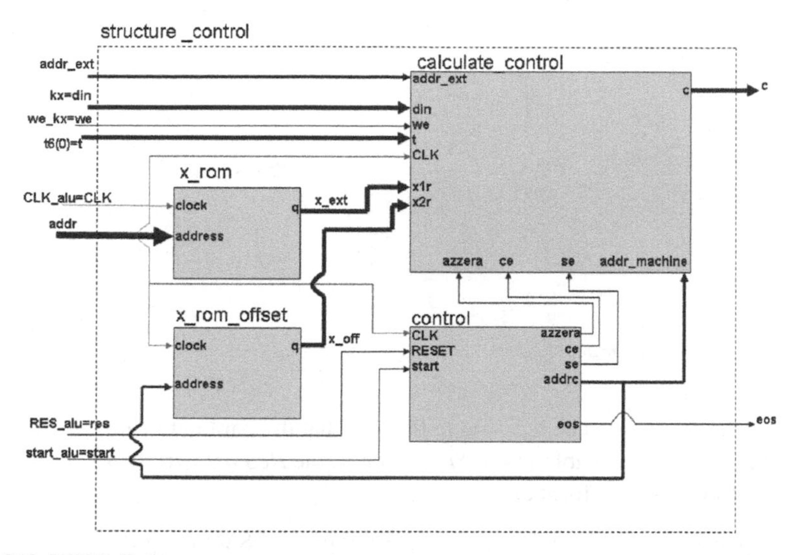

Fig. 6.13 VHDL Entity *structure_control*

The *structure_control_x* and *structure_control_y* blocks are two instances of the same VHDL entity named *structure_control*. The function of this entity is to calculate the control signals, while the k_{xi} and k_{yi} inputs represent the control gains corresponding to the reference i addressed with the input *addr_kx* and *addr_ky*, respectively. Figure 6.13 shows a block diagram of the entity for the x state variable. The reference signals are calculated off-line using (6.7), with x_{off} set to zero. The samples obtained are stored in the read-only memory (ROM) *x_rom*; the *x_rom_offset* ROM block stores the offset that must be added to the sample to obtain the desired reference signal; the *calculate_control* block computes the control gains and the *control* block is a finite state machine (FSM) that manages the activity of the entity.

Finally, in Fig. 6.14, the diagram of the FSM implemented is shown. The function of each state is reported in the following:

- *St_ini* sets the initial conditions, only at the startup
- *St_start* sets all the outputs to default values
- *St_sim* controls one step of simulation
- *St_mem* stores the produced samples into the *ram* block
- *St_eos* notifies the end of simulation
- *St_load, St_send, St_inc* together with the *Start_s* input and the *sx* output allow the NiosII processor to access correctly all the simulation parameters in memory.

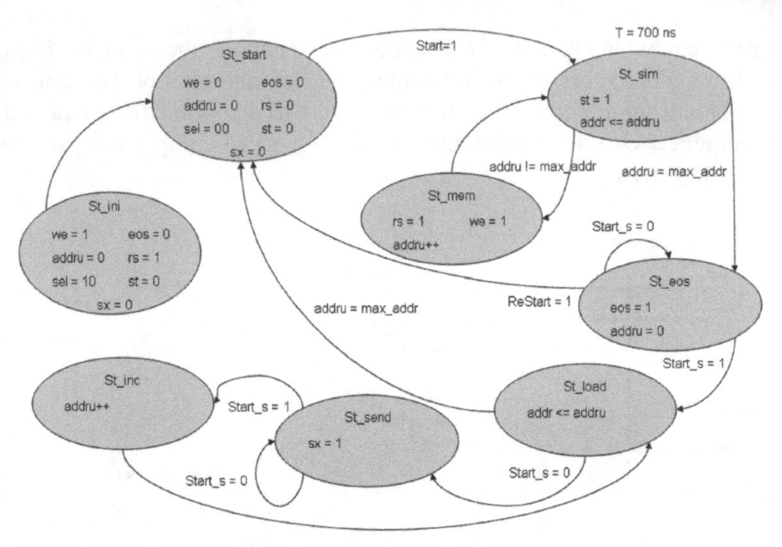

Fig. 6.14 VHDL Entity *sim_machine*

As this process is not needed (due to the fact that the parameters needed to identify the cycle are available in the *St_eos* state), the *ReStart* permits to skip it, and so to optimize the performances.

The inputs *Clk*, *Start*, *Reset* allow the control of the FSM activity.

Figure 6.15 represents an example of the evolution of the controlled multiscroll system implemented in FPGA. The hardware results are equivalent to the simulation ones, and so the fixed point algebra and the *sim_full* VHDL entity are suitable for implementing the WCC.

Table 6.2 SPARK board FPGA resources needed to implement the WCC.

Resource	Used
ALUTs	27 %
Memory	34 %
DSP	56 %

Finally, Table 6.2 shows the FPGA resources of the SPARK board used to implement the proposed control architecture.

6.5 Experiments

The proposed action-oriented perceptual architecture was tested both in a simulated environment and in a real arena. The simulation tool used to evaluate the system capabilities is called SPAN [8], and has been already used to compare the reactive

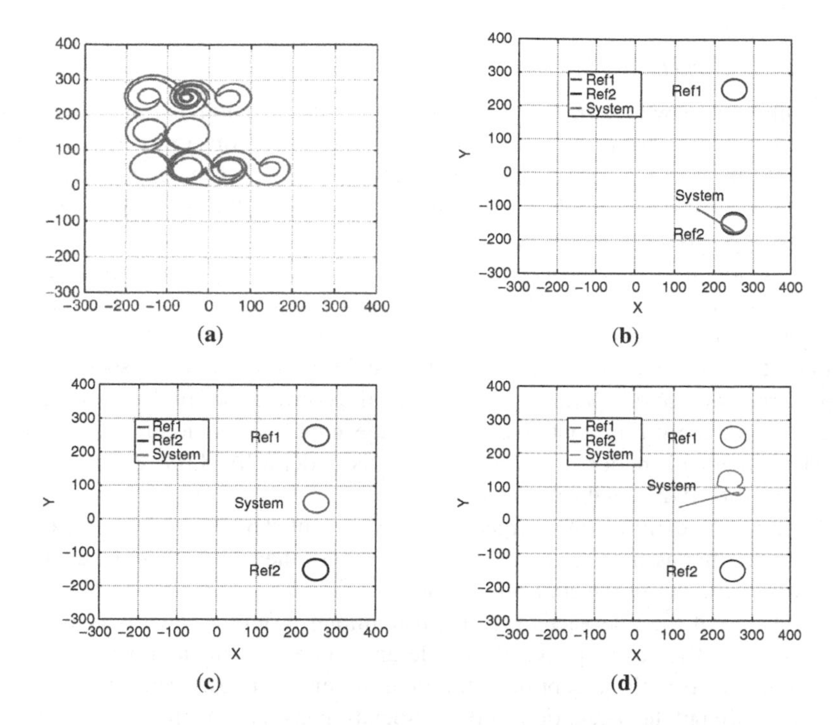

Fig. 6.15 An example of the evolution of the multiscroll system when controlled by reference dynamics associated with sensors. (**a**) When there are no input sensorial stimuli, the system shows a chaotic behavior; (**b**) when a single stimulus is perceived, the system converges to the reference cycle; (**c**),(**d**) when a different stimuli is perceived, the resulting cycle is placed between the references and the position depends on the gain value associated to the references

layer of the WCC approach (i.e., with a deterministic association between internal dynamics and corresponding robot actions) with more traditional approaches to navigation like the potential field (see [6] for more details). Here we want to demonstrate how a robot can learn to associate rewarding behaviors to the different environmental situations encountered, aiming at performing a given task that in the following experiments consists in retrieving targets avoiding obstacles.

6.5.1 Experimental Setup

During the learning phase, a growing set of emerged cycles, arisen in response to different environmental conditions, are associated to suitable actions through the MM algorithm. To evaluate the robot's performances in a quantitative way, the following parameters have been considered:

- P_{new}: cumulative number of new perceptual states that emerge during learning
- *Bumps*: cumulative number of collisions with obstacles

- *Explored Area*: new area covered (i.e., exploration capability)
- *Retrieved Targets*: number of targets retrieved in the environment

Initially each new pattern is associated with a random action, but the continuous emerging of such a pattern leads the action selection network to tune its weights to optimize the association between the perceptual state and the action to be performed. It is also desirable that new patterns occur only during the first learning steps (i.e., epochs). To guarantee the convergence of the algorithm, the learning process cannot be considered as ended while new patterns continue to emerge with a high frequency. Moreover, to solve the robot navigation problem, it is necessary that a pattern occurs several times, as the robot learns by trial and error.

As the term a_{s_q} gives information about the stability of the action associated with the pattern q, we use this parameter to evaluate the convergence of the learning process. The LRU algorithm (that manages the pattern table) was modified to consider a_{s_q}. The pattern q cannot be replaced if its a_{s_q} is under a fixed threshold (AS_{Learn}) that is determined during the design phase.

The code that identifies an emerged cycle is constituted by the four parameters x_q, y_q, $\bar{x}_q$, and $\bar{y}_q$ that assume continuous values, because they depend on the evolution of the state variables of the controlled system.

The choice of the tolerance to distinguish among different patterns is a crucial problem during the design phase. If the tolerance increases, the number of patterns representing the robot's perception of the environment decreases. Then, the learning time is reduced but the perception–action association is more rough. On the contrary, if the tolerance is reduced, the number of actions increases, producing a wider range of different solutions for the navigation task. In this way it is feasible to reach a better solution, but it is more time consuming. Table 6.3 shows the value of the most relevant parameters used during the learning phase. The target sensor equipped on the robot, based on cricket's phonotaxis, is not able to estimate accurately the distance between robot and sound source, for this reason the parameter h_D used in (6.11) has been not considered.

6.5.2 Experimental Results

The environment used to perform tests and comparisons is a $10 \times 10\,\mathrm{m}^2$ room with three obstacles and two targets. The simulated arena and the real arena are shown in Fig. 6.16. Both in simulation and in the real case, a set of five learning trials was performed, with the MM structure randomly initialized. The learning phase is maintained for 1,200 epochs (i.e., actions) that correspond to 40 min for the real robot and a few minutes in the virtual arena. Each robot simulation step (i.e., epoch) corresponds to a single robot action: this is determined simulating the dynamical system for 2,000 steps with an integration step equal to 0.1. These parameters guarantee the convergence of the multiscroll system to a stable attractor (i.e., a cycle) when external stimuli are perceived by the robot; otherwise a chaotic evolution is shown.

Table 6.3 Relevant parameters of the MM-like structure.

RF parameters		Learning parameters	
k_i for frontal distance sensors	15	a_s start value	0.6
k_i for lateral distance sensors	10	a_s decrement factor	0.01
h_A	10	Tolerance	8%

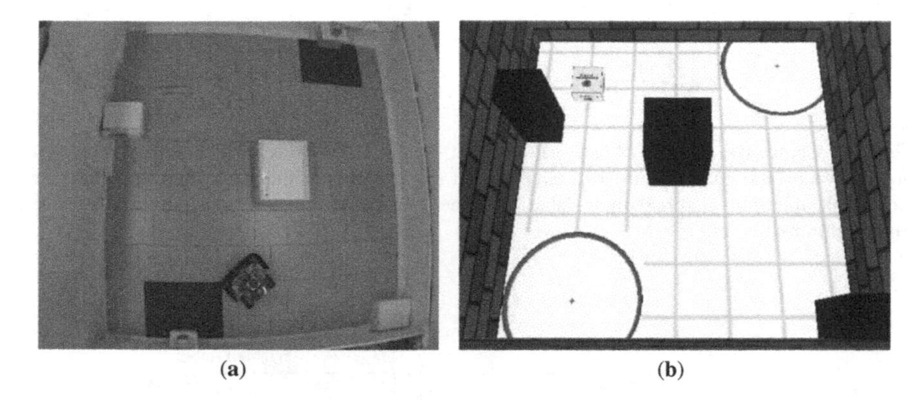

Fig. 6.16 3D view of the environment used during the learning phase in real (**a**) and simulated experiments (**b**). The dimensions of the environments are $10 \times 10\,\text{m}^2$

During the learning phase, a sequence of new patterns, in the form of cycles, emerges and the robot learns how to behave in the current situation. To evaluate the convergence of the learning phase in Fig. 6.17, the trends of the cumulative number of new patterns that arise (P_{new}) is shown in the case of the simulated agent and the roving robot. The learning process leads to a huge improvement of the robot behavior for the situation (i.e., perceptual state) that more often occurs, while some other patterns cannot be suitably learned if they seldom emerge.

After the first learning epoches, the number of new emerging patterns is very low and a steady state condition is reached after 100 epoches. These results were obtained adopting the learning parameters defined in Table 6.3. In particular, the total number of patterns that emerges, around 25–30, is directly related to the tolerance parameter that has been set to 8%. In other simulations carried out, increasing the tolerance factor to 15% reduces the number of emerging patterns to 20%. A reduced number of emerged patterns leads to the speed-up of the learning phase, but decreases the specialization of the robot actions with a consequent reduction of the performances.

a_s is an important parameter to evaluate the efficacy of the learning phase. The value of a_s is directly related to the stability of the association among perceptual state and robot action. The total number of emerged patterns is about 25–30 and

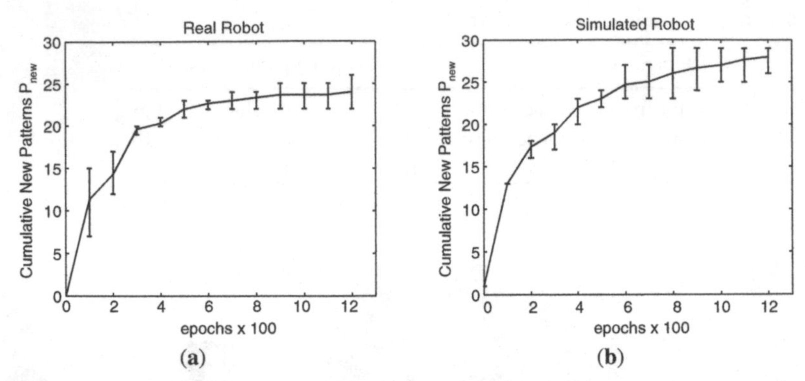

Fig. 6.17 Cumulative number of new patterns that emerge during the learning phase calculated in windows of 100 epochs in the real environment (**a**) and in the simulated arena (**b**). The *bars* indicate the minimum and maximum value while the *solid line* is the mean value of the set of five trials performed

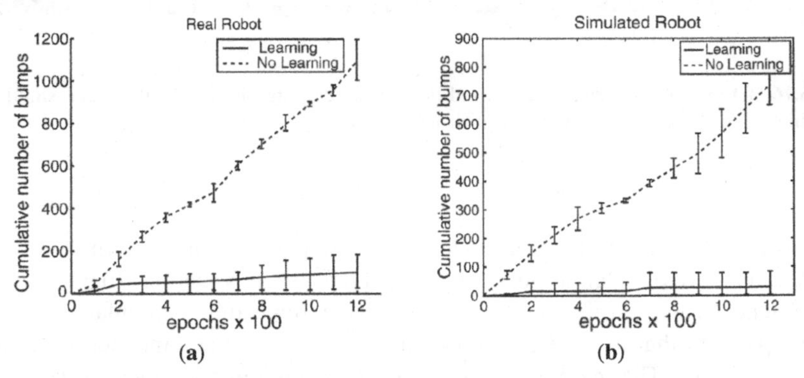

Fig. 6.18 Cumulative value of bumps calculated in windows of 100 epochs in the conditions of learning and no-learning for the real (**a**) and simulated (**b**) environments. The two trends span among the minimum and maximum value for each window

more than 60% of the patterns have an $a_s < 0.5$: this corresponds to more than 100 updates of the associated action following the indication of the reward function, and about the 25% of the patterns have an $a_s < 0.1$ that correspond to more than 500 updates.

The learning process guided by the reward function significantly improves the robot capabilities evaluated in terms of the number of bumps and target retrieved. In Figs. 6.18 and 6.19, the cumulative number of bumps and targets found is shown for the two learning cases, comparing the behavior of the system during learning with the same architecture when the learning is not activated. The results show that a significant difference in terms of performance is evident.

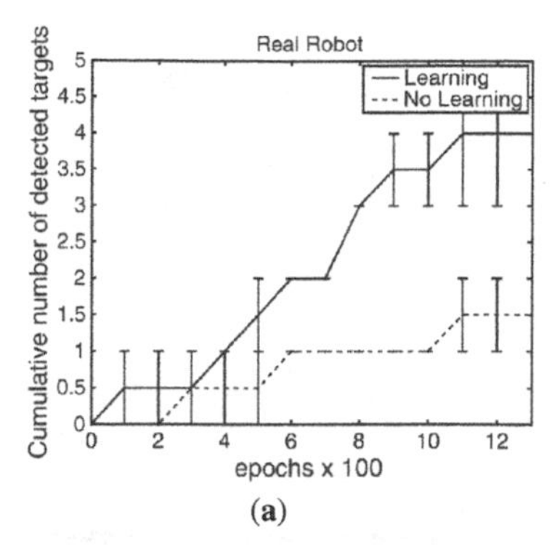

(a)

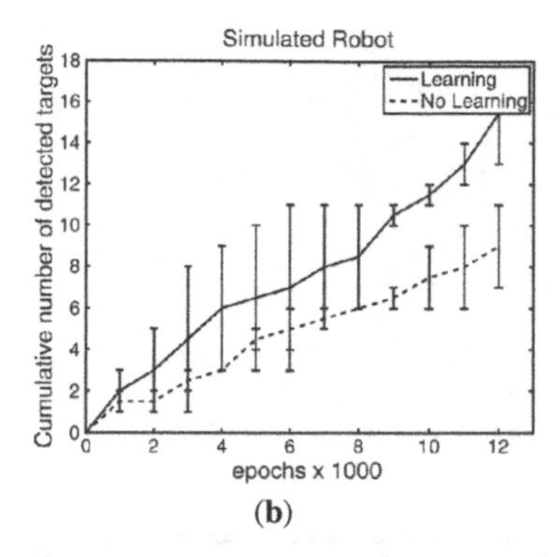

(b)

Fig. 6.19 Cumulative value of retrieved targets calculated in windows of 100 and 1,000 epochs in the conditions of learning and no-learning for the real (**a**) and simulated (**b**) environments. The two trends span among the minimum and maximum value for each window

As far as the avoidance behavior is concerned, the robot learns how to manage the information coming from distance sensors to avoid bumps. The improvement is evident both in simulation and in the real experiments.

Another important index to be considered is the number of targets retrieved. When a target is found by the robot, it is turned off until another target is reached. Because of the limited power supply, the robot experiment was limited to 1,200 epoches, while the simulation was extended to 12,000 epoches to better appreciate the improvement over a more relevant number of target retrieving (see Fig. 6.19). The number of target found by the real robot when the learning process does not work is very low, in fact the robot is often trapped by the obstacles placed in the arena. The WCC performances were also compared with other more traditional navigation algorithms showing similar results [6].

To understand how the robot behavior is modified during the learning process, in Fig. 6.20, the trajectories followed by the robot before and after learning are shown.

Cmparison between the performance with and without learning outlines the capability of the control system to create a suitable link between perception and action.

To further outline the results of the testing phase for a learned architecture, information on the association between perceptual patterns and corresponding actions is reported. In particular, in Fig. 6.21, the final emerged actions associated to the mostly used patterns are shown. Figure 6.21 shows the x-y phase plane of the multiscroll system together with the internal patterns emerged during a learning phase. For sake of clarity, each class is reported only with a marker, indicating its position (i.e., parameters x_q and y_q). The vector associated to each pattern shows module and phase of the corresponding action performed by the robot with respect to the x-axis that indicates frontal direction of the robot motion.

It is important to notice that, in this case, the robot through learning is able to define a series of actions needed in specific situations that can be adopted to accomplish the food retrieval task. The approach can be extended to other tasks,

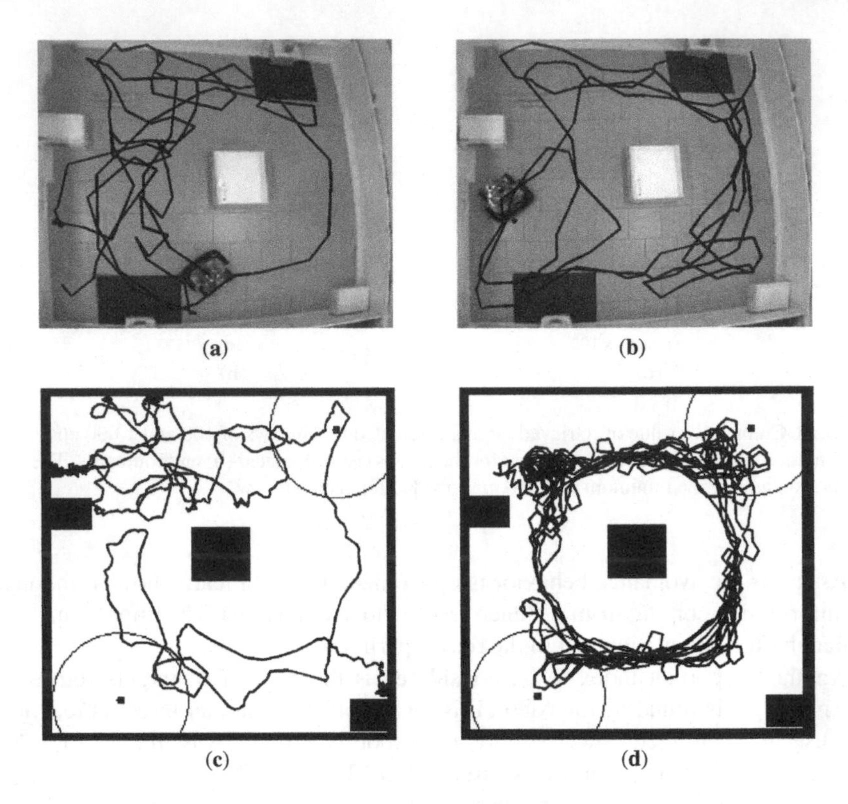

Fig. 6.20 Trajectories followed by the robot during a test in the learning environments. (**a**) and (**b**) Behavior of the robot without learning; (**c**) and (**d**) Trajectories followed after the learning phase

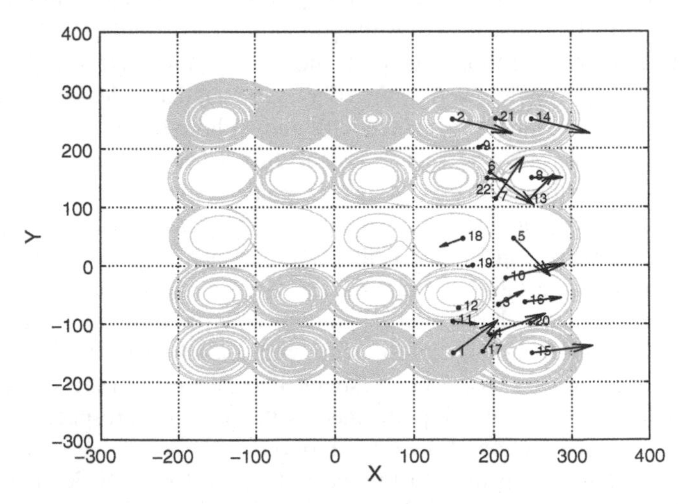

Fig. 6.21 *Action Map* created during the learning phase by the Rover II. Each vector indicates the phase action associated with the emerged pattern indicated only with its position in the phase plane (x_q, y_q)

considering other sensors and also different time scales, associating to the robot's internal state of not only a simple action but also a complete behavior.

Further experiments together with multimedia materials are available on the web [37].

6.6 Summary and Remarks

The introduced strategy is based on the WCC technique, with the addition of a plastically adaptable self learning layer. The technique is based on a state feedback approach and the control gains were here chosen to grant the controlled system stability under stimulation with reference signals (see [6] for details). The term *WCC* refers to a strategy that does not aim to the exact matching between the reference and the controlled signal; instead, the chaotic signal has to collapse to an orbit near the reference signal. At this point, the amplitude value of the control gains is related to the matching degree between the reference and the controlled signal. The control gain value could be therefore used as an additional choice to weight a kind of "degree of attention" that the learning system could pay to the corresponding sensor signal. If learning is used also to choose the control gains, at the end of the learning phase, the robot could be allowed to discard useless sensor signals and to "pay attention" to the important ones. This approach is currently under active investigation. Moreover, the strategy introduced in this chapter was applied to the apparently simple task of autonomous navigation learning. The clear advantages over the classical approaches, for example, related to the potential field, are that the control structure, based on the multiscroll system, is quite general. The fact that the results obtained are comparable with those of the potential field is relevant (see [6] for details). This means that a general approach to learning the sensing–perception–action cycle using the power of information embedding typical of chaotic systems, applied to a traditional task, succeeds in reaching the same results as a technique peculiarly designed to solve that task. The WCC-MM approach can in fact be applied to learn an arbitrary *action-map* or, in general, a *behavior map*. In particular, we exploit the rich information embedding capability of a chaotic system with a simple learning that gives a "meaning" to the embedded information (taking inspiration from [19], within the context of the robot action. This contextualization is decided through the reward function definition. According to the best of the authors' knowledge, it is the first time that the chaotic circuits and system theory, linked to a simple neural learning, is used to approach problems relevant in robot perception, even in the simple case of navigation. In fact, here navigation is treated as a perceptual task. Several other examples are currently under investigation to further generalize the approach and a considerable theoretical effort is being nowadays paid to include the strategy introduced here within a more general scheme for robot perception in unknown conditions and environments.

6.7 Conclusion

In this chapter, a new architecture used to fulfill the sensing–perception–action loop has been proposed. To evaluate the main characteristics of the proposed approach, a case study has been considered: navigation control of a roving robot in unknown environments. The method based on the control of a chaotic multiscroll system allows to synthesize a perceptual scheme in a compact form easy to be processed. Moreover, an action layer has been introduced to associate with the robot's internal states, new behaviors through an unsupervised learning driven by a reward function. The architecture has been implemented in an FPGA-based hardware designed to fulfill the flexibility needed by a cognitive architecture. A completely autonomous roving robot, equipped with a suite of sensors used to perform a food retrieval task, was tested, showing the improvement, in terms of robot capabilities, obtained through a reward-based learning.

As already envisaged, the chaotic dynamical system representing the core of the perceptual layer allows to extend the sensory system in a very simple and modular way. Further developments will include the introduction in the real robot of other kinds of sensors like visual, needed to improve the system capabilities to accomplish the assigned task.

References

1. Alba, L., Arena, P., De Fiore, S., Listan, J., Patané, L., Scordino, G., Webb, B.: Multi-sensory architectures for action-oriented perception. In: Proceedings of Microtechnologies for the New Millennium (SPIE'2007). Gran Canaria, Spain (2007)
2. Altera Corporation: Altera home page. http://www.altera.com.
3. Anafocus sl: Eye-ris producer home page. http://www.anafocus.com.
4. Aradi, I., Barna, G., Erdi, P.: Chaos and learning in the olfactory bulb. Int. J. Intell. Syst. **10**(1), 89–117 (1995)
5. Arena, P., Crucitti, P., Fortuna, L., Frasca, M., Lombardo, D., Patané, L.: Turing patterns in rd-cnns for the emergence of perceptual states in roving robots. Bifurcation Chaos **17**(1), 107–127 (2007)
6. Arena, P., De Fiore, S., Fortuna, L., Frasca, M., Patané, L., Vagliasindi, G.: Reactive navigation through multiscroll systems: from theory to real-time implementation. Autonomous Robots **25**(1–2), 123–146 (2007)
7. Arena, P., Fortuna, L., Frasca, M., Hulub, M.: Implementation and synchronization of 3×3 grid scroll chaotic circuits with analog programmable devices. Chaos **16**(1) (2006)
8. Arena, P., Fortuna, L., Frasca, M., Lo Turco, G., Patané, L., Russo, R.: A new simulation tool for actionoriented perception systems. In: Proceedings of 10th IEEE International Conference on Emerging Technologies and Factory Automation (EFTA'2005), pp. 19–22. Catania, Italy (2005)
9. Arena, P., Fortuna, L., Frasca, M., Lombardo, D., Patané, L.: Learning efference in cnns for perception-based navigation control. In: Proceedings of International Symposium on Nonlinear Theory and its Applications (NOLTA'2005), pp. 18–21. Bruges, Belgium (2005)
10. Arena, P., Fortuna, L., Frasca, M., Pasqualino, R., Patané, L.: Cnns and motor maps for bio-inspired collision avoidance in roving robots. In: 8th IEEE Int. Workshop on Cellular Neural Networks and their Applications (CNNA'2004). Budapest (2004)

11. Arena, P., Fortuna, L., Frasca, M., Patané, L.: Sensory feedback in CNN-based central pattern generators. Int. J. Neural Syst. **13(6)**, 349–362 (2003)
12. Arena, P., Fortuna, L., Frasca, M., Patané, L., Pavone, M.: Towards autonomous adaptive behavior in a bio-inspired cnn-controlled robot. In: Proceedings of IEEE International Symposium on Circuits and Systems (ISCAS'2006), pp. 21–24. Island of Kos, Grecia (2006)
13. Arena, P., Patané, L.: Spatial Temporal Patterns for Action Oriented Perception in Roving Robots. Springer, Berlin (2008)
14. Arkin, R.C.: Behavior-Based Robotics. MIT Press, Cambridge, MA (1998)
15. Boccaletti, S., Grebogi, C., Lai, Y.C., Mancini, H., Maza, D.: The control of chaos: theory and applications. Phys. Rep. **329**, 103–197 (2000)
16. Freeman, W.J.: The physiology of perception. Sci. Am. **264**(2), 78–85 (1991)
17. Freeman, W.J.: Characteristics of the synchronization of brain activity imposed by finite conduction velocities of axons. Bifurcation Chaos **10**(10), 2307–2322 (1999)
18. Freeman, W.J.: A neurobiological theory of meaning in perception. Part I: Information and meaning in nonconvergent and nonlocal brain dynamincs. Bifurcation Chaos **13**(9), 2493–2511 (2003)
19. Freeman, W.J.: How and why brains create meaning from sensory information. Bifurcation Chaos **14**(2), 515–530 (2004)
20. Freeman, W.J., Kozma, R.: Local-global interactions and the role of mesoscopic (intermediate-range) elements in brain dynamics. Behav. Brain Sci. **23**(3), 401 (2000)
21. Gutierrez-Osuna, R., Gutierrez-Galvez, A.: Habituation in the kiii olfactory model with chemical sensor arrays. IEEE Trans. Neural Netw. **14**(6), 1565–1568 (2003)
22. Harter, D., Kozma, R.: Chaotic neurodynamics for autonomous agents. IEEE Trans. Neural Netw. **16**(3), 565–579 (2005)
23. Kohonen, T.: Self-organized formation of topologically correct feature maps. Bio. Cybern. **43**(1), 59–69 (1972)
24. Kozma, R., Freeman, W.J.: Chaotic resonance – methods and applications for robust classification of noisy and variable patterns. Bifurcation Chaos **11**(6), 1607–1629 (2000)
25. Kuniyoshi, Y., Sangawa, S.: Emergence and development of motor behavior from neural-body coupling. In: Proc. IEEE Int. Conference on Robotics and Automation (ICRA'2007). Rome, Italy (2007)
26. Lapidus, L., Seinfeld, J.: Numerical Solution of Ordinary Differential Equations. Academic Press, New York (1971)
27. Lu, J., Chen, G., Yu, X., Leung, H.: Design and analysis of multiscroll chaotic attractors from saturated function series. IEEE T Circuits-I **51**(12), 2476–2490 (2004)
28. Makarov, V.A., Castellanos, N.P., Velarde, M.G.: Simple agents benefits only from simple brains. Trans. Eng. Comput. Tech. **15**, 25–30 (2006)
29. Manganaro, G., Arena, P., Fortuna, L.: Cellular Neural Networks: Chaos, Complexity and VLSI processing. Springer, Berlin (1999)
30. Murray, J.D.: Mathematical Biology. Springer, Berlin (1993)
31. Ott, E., Grebogi, C., Yorke, J.A.: Controlling chaos. Phys. Rev. Lett. **64**(11) (1990)
32. Pyragas, K.: Continuos control of chaos by self-controlling feedback. Phys. Lett. A **170**, 421–428 (1992)
33. Pyragas, K.: Predictable chaos in slightly pertirbed unpredictable chaotic systems. Phys. Lett. A **181**, 203–210 (1993)
34. Reeve, R., Webb, B.: New neural circuits for robot phonotaxis. Phil. Trans. R. Soc. A **361**, 2245–2266 (2002)
35. Ritter, H., Martinetz, T., Schulten, K.: Neural Computation and Self-Organizing Maps. Addison Wesley, Reading, MA (1992)
36. Skarda, C.A., Freeman, W.J.: How brains make chaos in order to make sense of the world. Behav. Brain Sci. **10**, 161–195 (1987)
37. SPARK Project: Spark eu project home page. `http://www.spark.diees.unict.it`
38. Steels, L., Brooks, R.A.: The Artificial Life Route to Artificial Intelligence: Building Embodied Situated Agents. Lawrence Erlbaum Associates, Hillsdale (1995)

39. Vernon, D., Metta, G., Sandini, G.: A survey of artificial cognitive systems: implications for the autonomous development of mental capabilities in computational agents. IEEE T. Evolut. Comput. **11**(2), 151–180 (2007)
40. Webb, B., Scutt, T.: A simple latency dependent spiking neuron model of cricket phonotaxis. Biol. Cybern. **82(3)**, 247–269 (2000)
41. Yalcin, M.E., Suykens, J.A.K., Vandewalle, J., Ozoguz, S.: Families of scroll grid attractors. Bifurcation Chaos **12**(1), 23–41 (2002)

Chapter 7
Nature-inspired Single-electron Computers

Tetsuya Asai and Takahide Oya

7.1 Introduction

A single-electron circuit is one that creates electronic functions by controlling movements of individual electrons [11]. The circuit uses tunneling junctions, each of which generally consists of two conducting materials facing each other very closely (statically, they are normal capacitors). Under a low-temperature environment, electron tunneling is governed by the physical phenomenon called the Coulomb blockade, where an electron does not tunnel through a junction if the tunneling increases the circuit's electrostatic energy (E_c). To comply with the Coulomb blockade, the capacitance of a tunneling junction must be sufficiently small; for example, if we use 1 pF of capacitance, E_c corresponds approximately to 1 mK in temperature (T). Generally, observing the Coulomb blockade in practical experimental environment (e.g., $T \sim 0.1$ K) is difficult because the blockade effect is disturbed by thermal fluctuations. Therefore, elemental devices of single-electron circuits, that is, tunneling junctions and capacitors, must be constructed in nanoscale (lower than a few tens of nanometers).

These intrinsic quantum behaviors may give us an insight in developing modern computing paradigms, including nature-inspired computing and quantum computing. However, if we employ conventional (deterministic) computing architectures, we need a fully worked-out plan for both computing and circuit architectures, for example, see [5, 10, 27, 29, 30]. Thermal noise tolerance is an important characteristic of single-electron computers, because the rate of random electron tunneling increases exponentially as the temperature increases. Several practical circuits have been developed by improving the process for fabricating ultra-low capacitance of tunneling junctions [29] and by using an error-compensation algorithm in the architecture [27].

On the other hand, one can easily observe robust, fault- and noise-tolerant systems in nature. For example, nature-inspired reaction-diffusion (RD) computers [3] can perform specific computing under natural environment without paying much

A. Adamatzky and M. Komosinski (eds.), *Artificial Life Models in Hardware*,

attention to, for example, device-size variations, thermal fluctuations, etc., as compared to present nanoscale semiconductor artifacts for both analog and digital computing. How can we incorporate such robust properties into single-electron circuits? One possible way is to build an electrically equivalent circuit that implements convenient natural systems, for example, RD computers. Complementary metal-oxide-semiconductor (CMOS) and single-electron circuits for image restoration and computation of a Voronoi diagram (VD) have already been proposed in [20, 28]. The other way is to learn from central nervous systems where neurons are fluctuated by thermal noises, as well as single-electron circuits. Neurons would utilize thermal noise to detect weak neuronal signals buried under the noise. The reappearance of neuronal behaviors and matching properties between neurons are really poor compared with semiconductor neural devices; nevertheless, our brains work robustly.

Constructing an electrical analog of natural or biological systems would enable us to generate artificial dynamics on a LSI chip and to develop novel information processing systems. This chapter briefly introduces recent topics on the development of single-electron circuits that perform nonclassical computation inspired by chemical or biological systems. In Sect. 7.2, a novel single-electron device for the computation of a VD is introduced. A cellular-automaton model of VD formation [1, 2] is employed to construct the device that consists of three layers of a 2D array of single-electron oscillators. In Sect. 7.3, a single-electron neural circuit for a robust synchrony detection among burst spikes is presented. A simple single-electron circuit for a single-layer nanodot array is designed for implementing depressing synapses for efficient synchrony detection. The circuit can be used as a unit element for spiking neural networks and its applications. Although the synapse circuit consists of only three single-electron oscillators, they emulate fundamental properties of depressing synapses. This work has been extended to utilize stochastic resonance between single-electron neurons for possible robust computation on single-electron circuits (Sect. 7.4). In Sect. 7.5, a novel semiconductor device in which electronic-analogue dendritic trees grow on multilayer single-electron circuits is introduced as an extreme example of artificial life on single-electron circuits.

7.2 A Single-electron Reaction-diffusion Device for Computation of a Voronoi Diagram

Computation of VD is one of the typical problems in computer science, and VDs are used in graphics, statistics, geography, and economics [14, 18]. The key feature of VD construction is the partition of two- or three-dimensional space on a sphere of influences generated from a given set of objects, points, or arbitrary geometrical shapes. This section introduces a novel single-electron device for the computation of a VD. A cellular-automaton model of VD formation [1, 2] is used to construct the device that consists of three layers of a 2D array of single-electron oscillators.

The authors and several colleagues have proposed to use single-electron reaction–diffusion (SE-RD) devices for VD computation [20]. The original SE-RD

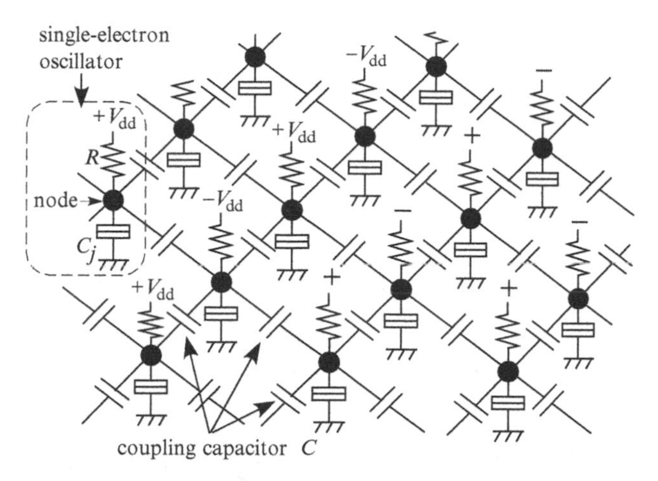

Fig. 7.1 Circuit configuration of single-electron reaction-diffusion device [23]

device consists of arrayed single-electron oscillators and can imitate the operation of chemical RD systems [23]. Figure 7.1 illustrates the original SE-RD device. The main component is a single-electron oscillator that consists of a tunneling junction C_j and a high resistance R connected in series at a node and biased by a positive voltage V_{dd} or a negative one $-V_{dd}$. It has voltage V_{node} of the node, and V_{node} shows the excitatory oscillation that is indispensable for imitating RD systems [23].

To compute a VD with RD systems, spatially localized waves that travel upon computing media at a constant speed are necessary [1, 2], that is, the wavefronts must be smooth and their speed must be constant. The original SE-RD device can generate nonlinear voltage waves. However, the device was not suitable for computing a VD because the wavefronts were not smooth, as shown in Fig. 7.2, and their spreading speeds were not constant. The stochastic tunneling of electrons at each oscillator is the reason why the waves cannot travel at a constant speed. To make the wavefronts smooth and the speed of the waves constant, new oscillators in which the tunneling probability is averaged are necessary. Therefore, the authors have developed new single-electron oscillators with multiple-tunneling junction (MTJ) as shown in Fig. 7.3. The MTJ oscillator consists of a MTJ C_m that has n tunneling junctions and a high resistance R connected in series at the node, and is biased by V_{dd}. It has voltage V_{node} that shows the excitatory oscillation like the original oscillator does. There are many tunneling junctions in the oscillator, and so the tunneling probability is averaged. Consequently, V_{node} changes smoothly as shown in Fig. 7.3b. The improved SE-RD device consists of these MTJ oscillators [20].

Adjacent oscillators have to be coupled with a capacitor for the voltage waves to travel on the MTJ device. Figure 7.4 shows simulation results of a one-dimensional chain of MTJ oscillators. In the figure, MTJ oscillators are denoted by A1, A2,..., with their nodes represented by closed circles that are connected to their adjacent oscillators through intermediary oscillators biased by a negative voltage $-V_{dd}$ (these are denoted by B1, B2, ..., with their nodes represented by open circles) and coupling

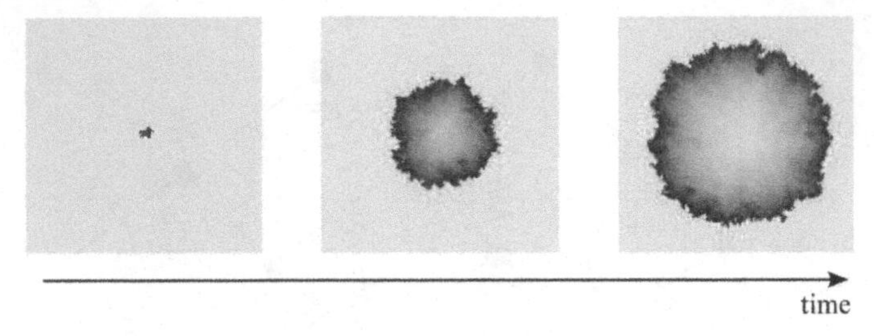

Fig. 7.2 Spatially localized voltage wave that is generated by the original SERD device (simulated). The device has 100 × 100 oscillators. Simulated with parameters: tunneling junction capacitance $C_j = 1$ aF, tunneling junction conductance = 1 μS, high resistance R = 137.5 MΩ, coupling capacitance $C = 1$ aF, bias voltage $V_{dd} = 16.5$ mV, and zero temperature [23]

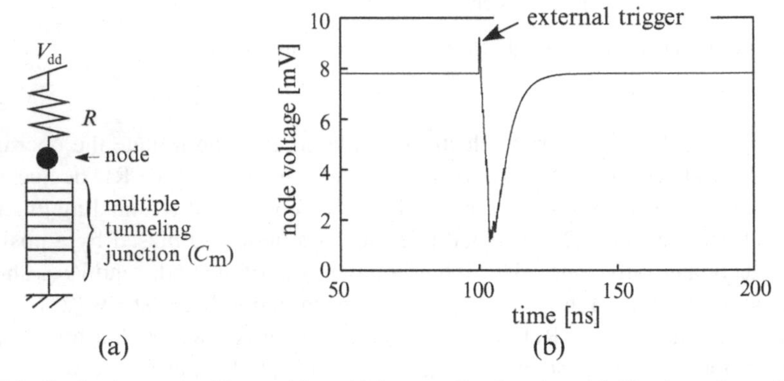

Fig. 7.3 Single-electron oscillator with multiple-tunneling junction. (**a**) Circuit configuration and (**b**) its operation (simulated). Simulated with parameters: tunneling junction capacitance C_m = 10 aF (500 aF/50 junctions), tunneling junction conductance = 5 μS, high resistance R = 20 GΩ, bias voltage $V_{dd} = 7.8$ mV, and zero temperature

capacitors C (Fig. 7.4a). When electron tunneling occurs in one oscillator in this structure, the node voltage of the oscillator decreases gently, and this induces electron tunneling in an adjacent intermediary oscillator. The induced tunneling changes the node voltage of the intermediary oscillator from low to high, and this further induces electron tunneling in an adjacent oscillator. Consequently, changes in node voltage that are caused by the electron tunneling are transmitted from one oscillator to another along the oscillator chain (Fig. 7.4b). Note that the voltage waves travel at almost constant speed because the tunneling probability is averaged over all oscillators.

A SE-RD device can be constructed by connecting MTJ oscillators into a network by means of intermediary oscillators and coupling capacitors, as shown in Fig. 7.5. Each oscillator is connected to its four adjacent oscillators by means of four intermediary oscillators and coupling capacitors. Nonlinear voltage waves travel on the device at a constant speed, as shown in Fig. 7.6. One can compute VDs by using

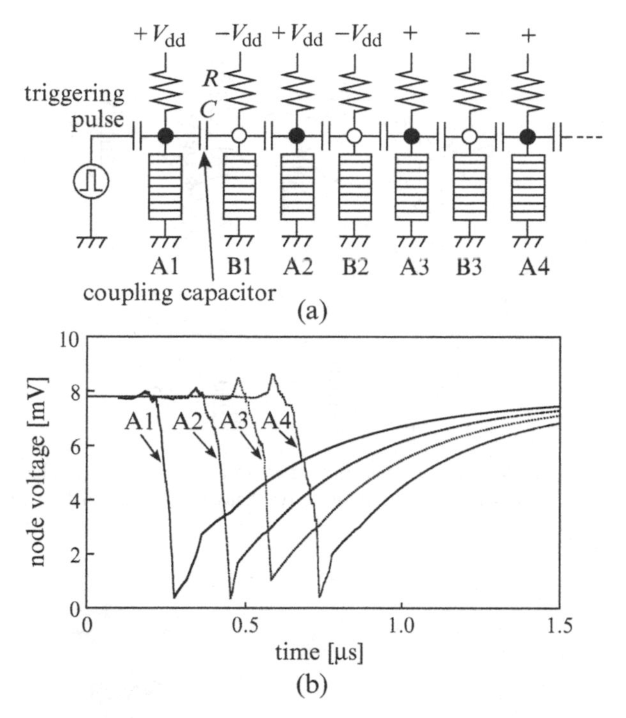

Fig. 7.4 One-dimensional chain of MTJ oscillators. (**a**) Circuit configuration and (**b**) its operation (simulated). Simulated with parameters: tunneling junction capacitance $C_m = 10\,\mathrm{aF}$ (500 aF/50 junctions), tunneling junction conductance = 5 μS, high resistance $R = 20\,\mathrm{G\Omega}$, coupling capacitance $C = 2.2\,\mathrm{aF}$, bias voltage $V_{dd} = 7.8\,\mathrm{mV}$, and zero temperature

the information on the collision points of the nonlinear waves. In [1, 2], a cell that connects eight adjacent cells changes its state according to the states of the adjacent cells. The cell state transition rule is

$$x^{t+1} = \begin{cases} \beta, & \text{if } x^t = \bullet \text{ and } 1 \le \sigma(x)^t \le 4, \\ \alpha, & \text{if } x^t = \beta \text{ and } 1 \le \sigma(x)^t \le 4, \\ x^t, & \text{otherwise,} \end{cases}$$

where x is the state of the middle cell, t is the time step, $\bullet$ is the resting cell, α is the colored precipitate, β is the reagent, and $\sigma(x)^t$ is the number of β cells in the eight adjacent cells. In this model, the collision points are memorized as the precipitate of reagents.

To apply this rule to single electron device, the authors used single-electron threshold detectors, specifically the single-electron boxes (SEB) for logic gate devices [19, 22]. The SEB consists of a single-electron trap (two identical tunneling junctions C_j connected in series, a capacitor C_L, and a bias voltage V_{dd}) as shown in Fig. 7.7a. This circuit has a hysteretic sawtooth function for V_{dd}, as shown in Fig. 7.7b. The authors made use of this characteristic for threshold operation.

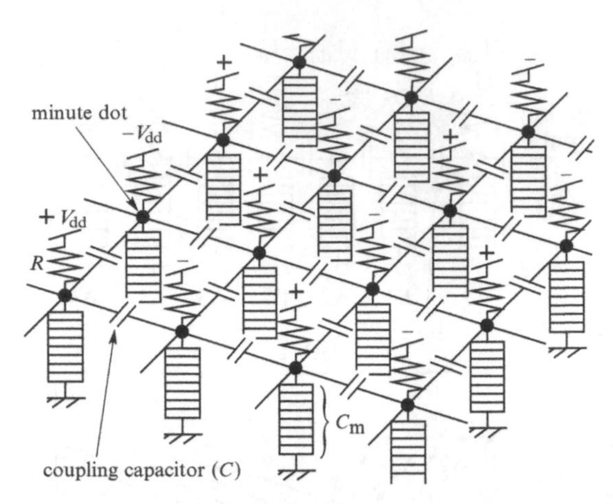

Fig. 7.5 Two-dimensional RD device consisting of network of MTJ single-electron oscillators. Each oscillator is connected with four neighboring oscillators by means of four intermediary oscillators and coupling capacitors

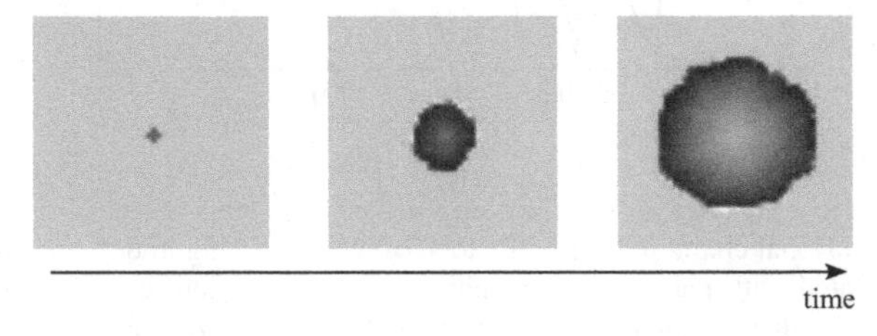

Fig. 7.6 Traveling nonlinear wave that is generated by the improved SE-RD device (simulated). 50×50 oscillators are placed in the device. This simulation used the same parameters as in Fig. 7.4b [20]

Let us consider the threshold operation for computing a VD based on the CA model, and assume the threshold value that is the number of β cells in the eight adjacent cells to be 4.5, that is, no electron tunneling occurs in the SEB when the node voltages of four or fewer adjacent oscillators are changed by electron tunneling in the oscillators. On the other hand, electron tunneling occurs in the SEB when the node voltages of five or more adjacent oscillators are changed. In addition, one can find the collision points by comparing the state of the center oscillator with the state of the SEB threshold detector. To compare the states, SEBs with the threshold set to 1.5 can be used, that is, no electron tunneling occurs in the SEB when electron tunneling occur in both the above SEB and the center oscillator.

Figure 7.8 shows a SE RD device with three layers for computing VDs. The top layer (Fig. 7.8a) is the MTJ device shown in Fig. 7.5. The middle layer (Fig. 7.8b) is the first logic layer of SEB threshold detectors. The SEB that is biased by the negative voltage $-V_{b1}$ (Fig. 7.8e) are directly placed under the oscillators biased by $+V_{dd}$ (oscillator 9 in Fig. 7.8) and connects to the eight adjacent oscillators of the top layer (oscillators 1–8 in Fig. 7.8d), that is, the SEB accepts eight signals from the

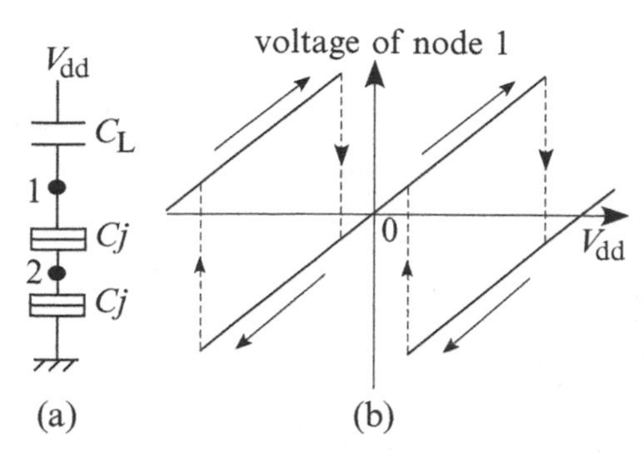

Fig. 7.7 Single-electron box. (**a**) Circuit configuration and (**b**) its operation [19,22]

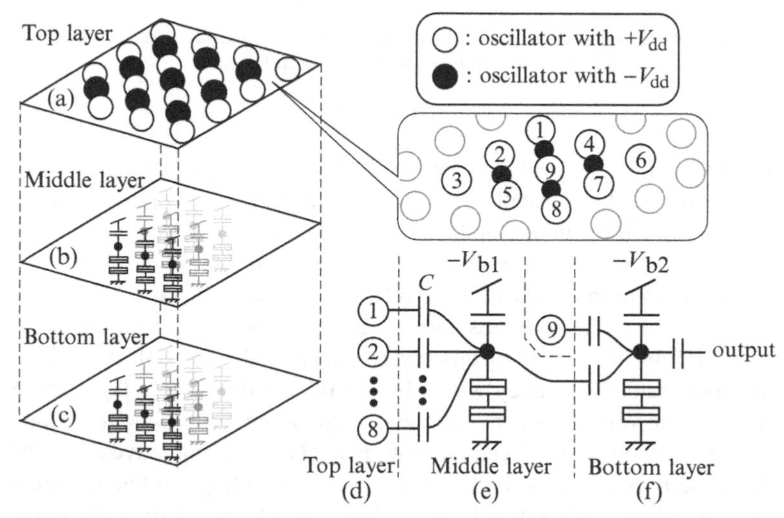

Fig. 7.8 SE RD device that has three layers for computing VDs. The top layer is the device shown in Fig. 7.5. The middle layer is the first logic layer with SEB threshold detectors. The bottom layer is the second logic layer and it produces the VD

eight oscillators as inputs. The bottom layer (Fig. 7.8c) is the second logic layer. The SEB that is biased by the negative voltage $-V_{b2}$ (Fig. 7.8f) connects to the oscillator 9 and the SEB in the second layer (Fig. 7.8e), that is, the second SEB accepts two signals from the oscillator 9 and the SEB in the second layer as inputs. The bottom layer produces the output, that is, its output is used to draw the VD.

Figures 7.9–7.11 show the simulated results. Figure 7.9 shows the density of node voltages on the top layer (bright and dark dots represent high and low voltages on the 2D device). Figures 7.10 and 7.11 show the voltages on the middle layer and on the bottom layer. In Fig. 7.10, "A" indicates the wavefront in the top layer, "B" indicates the wave-front in the middle layer, and "C" indicates collision points. In this simulation, three oscillators of the top layer were triggered as planar points for a VD. Nonlinear voltage waves traveled at a constant speed, and gave the data

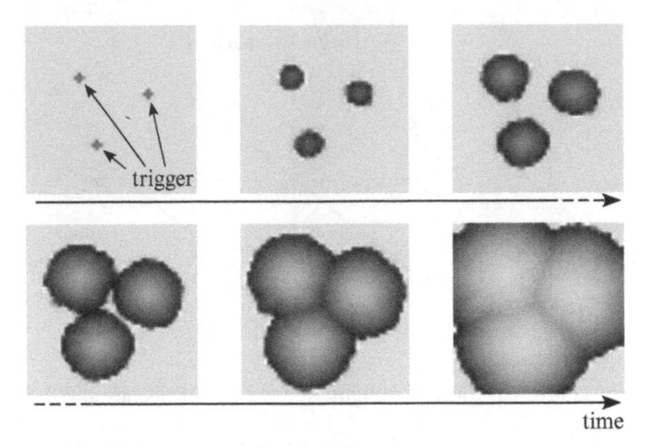

Fig. 7.9 Expanding circular pattern in the top layer of the device. Snapshots for six time steps. The simulation used the same parameters as in Fig. 7.4b for the simulation

to the middle and bottom layers. In the middle layer, the SEBs changed their node voltage when five or more oscillators of the upper eight oscillators changed their node voltage. Wavefronts in the top layer had four or fewer oscillators that changed their voltages. As a result, traveling nonlinear waves in this layer (B in Fig. 7.10) followed the waves in the top layer (A). When wave "A" collided with other waves in the top layer, the collision points had five or more oscillators that changed their voltages. Therefore, wave "B" in this layer overtook "A" and collided with other waves just like spanning a valley with a bridge (C). In the bottom layer, the SEBs changed their node voltages when both the voltage of the oscillators in the top layer and the SEBs in the middle layer are low. Namely, traveling waves that did not collide were memorized by the bottom layer as a high voltage. When the nonlinear waves of the top layer collided with each other, the voltages of the collision points in the top were low and the node voltages of the SEBs in the middle were high. As a result, the node voltages of the SEB in the bottom were kept low. Therefore, the bottom layer memorized the result of computing the VD (Fig. 7.12).

7.3 Neuronal Synchrony Detection on Single-electron Neural Networks

Synchrony detection between burst and nonburst spikes is known to be one functional example of depressing synapses. Kanazawa et al. demonstrated synchrony detection with CMOS depressing synapse circuits [12]. They found that the performance of a network with depressing synapses that discriminates between burst and random input spikes increases nonmonotonically as the static device mismatch is increased [4]. The authors have designed a single-electron depressing synapse and constructed the same network as in Kanazawa's study to develop

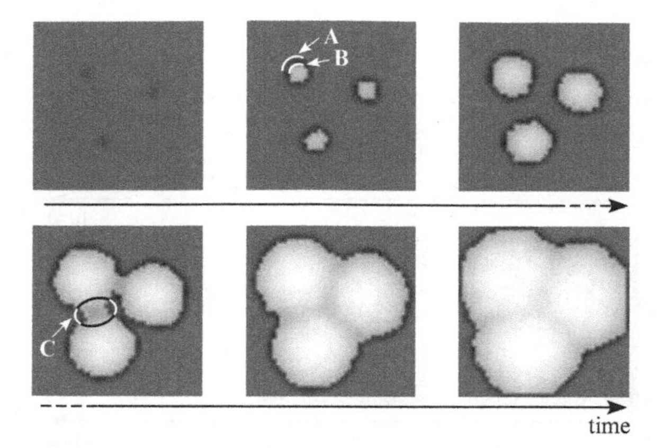

Fig. 7.10 Expanding circular pattern in the middle layer of the device. Snapshots for six time steps. Parameters: tunneling junction capacitance $C_j = 20\,\mathrm{aF}$, tunneling junction conductance = $5\,\mu\mathrm{S}$, bias capacitance $C_L = 10\,\mathrm{aF}$, coupling capacitance $C = 2.2\,\mathrm{aF}$, bias voltage $V_{b1} = 26.5\,\mathrm{mV}$, and zero temperature

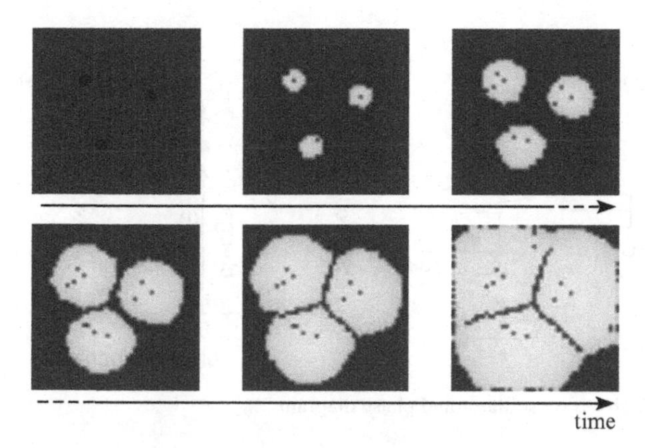

Fig. 7.11 Expanding circular pattern in the bottom layer of the device. Snapshots for six time steps. Traveling nonlinear waves in this layer construct a VD. Parameters are the same as in Fig. 7.10 without bias voltage $V_{b2} = 18.5\,\mathrm{mV}$

noise-tolerant single-electron circuits. This section shows the temperature characteristics, and explores possible architecture that enables single electron circuits to operate over absolute zero temperature.

The authors and several colleagues have proposed neuromorphic single-electron circuits for fundamental neural components in modern spiking neural networks [24]. Our aim was to implement artificial neural networks on a single or multilayer nanodot array. A unit circuit consists of a pair of single-electron oscillators. Using these unit circuits with coupling capacitors, we design a single-electron neuron circuit that consists of excitable axons and dendrites, excitatory and inhibitory synapses, and a soma. The authors have demonstrated an application of the neuron circuit

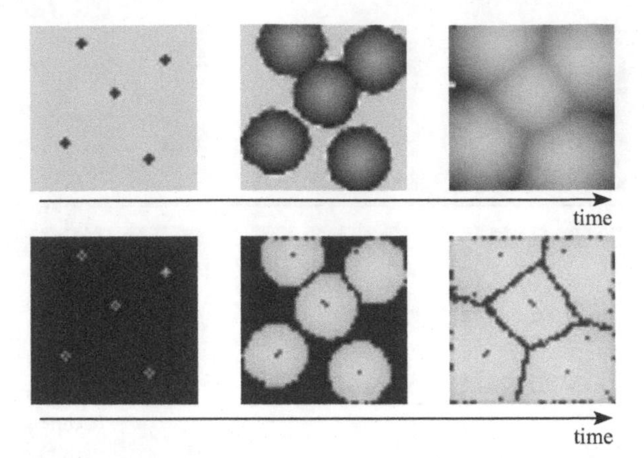

Fig. 7.12 Simulated results of VD computing with five planar points by using proposed device. Upper three snapshots show the voltage density of the top layer and the bottom three snapshots show the voltage density of the bottom layer

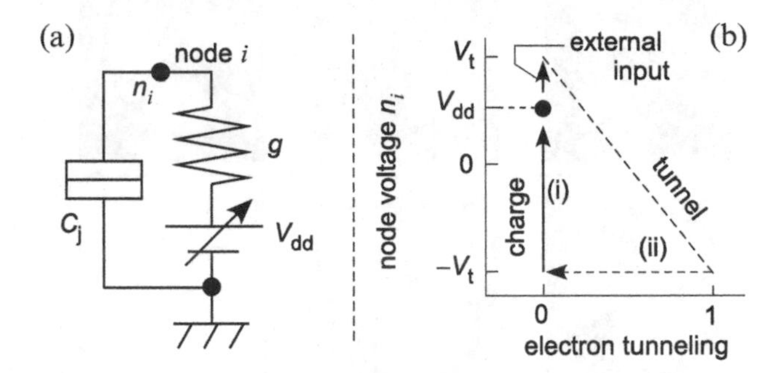

Fig. 7.13 Single-electron oscillator and phase diagram

in an inhibitory competitive neural network [24], where the neurons compete with each other in the temporal domain. However, we could observe expected neural competition at a very low temperature (≤ 0.1 K). In this section, we explore a possible solution to improve the performance in another application, that is, neuronal synchrony detection, by using the proposed single-electron depressing synapse.

To design a depressing synapse circuit, we use a pair of single-electron oscillators that were proposed for a spiking neuron circuit [24] and an excitable media [23]. As shown in Fig. 7.13a, one oscillator consists of a tunneling junction (C_j), a conductive device (g), and a bias voltage source (V_{dd}). The oscillator has an island node n_i where excess electrons are stored. Figure 7.13b is a nominal phase diagram of this circuit for positive V_{dd}. The vertical and horizontal axes represent node voltage n_i and a tunneling phenomenon [= 1 (when an electron tunnels), 0 (else)] at C_j. Note that trajectories between the tunneling phenomenon (0 and 1) in the figure do not have any quantitative physical meaning, but they have been used only to explain this circuit's operation. Let us assume that $V_{dd} < e/2C_j$ ($\equiv V_T$: tunneling threshold voltage of junction C_j). Because tunneling junction C_j is charged by V_{dd} [(i) in Fig. 7.13b], the circuit is stable when $n_i = V_{dd}$. Under this resting condition, if n_i is further increased by an external input and exceeds V_T, an electron tunnels from the ground to node i through junction C_j, which results in a sudden decrease in n_i from V_T to $-V_T$ [(ii) in Fig. 7.13b]. Then, V_{dd} starts charging C_j, and the circuit becomes stable again [(i) in Fig. 7.13b].

Note that there is a time lag from when the junction voltage exceeds V_T to when tunneling actually occurs. On can utilize this "monostable" (excitable) oscillatory property to produce the depressing characteristics of the synapses, that is, we regard an array of oscillators as a depressing synapse because input spike trains are depressed by each neuron operating in its refractory period. Therefore, we can use an array of single-electron oscillators to construct the single-electron depressing synapse (SEDS) shown in Fig. 7.14. It should be noted that the term of the refractory period increases as the values of g_{Na} and g_K increase [23].

A neuromorphic relationship exists between the proposed SEDS and electronic Hodgkin–Huxley (H-H) models: (1) a tunneling junction (C_j) corresponds to a membrane capacitance and voltage-controlled gates in the H-H models, (2) nonlinear chemical reactions between Na^+ and K^+ can be mediated by a coupling capacitance (C) because of the neuron's dielectric inside the soma.

Let us observe the depressing properties of a single SEDS through numerical simulations. Typical parameter values for the single-electron circuit [23], except for $g_{Na}(= g_K) = 5, 2.5$, and $1\ \mu$S were used. Figure 7.15 shows synaptic conductivities (approximately the number of post-synaptic spikes) for inter-spike intervals (ISIs) of input spike trains. As the ISI increases, the conductivity increases because each SEDS can easily be recovered from its depressed (refractory) period as the ISI increase. Because the depressed period increases as g_{Na} and g_K increase, the SEDS's conductivity for increasing ISIs decreases significantly.

Applications of SEDSs to synchrony detection has been demonstrated. We used a typical functional example of depressing synapses proposed by Senn [26]. He showed that an easy way to extract coherence information between cortical neurons is by projecting spike trains through depressing synapses onto a post-synaptic neuron [26]. We demonstrate it here by using single-electron synapse circuits.

Let us assume a simple circuit as shown in Fig. 7.16. The circuit is designed based on the construction of Senn's neural network. The bottom right part represents a post-synaptic neuron and the left part represents its dendrite with our synapse circuits. The post-synaptic neuron consists of a membrane capacitance (C_m) and a leak conductance (g_m). In this study, we omit a threshold (V_{th}) detector from the post-synaptic neuron circuit, that is, the post-synaptic neuron circuit never fires. The post-synaptic neuron accepts spike inputs from excitatory neurons through depressing synapses. If the post-synaptic neuron circuit has a firing function, it outputs a spike when its EPSP $< V_{th}$, and resets the EPSP after the firing. In this setup, the average values of the EPSP increase in proportion to the number of pre-synaptic active neurons. Therefore, it can detect the number of pre-synaptic active neurons

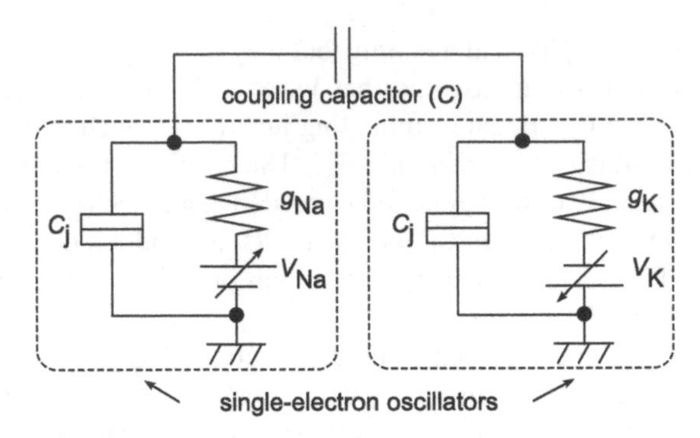

Fig. 7.14 Depressing synapse circuit with single-electron oscillator

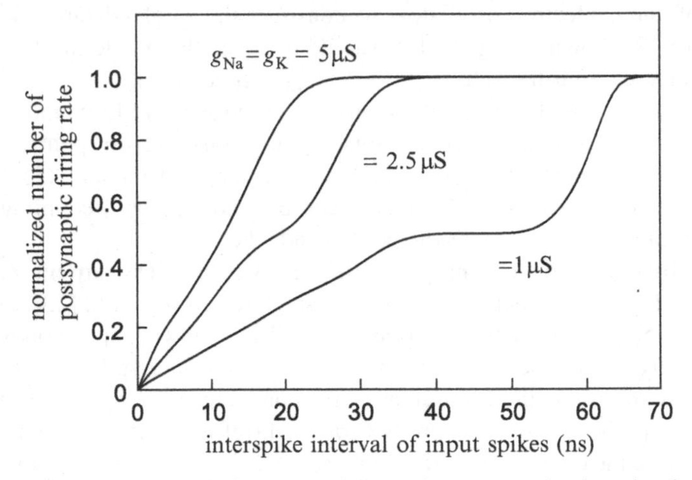

Fig. 7.15 Changes in postsynaptic firing rate of depressing synapse circuit against interspike interval of input spikes

by setting the appropriate threshold V_{th} corresponding to the number of active neurons. On the other hand, the EPSP also increases in proportion to the firing rate of spiking neurons. Therefore, the performance needed to discriminate the number of pre-synaptic active neurons largely deteriorates if the firing rate is not a constant value. During a burst input, the output current of the depressing synapse circuit that flows via a conductance (g') rapidly decreases for successive spikes due to the refractory properties of the single-electron oscillator. But during a nonbursting period, the oscillator has time to be in a resting period, and these results in a strong EPSP at the onset of the next burst. If we compare this dynamic response with that for a nondepressed synapse evoking the same EPSP on average, the depressed synapse will have a larger response at the burst onset and a smaller response toward the end of the burst.

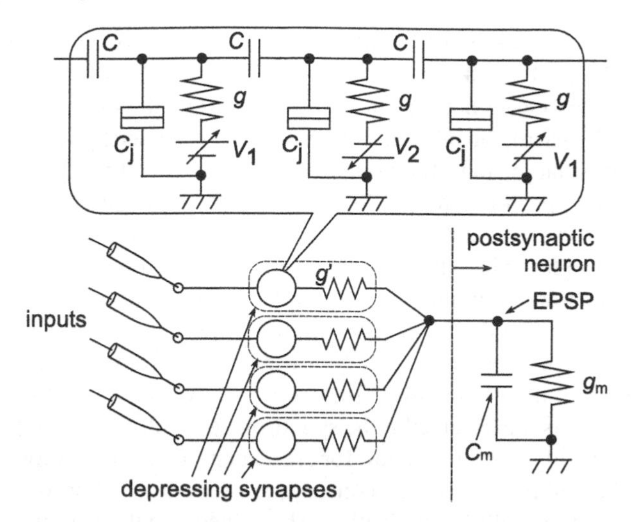

Fig. 7.16 Circuit configuration of depressing synapses with postsynaptic neuron circuit

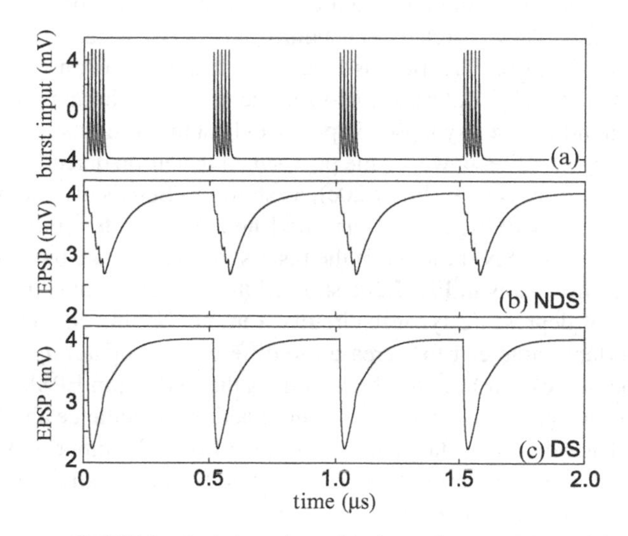

Fig. 7.17 Responses of EPSP for single burst input (**a**) via nondepressed (**b**) and depressed synapse circuit (**c**)

Figure 7.17 shows the response of the EPSP with bursting inputs for (a) a non-depressed synapse (b) and a depressed synapse circuit (c). The results ensure that the EPSP caused by the depressed synapse circuit has a larger response at the burst onset, as compared with a nondepressed synapse circuit.

Let us see that the depressing synapse circuit can detect the synchrony in the burst times. We used two bursting neurons as the input of the post-synaptic neuron that receives the burst inputs through depressed or nondepressed synapses.

Figures 7.18 and 7.19 show the results. When the input bursts were not synchronized (Fig. 7.6a, b), the peak EPSPs evoked by nondepressed [Fig. 7.18c] and depressed synapses (Fig. 7.18d) were both around 3 mV. But, when the input bursts were synchronized (Fig. 7.19a, b), the peak EPSPs evoked by depressed synapses (Fig. 7.19d) were significantly larger than the nondepressed synapses (Fig. 7.19c). Therefore, after defining an appropriate threshold V_{th} of the post-synaptic neuron, for example, $V_{th} = 1.5$ mV in the experiments, the post-synaptic neuron with the depressing synapse circuit can fire when the burst inputs are synchronized.

Figure 7.20a shows a simulated result of 100 neurons driven by random spike trains. According to Senn's report [26], there is experimental evidence to assume that before and during the tone, auditory cortical neurons fire in short bursts, with bursts of three to four spikes at 40–50 ms, repeated every 200–250 ms. During the tone, the burst onsets are assumed to be synchronized within the groups of 70 neurons that are randomly assembled anew for each burst. In our simulations, the overall firing rate of the population remains constant, apart from the short onset and offset of the tone when most cells burst together. This is because the bursting times of the groups alternate during the on-going tone (see Fig. 7.20b).

The neurons respond at the onset and offset by applying a tone stimulus (1.5–4.0 μs in Fig. 7.20). They correlate their bursts only between randomly assembled subgroups during the stimulus. Because the mean firing rate is on the background level during the tone (Fig. 7.20b), a post-synaptic neuron gathering the input spike trains through nondepressed synapses responds only at the stimulus onset and offset. With a depressing synapse, however, the post-synaptic neuron detects the correlated bursts, and then it fires as well (Fig. 7.20c), as shown in Senn's original work.

The difference in EPSP between burst and nonburst inputs represents the network's signal-to-noise (SN) ratio when the task is to discriminate burst spikes from nonburst ones. The results in Fig. 7.20c showed that it was around 1 mV. Note that the parameters of depressing synapse circuits were not optimized well. So what is the most important parameter to increase the difference? Apparently, it is the time constant of the depression because it determines the maximum EPSP, as shown in Fig. 7.19d. The constant is proportional to the junction capacitance and the channel conductance. The other important parameter is the ISI of the input bursting spikes. In the aforementioned simulations, the authors used typical bursting inputs that can easily be generated by external spike generators.

To consider the noise tolerance, we examine Monte-Carlo simulations for the network circuit with typical parameter sets. What we want here is to examine the quantitative difference between the original model [4] and the proposed single-electron circuit to optimize the performance. As described, the performance of the discrimination strongly depends on the SN ratio between the burst and the nonburst spike inputs. Increasing temperature results in an increase in the averaged EPSP. Our interest here is whether the SN ratio is constant or not for increasing the temperature. Of course, we need to recalculate an appropriate threshold for the discrimination. The following shows that the performance is definitely increased by increasing the temperature; however, all of the parameter sets are not optimized.

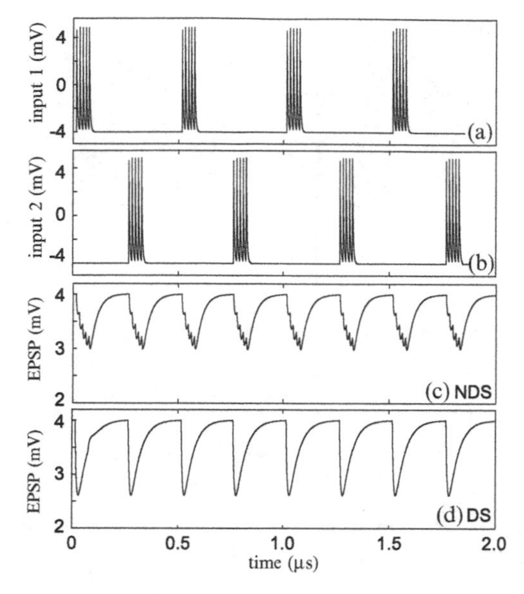

Fig. 7.18 Responses of EPSP for asynchronous burst input [(**a**) and (**b**)] via nondepressed (**c**) and depressed synapse circuit (**d**)

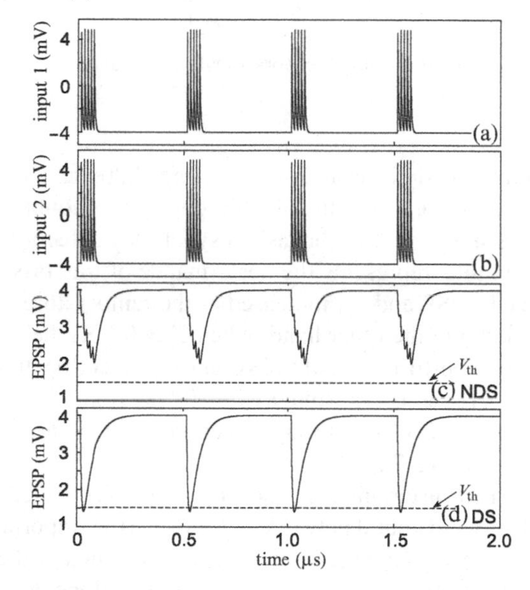

Fig. 7.19 Responses of EPSP for synchronous burst input [(**a**) and (**b**)] via nondepressed (**c**) and depressed synapse circuit (**d**)

To investigate the noise tolerance of Senn's network with our circuits, the 100-neuron network was simulated. To evaluate the noise tolerance, we calculate the difference between the averaged EPSP for the bursting and nonbursting periods and

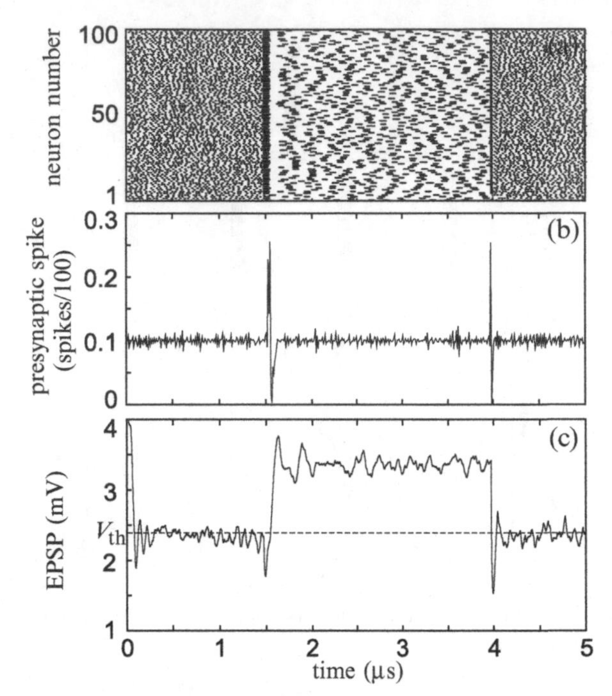

Fig. 7.20 Simulation results of 100-neuron network simulated by random spike trains through our depressing synapses

the threshold V_{th} that was defined as 2.4 mV in Fig. 7.20c as shown in Fig. 7.21. Ideally, the post-synaptic neuron must not fire during the nonbursting period but rather during the bursting period for the task of synchrony detection. The difference between the numbers thus represents the performance of this task. The difference between the averaged EPSP and V_{th} increased as the temperature increased during the nonbursting period. On the other hand, when $T > 0.5$ K, the difference started increasing. Namely, the performance of the synchrony detection did not change significantly due to an increase in T as long as $T < 0.5$ K. Remarkably, the difference (approximately performance of synchrony detection) changed nonmonotonically as T increased, as shown in Fig. 7.22.

The post-synaptic neuron circuit does not yet have any firing mechanism, because the circuit is a feed-forward neural network and because the important value for discriminating the burst spikes from the nonburst spikes is whether the EPSP is lower than the threshold or not. However, for the visualization alone, a significant difference is evident between the original model [4] and our circuit in Fig. 7.20c, where the EPSP is oppositely represented due to the lack of firing (discharging the membrane capacitance). The results shown in Fig. 7.22 indicated that the performance increased up to 0.5 K in the simulations when the temperature was increased. In our previous work [24], the maximum temperature for desired competitive operation was 0.1 K. Needless to say, the temperature is not enough to operate at room temperature. However, the phenomena where the performance increases monotonically

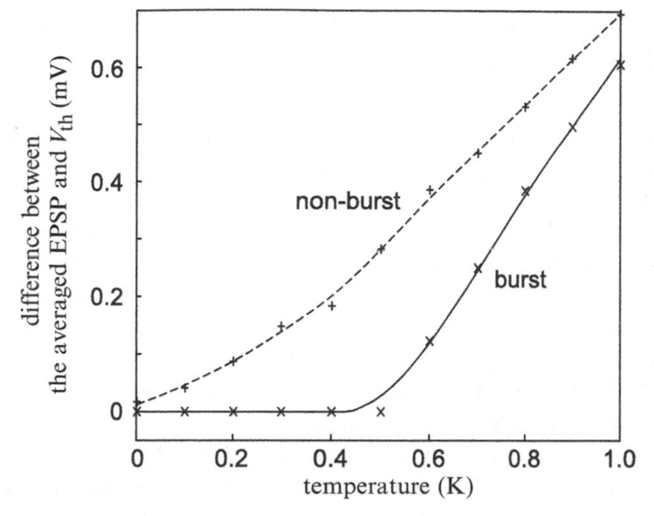

Fig. 7.21 Changes in the difference between the averaged EPSP and the threshold V_{th} during bursting and nonbursting period as a function of temperature

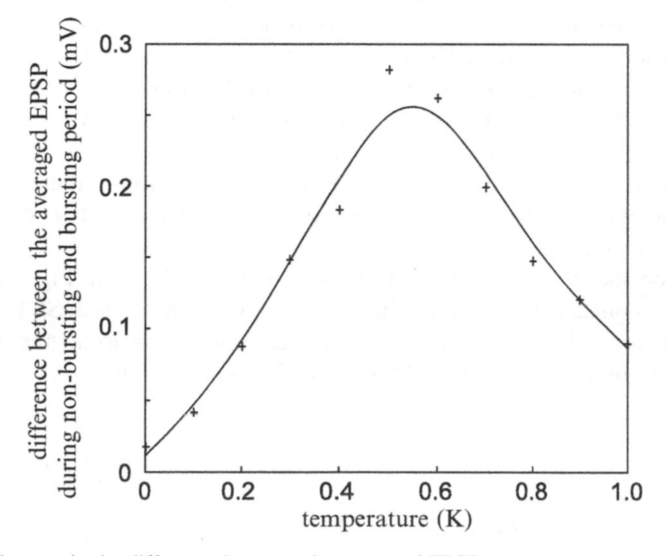

Fig. 7.22 Changes in the difference between the averaged EPSP during bursting and nonbursting period as a function of temperature

as the temperature increases have different physical meanings for conventional stochastic resonance. We are currently optimizing the device and environmental parameters to make the phenomena clear and to optimize these parameters.

7.4 Stochastic Resonance Among Single-Electron Neurons on Schottky Wrap-Gate Devices

Neuromorphic computing based on single-electron circuit technology has become widely noticed because of the recent claim about its massively increased computational efficiency and its increasing relevance between computer technology and nanotechnology. Its impact will be strongly felt when single-electron circuits based on a fault- and noise-tolerant neural structure are able to operate in a room-temperature environment. To fabricate such robust single-electron devices, the authors investigated stochastic resonance (cf. [8]) in an ensemble of single-electron boxes (SEB) [21]. We employed a single-electron transistor (SET) on a schottky wrap-gate (WPG) device [13], instead of a SEB, as a neuron, and examined statistical results of the network by numerical simulation.

The reason why we employ WPG-SETs instead of SEBs is that SETs have a switching characteristic as CMOS transistors do. A general SET consists of a capacitor (C) for an input terminal and two tunneling junctions (C_js) as shown in Fig. 7.23. A SET has three terminals, and one can connect controllable voltage sources to the terminals. Now let us connect terminals with bias V_d, input V_g, and ground, respectively, and control the switching characteristic by controlling the voltage sources. When the SET is in operation, an electron can tunnel through two C_js (between ground and a node, or between a node and V_d) in a low-temperature environment, because electron tunneling is governed by the physical phenomenon called the Coulomb blockade effect. In addition, one can easily observe the operations of practical SET devices. However, we must also be careful when we use single-electron circuits in a high-temperature environment. The reason is that the electrons randomly tunnel through C_js because the Coulomb blockade effect is disturbed by thermal fluctuations.

Let us consider SR among N SETs in a network, as shown in Fig. 7.23. When SETs are not connected with each other, electron tunneling in each SET's junction occurs independently. As in [8], we apply a common input to all the SETs and calculate the sum of outputs of the SETs. For simplicity, we apply a common input

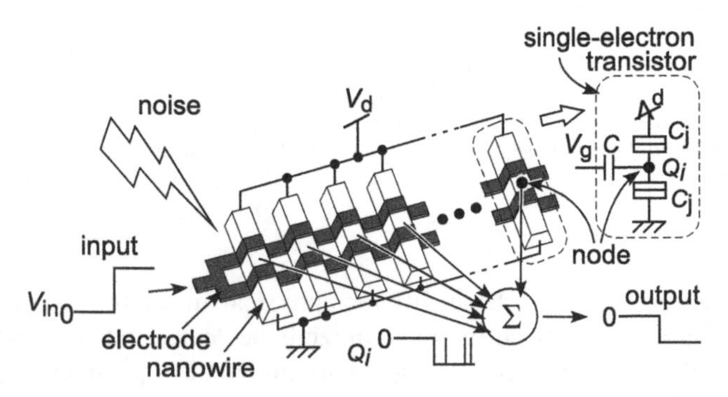

Fig. 7.23 Schematic image of SET array

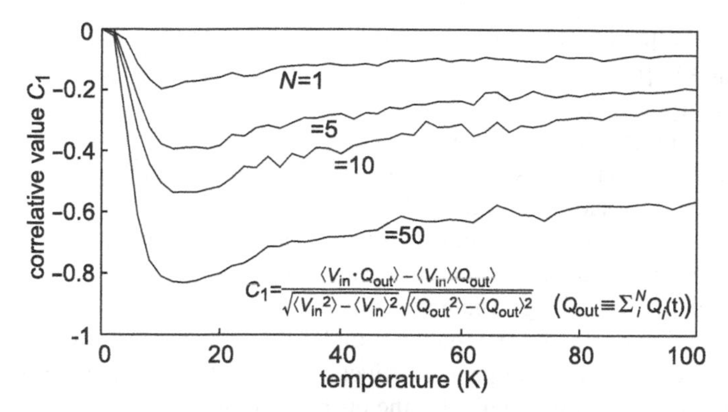

Fig. 7.24 Stochastic resonance in ensemble of single-electron transistors

voltage V_{in} to all the SETs as V_g, and do not consider practical circuits that calculate the sum of the changes in each node's electric charge Q_i. The ith SET's charge (Q_i) was increased with the input voltage, while the magnitude of the input was set to a very low value so that no electron would tunnel through C_js. Under this condition, increasing the magnitude of thermal noise (temperature) enables electrons to tunnel through each C_j.

Figure 7.24 shows simulation results of an ensemble of SETs for $N = 1, 5, 10$, and 50. The temperature was increased from 0 to 100 K and correlation values (C_1) between the input voltages and the summed output were calculated. The results showed characteristic signatures of SR-type behavior: a rapid rise to a peak, and then a decrease at high temperatures. One could observe that the magnitude of $|C_1|$ increased as N increased, as expected. The resonant temperatures were approximately 10 K for all the values of N. In addition, C_1 took a large value -0.6 at 100 K when $N = 50$, and increased as N increased. According to [8], the correlation value should become almost 1.0 when $N = 1{,}000$. The results indicate that when one employs such an SR network in single-electron circuits, it certainly acts as a transmission line that can cancel noises on the line, as well as, cancel the devices' intrinsic noises.

7.5 Single-electron Circuits Performing Dendritic Pattern Formation with Nature-inspired Cellular Automata

Ordered complex patterns can easily be observed everywhere in the natural world. Among these, bifurcated and branched patterns formed in open systems often serve as a basis for advanced functional structures. Indeed, these structures are essential for performing particular computational tasks in nature, for example, structures of a neuron's dendritic tree are responsible for various intelligent computing tasks. Recent advances in neuroscience have revealed that fundamental roles of these dendritic trees include not only the transmission of neuronal signals but also functional computation utilizing multiple properties of membranes and spines (early works can be found in [6, 15]). To incorporate the functions performed by dendritic trees into neuromorphic hardware, the authors and several colleagues developed a single-electron circuit that self-organizes spatial dendritic patterns on a multilayer nanodot array. As the first step, let us see a cellular automaton (CA) model based on a behavioral model of bacteria colonies [17].

It is difficult to implement a huge amount of physical wiring, that is, axons and dendrites, on a 2D semiconductor chip because the wiring is fabricated by stacking several layers of only wiring. Again we here use single-electron circuits, which are believed to have potential for next-generation VLSIs, to increase the wiring density. We also actively incorporate quantum effects and sensitivity to thermal noise into the design of compact unit circuits for the proposed CA.

One of the features of neurons is the complexity of their forms, such as tree-like, branching dendritic form. The details of dendritic pattern formation in neural systems have been mainly studied from the viewpoint of molecular biology, rather than that of general physics. This kind of branching pattern is also observed in many other systems, including trees, crystal growth, protoplasmic streaming tubes of slime molds, bacterial colonies, etc. Because of generality of these kinds of patterns, several models have been proposed to describe complex branching patterns [7, 16, 32]. One of the best-known simple models is the diffusion limited aggregation (DLA) model [32]. Another well-known type of model is the RD model for the pattern formation of bacterial colonies, which exhibit more diverse patterns than the DLA model [16]. Two methods are used to describe RD systems; one is based on partial differential equations (PDE) [31] and the other on discretized CA [9]. Space, time, and state variables are generally discrete in the CA model, whereas they are continuous in the PDE representation of the RD system. Here we employ CA representation of RD dynamics based on bacterial colony pattern formation to represent dendritic patterns because of the variety of patterns available and its expansion for device applications.

The skeleton of the RD pattern formation model of a bacterial colony consists of movement/schism of bacteria, diffusion of nutrients, and consumption of nutrients by the bacteria. In the model, the dynamics are described as "reaction," the

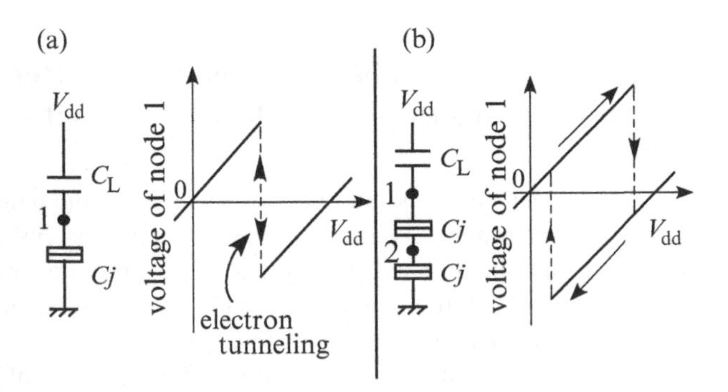

Fig. 7.25 Single-electron devices. (**a**) single-electron box, and (**b**) single-electron memory circuit

relationship of the bacteria and the nutrient, and "diffusion," which averages the bacteria or nutrient in the neighborhood. In CA models, the targeted space is divided into discrete areas of named cells. The time evolution of the state of each cell is decided by simple inner- and inter-cell rules, and a dynamic pattern or whole structure is generated from these local interaction rules. The algorithm of our CA model is as follows. Three variables are used to describe the state of the system in each cell; active bacteria (activator) a, inactive bacteria (inactivator, a trace of activator) w, and nutrient (substrate) f. The variables a and w take the digit value $\{0,1\}$, and f takes a digit or multivalue. When there are both a sufficient number of activators in the cell's neighborhood and sufficient substrate in the cell, the state of the activators becomes 1 ($a : 0 \rightarrow 1$), and substrate f is depleted in the next step. After ($a \rightarrow 1$), when the substrate in the cell itself is less than the threshold value as a result of depletion, the activator can no longer sustain an active state; the state of the activators becomes 0 ($a \rightarrow 0$), and the state of the inactivator becomes 1 ($w \rightarrow 1$). The substrate diffuses constantly with fluctuation. When there is insufficient substrate, the activators try to take up the limited substrate. As a result, the cluster of activators divides into several clusters, and a branching pattern appears as the cluster of cells, where inactivator $w = 1$.

To imitate the diffusion of the consumed substance with fluctuation in the model, we use an SE-RD device [23], SEBs, and SEM circuits [22] to implement the CA rules. A typical single-electron circuit consists of tunneling junctions, resistances, and capacitors. A tunneling junction that is similar to a capacitor is the main component of a single-electron circuit. In a junction, a quantum effect occurs. A point of difference between tunneling junctions and normal capacitors is that the two conductors of the junction face each other very closely. The junction has a threshold voltage value for the generation of a quantum effect that is electron tunneling. A single-electron passes through the junction when the junction potential is over the threshold voltage, and the potential of the junction changes suddenly. The tunneling event has a probability of occurring, given by

$$P(E) \sim \frac{1}{1 - \exp[-\Delta E/(k_{\mathrm{B}}T)]} \tag{7.1}$$

where P is the tunneling probability, E is the charging energy, k_{B} is the Boltzmann constant, and T is the temperature. The equation has a temperature factor. Therefore, the tunneling probability changes with the temperature. We would like to utilize this physical phenomenon to implement fluctuation in the diffusional operation of the CA model. The circuit configuration and example operations of an SE-RD device have already been shown in Figs. 7.1 and 7.2, respectively.

We use SEBs to change the input signals to binary signals. The left of Fig. 7.25a shows the circuit configuration of an SEB. It consists of a tunneling junction, bias capacitor, and bias voltage source. The right of Fig. 7.25a shows a sample operation. The SEB shows a positive voltage (logical 1) when no electron tunneling occurs, and a negative one (logical 0) when electron tunneling occurs. We also use SEMs as memory devices because they have a hysteretic function as a function of the input voltage. The left of Fig. 7.25b shows the circuit configuration of an SEM. It consists of two tunneling junctions, a bias capacitor, and a bias voltage source in series. The

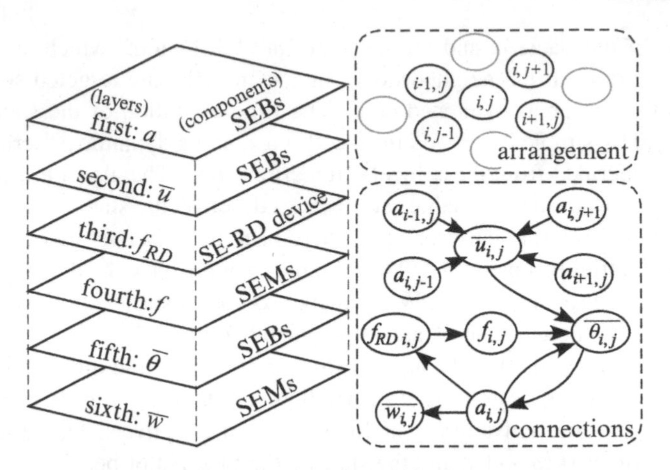

Fig. 7.26 Single-electron device for generation of dendritic patterns

right of Fig. 7.25b shows a sample operation [22]. The SEBs and SEMs play the role of active and inactive bacteria, respectively.

We design a new single-electron device based on the CA model, the SE-RD device, the SEBs, and the SEMs. The device consists of six layers (Fig. 7.26). We add two factors "$u_{i,j}$" and "$\theta_{i,j}$" to the device as supplementary functions. The first layer consists of arrayed SEBs that implements the "$a_{i,j}$" of the model. The second also consists of arrayed SEBs that implement the added factor "$u_{i,j}$." The third is the SE-RD device, and the fourth consists of arrayed SEMs. The third layer implements "$f_{i,j}$" in collaboration with the fourth layer. In this device, f takes a digit value. The fifth layer consists of arrayed SEBs that implement the added factor "$\theta_{i,j}$," and the sixth consists of arrayed SEMs that implement the "$w_{i,j}$" of the model. The unit circuits of each layer are assumed to be cells of the CA. The operations of each factor are represented by the following equations:

$$a_{i,j} = \begin{cases} 1 \text{ (if } \theta_{i,j} = 1), \\ 0 \text{ (otherwise)}, \end{cases} \tag{7.2}$$

$$u_{i,j} = \begin{cases} 1 \text{ (if } a_{i-1,j} + a_{i+1,j} + a_{i,j-1} + a_{i,j+1} = 1), \\ 0 \text{ (otherwise)}, \end{cases} \tag{7.3}$$

$$f_{\mathrm{RD}\ i,j} = \begin{cases} \mathrm{TN} \text{ (if } a_{i,j} = 1 \text{ or neighbor } f_{\mathrm{RD}} = \mathrm{TN}), \\ \mathrm{NT} \text{ (otherwise)}, \end{cases} \tag{7.4}$$

$$f_{i,j} = \begin{cases} 0 \text{ (if } f_{\mathrm{RD}\ i,j} = \mathrm{NT} \text{ or } f_{i,j} = 0), \\ 1 \text{ (otherwise)}, \end{cases} \tag{7.5}$$

$$\theta_{i,j} = \begin{cases} 1 \text{ (if } \overline{a_{i,j}} \cdot u_{i,j} \cdot f_{i,j} = 1), \\ 0 \text{ (otherwise)}, \end{cases} \tag{7.6}$$

$$w_{i,j} = \begin{cases} 1 \text{ (if } a_{i,j} + w_{i,j} = 1), \\ 0 \text{ (otherwise)}, \end{cases} \tag{7.7}$$

where "TN" and "NT" represent "tunneling" and "no tunneling," respectively.

In the simulation, the device had 200×200 elements in each layer. Figure 7.27 shows sample operations of the elements in each layer. In Fig. 7.27a, $a_{i,j}$ maintained a logical 0 state (negative voltage) until $\theta_{i,j}$ changed its 0 state to 1 (positive voltage). $u_{i,j}$ kept 0 state until neighbor a changed its 0 state to 1. $f_{i,j}$ ($f_{RD\ i,j}$) maintained 1 state until $a_{i,j}$ changed its 0 state to 1 or electron tunneling occurred in the neighboring f_{RD} and kept 0 state after it changed its state from 1 to 0. $w_{i,j}$ kept 1 state until $a_{i,j}$ changed its 0 state to 1 and kept 0 state after it changed its state from 1 to 0. Thus, the device implemented the CA model. However, operating errors sometimes occurred because of the tunneling probability in the device. Figure 7.27b shows sample operations. In the figure, p and q represent different points from i and j. Neighbors a, $u_{p,q}$, and $f_{p,q}$ operated well as did a, u, and f in Fig. 7.27a. $\theta_{p,q}$, however, showed a failed operation because of the tunneling probability. As a result, $a_{p,q}$ and $w_{p,q}$ could not change their state. This failed operation works as the diffusion of the consumed nutrient with fluctuation in the model. Figure 7.28 shows the results of a two-dimensional simulation. In the simulation, a spatio-temporal pattern was formed on the sixth (w) layer. The pattern grew from a planar point, but some parts of the growing points stopped because of both the correct and failed operations in each layer. As a result, a dendritic pattern appeared. This dendritic pattern will change with every simulation because of the tunneling probability.

7.6 Summary and Future Works

This chapter introduced four nature- or bio-inspired single-electron circuits for non-classical computation, that is, computation of a Voronoi diagram (VD), burst or nonburst spike detection, weak signal transmission based on stochastic resonance (SR) under noisy environment, and dendritic pattern generation based on an artificial model of bacteria colonies.

First, a SE-RD devices for computing a VD was introduced. The novel SE-RD device consists of three layers. The top layer is an improved SE-RD device in which nonlinear voltage waves are generated and travel, and the middle and bottom layer are threshold detectors. The operations of the middle and bottom layer are based on the CA model [1, 2]. The bottom layer outputs the results of computing a VD by using data from the top and middle layers.

Second, a single-electron depressing synapse and its characteristics was introduced for considering possible applications on noise-tolerant synchrony detection. Previous works on CMOS VLSI showed that the network had great noise-tolerant ability for static noise embedded as device (threshold) mismatches of MOSFETs [4]. We expanded this notion to dynamic ones that are usually a common problem in the area of single-electron circuits. The results showed that the performance is greatly increased by increasing the temperature until $T \leq 0.5$ K [25]. However, all the parameter sets are not optimized. The performance is apparently sensitive to the time constant of a single-electron oscillator and interspike intervals of input burst spikes. Our next goal is an appropriate theory for the emergence of the noise tolerance and

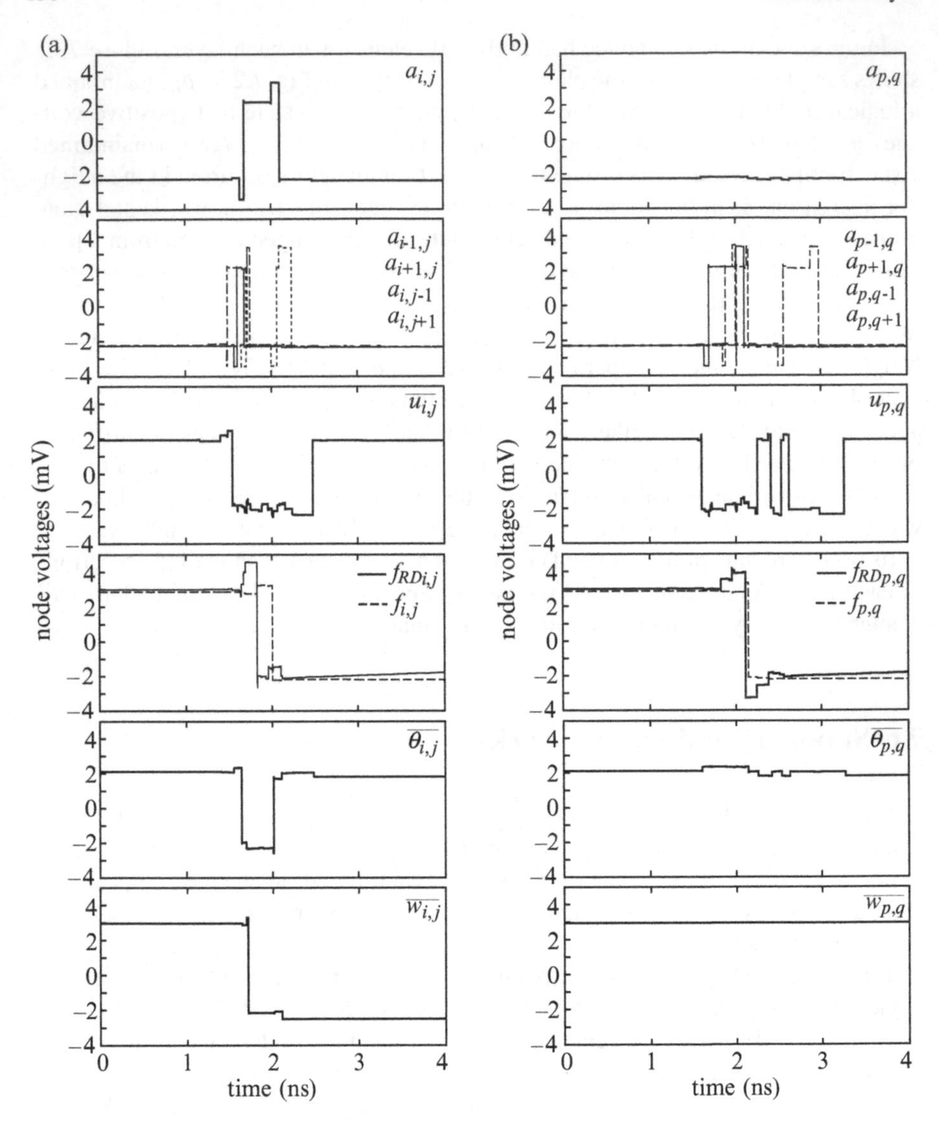

Fig. 7.27 Simulation results. (**a**) correct operations, (**b**) failed operations

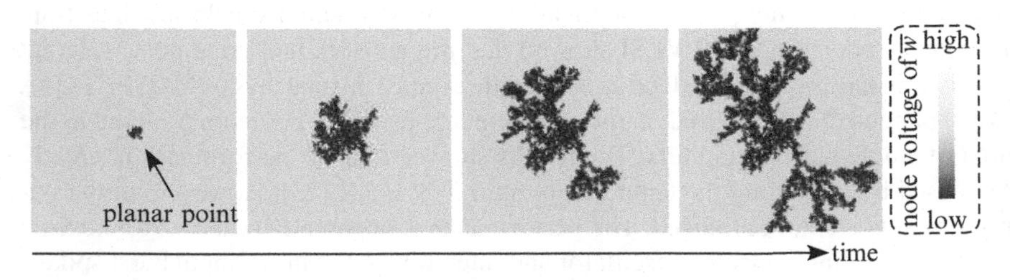

Fig. 7.28 Result of two-dimensional simulation of sixth layer

optimization of these parameters to explore the possible development of fault and noise-tolerant single electron computing devices.

Third, SR in an ensemble of single-electron neuromorphic devices was introduced. Recently, the authors and several colleagues have proposed a single-electron competitive neural network based on SR in an ensemble of single-electron boxes that can operate at room temperature [21]. Using realistic physical parameters, we confirmed the SR behavior of single-electron boxes. The resonant temperature was 20 K, independent of the number of boxes (N).

Finally, novel single-electron circuits for forming dendritic patterns were introduced. To construct the proposed device, we designed a six-layer single-electron circuit with nature-inspired cellular automata. The utilized cellular automaton had three factors. In the device, the top, second, and fifth layers consist of arrayed single-electron boxes, the fourth and sixth layers consist of arrayed single-electron memory circuits, and the third layer is an SE-RD device. Each layer described each factor of the CA rule with randomness. As a result, the device formed dendritic patterns in the sixth layer. These patterns will change with variations in the circuit parameters or temperature environment.

Recent progress in nanotechnology has certainly accelerated by advances in nanoscale processing, for example, elemental logic gates and memory cells for single-electron LSIs have been proposed in the literature, and significant reports of their fabrication have appeared. However, many problems concerning both static and dynamic "noises" still exist for practical use of single-electron circuits. Future works in this field is thus to find a clever way to cancel (or exploit if possible) the effects of thermal fluctuations in terms of circuit architecture, instead of improving nanoscale device fabrication technologies. Recently, neuromorphic computing based on single-electron circuits is gaining prominence because of its massively increased computational efficiency and the increasing relevance of computer technology and nanotechnology. The maximum impact of these technologies will be strongly felt when single-electron circuits based on fault- and noise-tolerant neural structures can operate at room temperature.

References

1. Adamatzky, A.: Reaction–diffusion algorithm for constructing discrete generalized voronoi diagram. Neural Networks World **6**, 635–643 (1994)
2. Adamatzky, A.: Voronoi-like partition of lattice in cellular automata. Math. Comput. Modelling 51–66 (1996)
3. Adamatzky, A., De-Lacy-Costello, B., Asai, T.: Reaction Diffusion Computers. Elsevier, Amsterdam (2005)
4. Asai, T., Kanazawa, Y., Hirose, T., Amemiya, Y.: A mos circuit for depressing synapse and its application to contrast-invariant pattern classification and synchrony detection. In: Proc. 2004 Int. Joint Conf. Neural Networksn, p. W107 (2004)
5. Benioff, P.: Quantum mechanical models of turing machines that dissipate no energy. Phys. Rev. Lett. **48**(23), 1581–1585 (1982)
6. Blackwell, K.T., Vogl, T.P., Alkon, D.L.: Pattern matching in a model of dendritic spines. Network Comput. Neural Syst. **9**(1), 107–121 (1998)

7. Carmeliet, P., Tessier-Lavigne, M.: Common mechanisms of nerve and blood vessel wiring. Nature **436**(7048), 193–200 (2005)
8. Collins, J.J., Chow, C.C., Imhoff, T.T.: Stochastic resonance without tuning. Nature **376**(6537), 236–238 (1995)
9. Gerhardt, M., Schuster, H., Tyson, J.J.: Cellular automaton model of excitable media II: Curvature, dispersion, rotating waves and meandering waves. Physica D **46**(3), 392–415 (1990)
10. Gershenfeld, N.A.: Bulk spin-resonance quantum computation. Science **275**(17), 350–356 (1997)
11. Gravert, H., Devoret, M.H.: Single Charge Tunneling – Coulomb Blockade Phenomena in Nanostructures. Plenum, New York (1992)
12. Kanazawa, Y., Asai, T., Ikebe, M., Amemiya, Y.: A novel cmos circuit for depressing synapse and its application to contrast-invariant pattern classification and synchrony detection. Int. J. Robotics Automation **19**(4), 206–212 (2004)
13. Kasai, S., Jinushi, K., Tomozawa, H., Hasegawa, H.: Fabrication and characterization of gaas single electron devices having single and multiple dots based on schottky in-plane-gate and wrap-gate control of two-dimensional electron gas. Jpn. J. Appl. Phys. **36**(3B), 1678–1685 (1997)
14. Klein, R.: Concrete and Abstract Voronoi Diagrams. Springer, Berlin (1990)
15. Mel, B.W.: Nmda-based pattern discrimination in a modeled cortical neuron. Neural Comput. **4**(4), 502–516 (1992)
16. Mimura, M., Sakaguchi, H., Matsushita, M.: Reaction–diffusion modelling of bacterial colony patterns. Physica A **282**(1–2), 283–303 (2000)
17. Motoike, I.N.: Simple modeling of branching pattern formation in a reaction diffusion system with cellular automaton. J. Phys. Soc. Jpn **76**(3), 034,002 (2007)
18. Okabe, A., Boots, B., Sugihara, K., Chiu, S.N.: Spatial Tesselations: Concepts and Applications of Voronoi diagrams. Wiley, Chichester (2000)
19. Oya, T., Asai, T., Amemiya, Y.: Single-electrron logic device with simple structure. Elec. Lett. **39**(13), 965–967 (2003)
20. Oya, T., Asai, T., Amemiya, Y.: A single-electron reaction-diffusion device for computation of a voronoi diagram. Int. J. Unconventional Comput. **3**(4), 271–284 (2007)
21. Oya, T., Asai, T., Amemiya, Y.: Stochastic resonance in an ensemble of single-electron neuromorphic devices and its application to competitive neural networks. Chaos, Solitons and Fractals **32**(2), 855–861 (2007)
22. Oya, T., Asai, T., Fukui, T., Amemiya, Y.: A majority-logic device using an irreversible single-electron box. IEEE Trans. Nanotech. **2**(1), 15–22 (2003)
23. Oya, T., Asai, T., Fukui, T., Amemiya, Y.: Reaction-diffusion systems consisting of single-electron circuits. Int. J. Unconventional Comput. **1**(2), 177–194 (2005)
24. Oya, T., Asai, T., Kagaya, R., Hirose, T., Amemiya, Y.: Neuromorphic single-electron circuit and its application to temporal-domain neural competition. In: 2004 Int. Symp. Nonlinear Theory and its Application, pp. 235–239 (2004)
25. Oya, T., Asai, T., Kagaya, R., Hirose, T., Amemiya, Y.: Neuronal synchrony detection on signle-electron neural network. Chaos, Solitons and Fractals **27**(4), 887–894 (2006)
26. Senn, W., Segev, I., Tsodyks, M.: Reading neuronal synchrony with depressing synapses. Neural Comput. **10**(4), 815–819 (1998)
27. Shor, P.W.: Scheme for reducing decoherence in quantum computer memory. Phys. Rev. A **52**(4), 2493–2496 (1995)
28. Suzuki, Y., Takayama, T., Motoike, I.N., Asai, T.: Striped and spotted pattern generation on reaction-diffusion cellular automata – Theory and lsi implementation. Int. J. Unconventional Comput. **3**(1), 1–13 (2007)
29. Takahashi, Y., Nagase, M., Namatsu, H., Kurihara, K., Iwadate, K., Nakajima, K., Horiguchi, S., Murase, K., Tabe, M.: Fabrication technique for si single-electron transistor operating at room-temperature. Elec. Lett. **31**(2), 136–137 (1995)
30. Tucker, J.R.: Complementary digital logic based on the coulomb blockade. J. Appl. Phys. **72**(9), 4399–4413 (1992)

31. Turing, A.M.: The chemical basis of morphogenesis. Phil. Trans. R. Soc. Lond. **237**(641), 37–72 (1952)
32. Witten, T.A., Sander, L.M.: Diffusion-limited aggregation, a kinetic critical phenomenon. Phys. Rev. Lett. **47**(19), 1400–1403 (1981)

Chapter 8
Tribolon: Water-Based Self-Assembly Robots

Shuhei Miyashita, Max Lungarella, and Rolf Pfeifer

Self-assembly is a process through which an organized structure spontaneously forms from simple parts. This process is ubiquitous in nature, and its amazing power is documented by many fascinating instances operating at various spatial scales. Despite its crucial importance, little is known about the mechanisms underlying self-assembly and not much effort has been devoted to abstract higher level design principles. Taking inspiration from biological examples of self-assembly, we designed and built a series of modular robotic systems consisting of centimeter size autonomous plastic tiles capable of aggregation on the surface of water. According to the characteristics of the modules composing them, the systems were classified as "passive," "active," and "connectable." We conducted experiments specifically aimed demonstrating the power of behavioral representation of each system with respect to the level of autonomy of its components. We focused mainly on the effect of the morphology (here shape) of the modules, in particular on the yield of the self-assembly process.

8.1 Introduction

Since the breakthrough discovery of DNA by James Watson and Francis Crick more than 50 years ago, there have been enormous advances in our understanding of the material bases of life. Despite the progress, the emergence of life from the local interactions of components (such as molecules, proteins, and cells) remains one of the big mysteries of modern science. Most discussions about life still gravitate around in-depth descriptions and modeling of the interactions between molecules and cells (e.g., through steady-state representations such as reaction pathways). The actual dynamical processes underlying these interactions, however, remain mostly unknown.

The main assumption of this chapter is that by taking a closer look at the region separating life from nonlife, it may be possible to get a better understanding of how

A. Adamatzky and M. Komosinski (eds.), *Artificial Life Models in Hardware*,

the interactions between nonliving entities (such as atoms and molecules) can give rise to life. For instance, the formation of the complex symmetrical protein shells of spherical viruses is a well-studied example of self-assembly. The shell of the T4 bacteriophage (so-called because it infects bacteria) is composed of hundreds of parts and it is not plausible to assume that the instructions for its construction are only contained in the virus' genetic material. This virus consists of about 70 different kinds of proteins and exploits the metabolism of the host cell (e.g., *E. Coli*) to generate around 100 of itself in approximately 40 min. Even more surprising is the fact that if the right kinds of proteins are mixed, the virus can be synthesized in vitro [1, 2], which is truly remarkable for such a complex system.

Although the discussion of whether viruses are living things or not has been controversial ever since, typically, viruses are considered to be nonliving entities, because they cannot reproduce without the help of another organism. As the research documented above shows, the rules that guide the interaction at a local level are simple; the interactions of a large number of entities can lead to the emergence of complex structures, through a process of self-assembly.

8.2 Self-Assembly Robots

Manufacturing technologies and industries heavily rely on robots. For macroscopic objects, industrial robots are not only economical but are also reliable, fast, and accurate. They are widely used in production processes (i.e., pick-and-place car manufacturing robots) and other situations where their environment as well as their task are well defined and no unexpected situations occur. However, as the assembled objects become more complex, conventional engineering technologies hit on a complexity barrier, which entails lower yields and higher fabrication costs.

Recent advances in robotics have, therefore, pointed out the importance of self-assembly for building complex objects, aiming at exploiting the obvious advantages of living organisms, such as their capability to self-configure or self-heal. Modular robots – autonomous machines consisting of typically homogeneous building blocks – promise a viable solution because they have the ability to be highly versatile. For instance, at least ideally, they can reconfigure and adapt their shape according to a given task-environment. Many attempts have been made to realize self-assembling and self-reconfigurable systems. Work has been mainly focused on the design and construction of basic building blocks of a typically small repertoire, with docking interfaces that allow transfer of mechanical forces and moments, and electrical power, and that can also be used for data communication.

The essential issues of how to develop cellular (modular) robotic systems were already described 20 years ago by Fukuda and his collaborators [3, 4]. This work was followed by a more detailed analysis of the kinematics and the metamorphic design of a modular robot inspired by amoebae [5]. Several different types of modular robotic systems, which could morph into desired target configurations (e.g., a snake or a quadruped), were designed by Murata et al. [6–8]. Locomotion of a reconfigurable modular robot was investigated by Yim [9]. Kotay et al. developed a robotic module which could aggregate as active three-dimensional structures that could move and change shape [10]. Another self-reconfigurable robot was proposed by Rus et al. [11]. This robot could form aggregates by expanding and contracting its shape. A self-contained modular system (it included its own processor, power supply, communication system, sensors, and actuators) was designed by Castano et al. [12]. These modules were designed to work in groups as part of a large configuration. A similar approach can be seen in the work presented by Jorgensen et al. [13], which consists of several fully self-contained robot modules. An autonomous modular robot capable of physical self-reproduction using a set of cubes was created by Zykov et al. [14].

All the robots described above can rearrange the connectivity of their structural units to create new topologies to accomplish diverse tasks. Because the units move or are directly manipulated into their target locations through deliberate active motion, the modular system is called "deterministically self-reconfigurable." The implication is that the exact location of the unit is known all the time, or needs to be calculated at run time.

By contrast, self-assembly systems are "stochastically self-reconfigurable" implying that (1) there are uncertainties in the knowledge of the modules' location (the exact location is known only when the unit docks to the main structure) and (2) the modules have only limited (or no) computational (deliberative) abilities. Pioneering experiments on artificial self-replication were conducted by Lionel and Roger Penrose almost 50 years ago [15]. They presented a mechanical model of natural self-replication in a stochastic environment. Followed by Hosokawa's work [16, 17], speculations about the clustering pattern of passive elements were conducted. The Whitesides group revealed different types of self-assembly at small scales [18–21]. Notable ideas about conformational switch were proposed by Saitou [22].

To date, a few self-reconfigurable modular robots relying on stochastic self-assembly have been built (White et al. [23, 24]; Shimizu et al. [25]; Bishop et al. [26]; Griffith et al. [27]). Although in all these systems the units interact asynchronously and concurrently, a certain amount of state-based control is still required for the modules to move, communicate, and dock. Such docking/undocking is one of the main challenges towards the realization of self-assembly systems. Except from molecular self-assembly (see following paragraph), to our knowledge there have not yet been any attempts at exploring stochastic macroscopic self-assembly systems where the parts have only limited (or no) computational (i.e., deliberative) abilities.

It is plausible to assume that self-assembly can also lead to innovation in applications such as macroscale multirobot coordination and manufacturing technologies for microscale devices. By taking tools and methods from nature, many inroads have already been made to utilize self-assembly for the fabrication of structures at molecular scales [28–32]. While they are designing DNA for the elements, Yokoyama assigned molecules for the assembly tasks [33].

8.2.1 The "ABC Problem"

For modular systems smaller than a few centimeters, there are three fundamental problems that still await a solution. These problems relate to actuator, battery (or power in general), and connector technology (henceforth, referred to as the "ABC problem"). When designing systems where a high quantity of modules of small size is desired, solutions for these problems are of particular relevance. First, actuation endows the parts with the ability to move and reconfigure. A common solution is to use electrical servo motors. These actuators, however, are typically big and heavy. Other means of actuation have also been proposed, e.g., pneumatic actuators. Although they are lightweight, they require a source of compressed air (e.g., a compressor). The second problem is concerned with providing power to the actuator(s). A typical solution is to use batteries. Batteries, however, are problematic, because they are only able to provide power for a limited period of time. Furthermore, their initial charge may vary which leads to heterogeneously actuated modules. Another popular solution involves propagating current through the binding locations. Unfortunately, this solution has the drawback that the alignment of the connecting points has to be very precise. In addition, such modules cannot segregate from the main structure, which prohibits this way of powering for mobile type robots. The third problem is the connection mechanism enabling the modular parts to dock to each other. Binding is crucial for reorganization and for a desired structure to hold. The most common ways of binding are magnets and mechanical latches.

There is a strong interdependency between these issues. The requirements of the connection mechanism as well as the actuator are partly determined by the weight of each module. The heavier the modules, the more force needs to be applied to the binding location. In addition, the actuators have to apply larger torques to displace the modules. The use of more powerful components in general leads to even heavier modules. Also, the power consumption increases as a result of stronger connection mechanisms and actuators. Surprisingly, small size and weight reduction of modular parts is not a good way to solve this problem, because not only does the power/weight ratio of the most common actuators decrease with a reduction in size, but also so does the strength/weight ratio of common connectors. This implies that the most common ways of actuating, powering, and connecting modular robots cannot be applied to small-sized entities. It follows that novel solutions to the ABC problem are necessary to make progress in small-scale modular robotics.

8.3 Tribolon: Water-Based Self-Assembly Robots

In this section, we introduce our three water-based self-assembly systems called Tribolon[1]: a "passive" system, an "active" system, and a "connectable" system (in Sects. 8.3.1, 8.3.2, and 8.3.3, respectively). The first two systems float on water, whereas the third one is immersed in water.

[1] The name is derived from Tribology. Movies and pictures are available from `http://shuhei.net/`

8.3.1 Passive Tile Model

The first prototype of our self-assembly robot was developed in 2005. The model consists of flexible plastic tiles (base plate) floating in a water tank (Fig. 8.1). Figure 8.1 a shows one of the proposed floating tiles (0.2 g, 22 × 22 × 2 mm). We constructed three different types of tiles: squares, circles, and squares with rounded corners (rounded squares). A vertically oriented magnet was attached to the bottom of each base plate such that either the North pole or the South pole was directed upwards (depending on the color of the elements). To produce turbulence on the water surface and to provide the system with the energy necessary to drive the self-assembly process, we attached two vibration motors to the sides of the water tank.

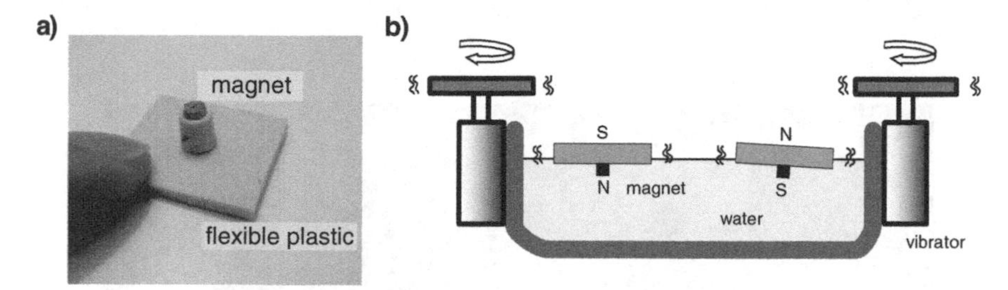

Fig. 8.1 The experimental setup for assembly of passive tiles. (**a**) A vertically oriented magnet was attached to the bottom of each base plate. (**b**) The stirring of water generates random, fluctuating forces providing the system with the necessary energy for assembly

For the experiments, we built 30 blue tiles (N) and 30 pink tiles (S) for each shape. In what follows, we summarize the results of our experiments as a function of the type of shape used.

- *Square* (Fig. 8.2 a): The tiles change their relative positions by sliding along their edges. The connection between two tiles is strong, that is, a relatively large amount of energy is required to break the connection. The clusters formed are stable and rather static, that is, the modules tend to stay in the same configuration for a prolonged period of time.
- *Circle* (Fig. 8.2 b): The tiles easily change their relative positions by moving along the edges of tile pairs. In most cases, several small groups are formed. The speed of aggregation is high, but the structures created are unstable and easily disintegrate due to turbulence; the results are configurations which wax and wane.
- *Rounded square* (Fig. 8.2 c): This type of tile combines the positive characteristics of the squared and the circular tile; that is, stable connections and flexible change of position. The lattice structure is reached rapidly and is stable for a prolonged period of time.

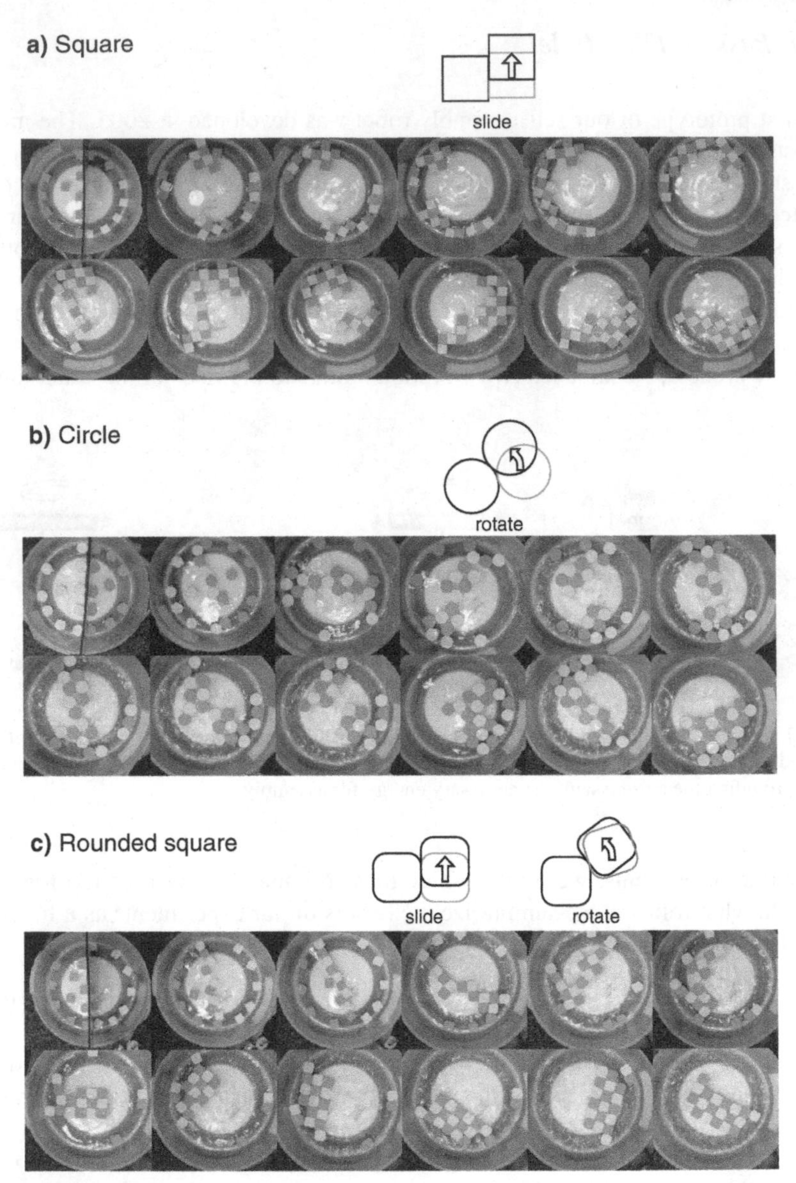

Fig. 8.2 Each shape displays a distinct behavioral pattern. The larger the curvature of the corners of the tiles, the faster the aggregation speed and the less stable the final configuration

As can be inferred from the above summary, each shape displays a distinct behavioral pattern. The larger the curvature of the corners of the tiles, the faster the aggregation speed and the more unstable the final configuration. This study effectively demonstrates how a small change of (local) morphology can induce a change of behavior affecting the global configuration of the system.

8.3.2 *Self-propelled Model: Tiles with Vibration Motors*

The word self-assembly implies that the involved elements or parts assemble spontaneously without external intervention (i.e., they are autonomous). It follows that the ability to "locomote" (i.e., to move around) is an important prerequisite for self-assembly. To address this requirement, we equipped each module with a vibration motor (pager motor).

8.3.2.1 Experimental Setup

Each module was built with flexible plastic. A small vibration motor was positioned on top of the tiles, and a permanent magnet was attached to the module's bottom surface (Fig. 8.3 a). The magnet was made of Neodymium, which had a surface magnetic flux density of 1.3 T. Because the actuating system had to be small and controllable, we chose vibration motors. This allowed the modules to move around in their environment like Brownian particles. Rather than using batteries for each module, we opted to supply electricity through a pantograph that draws current from a metal ceiling (aluminum) (Fig. 8.3 b) (a similar technique is used by cars in amusement parks, Fig. 8.3 c). This solution has several advantages: (1) all modules receive the same power, which is necessary because all modules should be identical and (2) the modules can be very lightweight. When an electrical potential was applied to the ceiling plate, current flowed through the pantograph to the vibration motor returning to ground via platinum electrodes immersed in the water. Salt, $83.3\,\mathrm{g\,l^{-1}}$ was added to the water to make it conductive (electrolyte). The weight of each module was approximately 2.8 g with a size of about 3 cm. As depicted in Fig. 8.3 b, the modules could tilt, inducing rather large fluctuations in the current flowing through the motors.

8.3.2.2 Experiment 1: The Ability to Express Various Behaviors

Experiments carried out with various types of modules are shown in Fig. 8.4. We used different types of primitive shapes, such as circles or triangles, and demonstrated their ability to express various behaviors in each combination. Three modular systems were powered via the ceiling (c, e, f), while the others were powered through a wire (a, b, d, g). Some of the modular configurations were initialized manually (a, b, d, g). Moreover, not all the modules contained a vibration motor (only the modules with a letter "V").

- *Reel* (Fig. 8.4 a): A vibrating module can reel a string of modules. Circular modules whose magnetic North points upwards and modules whose magnetic North points downwards rotate in opposite directions.
- *Gear* (Fig. 8.4 b): This example shows how far the rotational movement generated by the vibrating modules can be transferred.

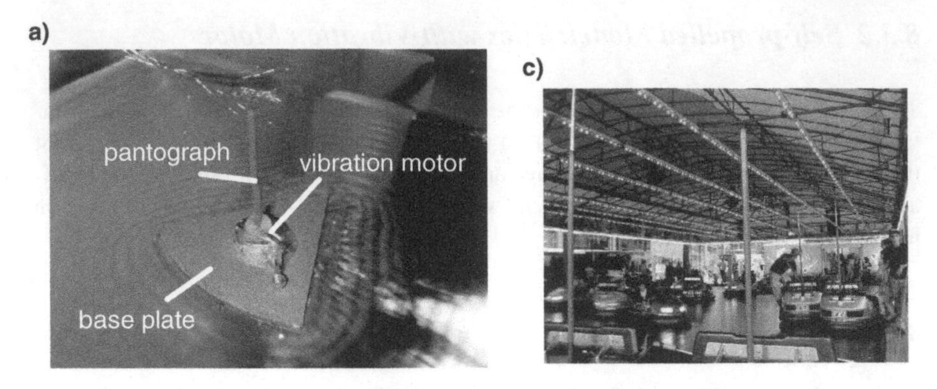

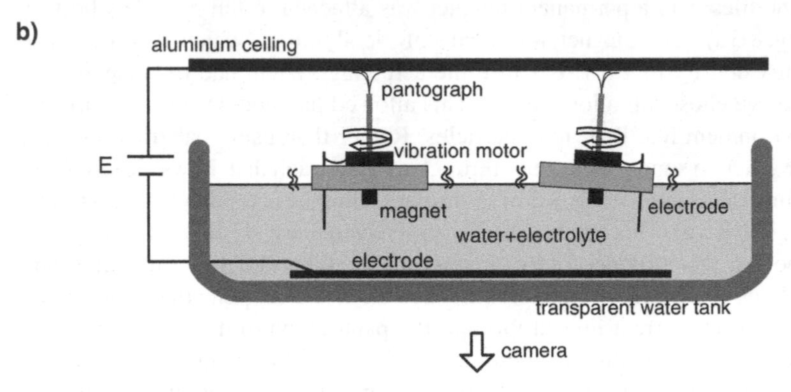

Fig. 8.3 Experimental setup. (**a**) Snapshot of a Tribolon module. (**b**) Illustration of the experimental environment with two modules. (**c**) Bumping cars in an amusement park, which employ a similar technique

- *Bike* (Fig. 8.4 c): A triangular vibrating module follows a wall. The addition of two circular modules close to it, turns them into "wheels" and the whole cluster starts following the wall.
- *Planets* (Fig. 8.4 d): A vibrating module placed manually attracts a small module through a magnetic force. The movement of the big module forces the small module to rotate around itself (rotation and revolution).
- *Dimer* (Fig. 8.4 e): All the three modules contain a vibration motor. Two modules (S) "chase" one module (N). After a while, one module reaches a stable position and the other modules stops the chase.
- *Lattice* (Fig. 8.4 f): All modules contain a vibration motor. Only the circular modules rotate, while the square ones stop moving.
- *Rotation* (Fig. 8.4 g): A triangular module can rotate around a circular module following its edge. The circular modules slide their positions and push the edge of the triangle module and generates a rotational movement.

It has to be mentioned that the reactions responsible for the structuring of the system act at a strictly local level, that is, all behaviors described above emerge spontaneously and autonomously through a decentralized process. Thus, by contrast to traditional manufacturing processes, less control is required for the assembly. The process and the configurations are also more robust against unforeseen damage, implying that the system is largely capable to recover from failure and external disturbances.

In addition, our simple and cheap experimental setup makes it suitable as an educational tool. It not only helps people grasp difficult concepts such as "open, nonequilibrium dynamical system," but also provides a starting point for thinking about natural systems.

8.3.2.3 Experiment 2: Circular Section Shape Module

A "shape" can be described by indicating angles and the length of the edges forming them. One of the inevitable problems when we discuss about shapes is that a change of one parameter can lead to changes in another parameter, which makes it hard to discuss the implications of a single parameter change. To study this issue, we designed a set of pie-shaped modules (wedges) spanning an angle of α degrees. A permanent magnet was attached, oriented orthogonally to the module's main axis (Fig. 8.5 a). Snapshots taken from three experiments using such six modules (for $\alpha = 60°$) are visualized in Fig. 8.6.

In each experiment, a different (constant) electric potential was applied between the ceiling plate and the bottom submerged electrode causing the Tribolon modules to aggregate in various ways. In the experiment reproduced in Fig. 8.6a, we applied a potential of $E = 7$ V. The modules first moved along random paths vaguely reminiscent of Brownian motion. After some time ($\approx$9 s), through magnetic attraction, some of the modules were pulled to each other forming two-unit clusters (denoted by X_2; X_i stands for the state of a cluster consisting of i modules). These clusters further combined to generate a four-unit cluster (X_4), then a five-unit cluster (X_5), and eventually a six-unit cluster (X_6) (sequential self-assembly; Fig. 8.6). Once this final state was reached, the whole circular structure started to rotate counter-clockwise at an almost constant speed, forming a propeller-like structure. A plausible explanation for this "higher level functionality" can be found in the intermittency of the contact of the pantograph with the aluminum ceiling (due to the stable configuration of the six-unit cluster on the water surface; see Fig. 8.6 b), which in turn lead to a pulsed current flow.

In the snapshots reproduced in Fig. 8.6 b, the potential was set to $E = 8$ V. As a result of the higher potential, the motors vibrated at a higher frequency, increasing the likelihood of segregation of clusters while decreasing the likelihood of aggregation. Most of the time, all types of clusters disintegrated shortly after formation, exception made for the six-unit cluster (X_6) which, due to its symmetry, proved to be a stable structure. It is important to note that the formation of the six-unit cluster at $T = 98$ s was accidental (one-shot self-assembly). This tendency, suppressing intermediate states, is thought to be a potential solution to the yield problem (see Sect. 8.3.2.4).

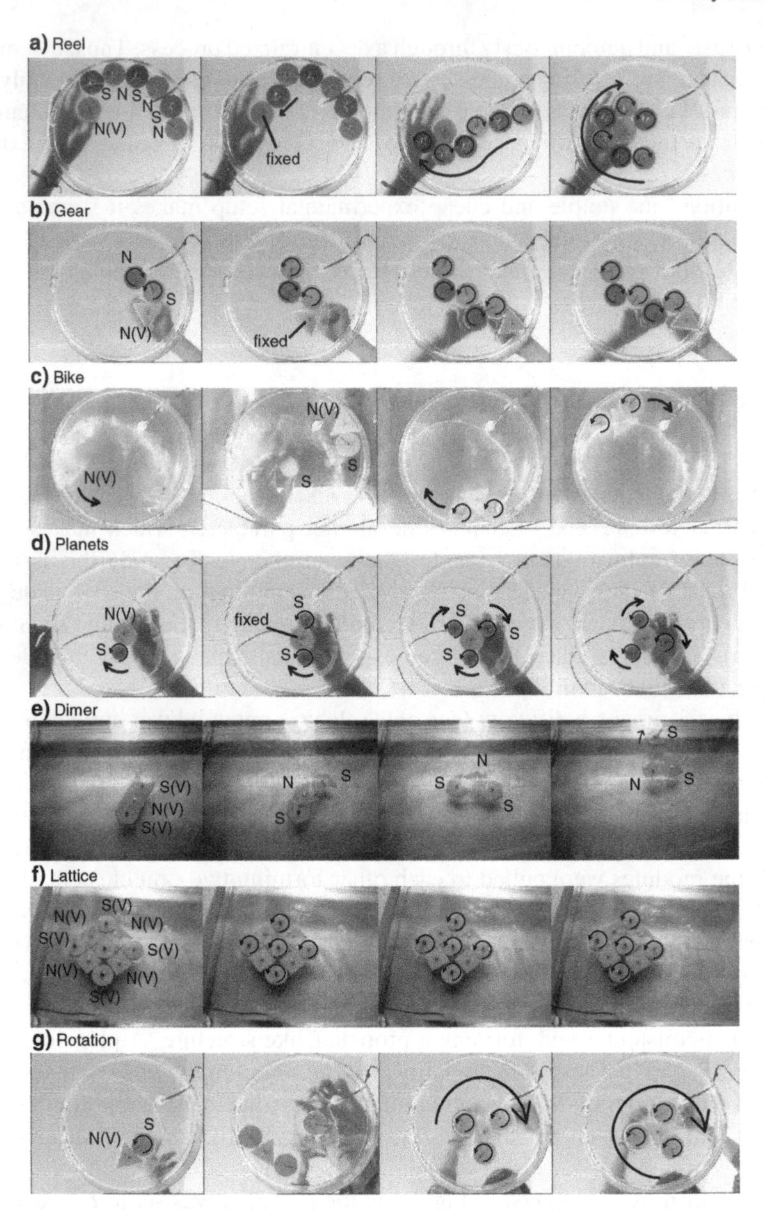

Fig. 8.4 The ability to express various behaviors. Three modular systems were powered via the ceiling (**c, e, f**), while the others were powered through a wire (**a, b, d, g**). Some of the modular configurations were initialized manually (**a, b, d, g**); not all the modules contained a vibration motor (only the modules with a letter "*V*")

The snapshots in Fig. 8.6 c were obtained by applying a potential of $E = 9$ V. This potential induced the vibration motors to vibrate even faster making the formation of a six-unit cluster unlikely. In fact, even for prolonged experiments no clusters

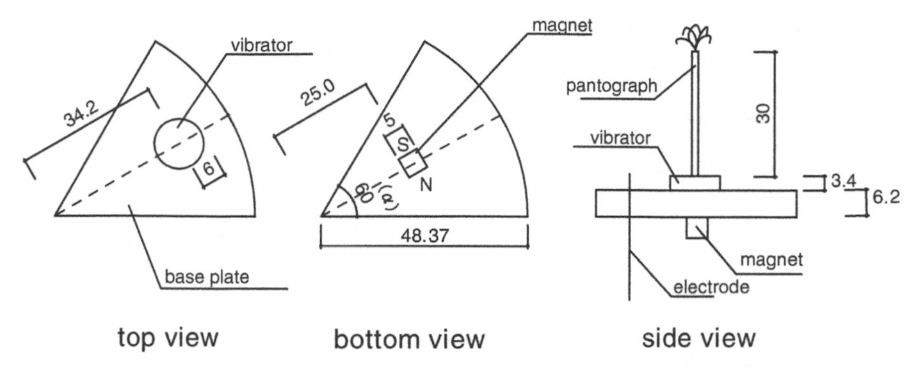

Fig. 8.5 Schematic representation of a Tribolon module (units: (mm), (degrees)). Each module weighs approximately 2.8 g and covers an area of 12.25 cm^2. The angle spanned by the circular sector is $\alpha = 60°$

could be observed (random movements). We confirmed this result by initializing the experiment with modules arranged in a circular configuration consisting of six units (the desired configuration); as expected, the cluster was unstable and disaggregated shortly after the start of the experiment.

8.3.2.4 Yield Problem

The problem of producing a desired configuration in large quantities (while avoiding incorrect assemblies) is known as the "yield problem" and has been studied in the context of biological and nonbiological self-assembly systems [16,17]. For example, let us assume that the self-assembly process is initialized with 12 modules – each one a pie-shaped wedge spanning 60° – with the objective to form two complete circles consisting of six modules each. In fact, the likelihood that the system actually settles into the desired configuration (a circle) is rather low (Fig. 8.7). In this respect, suppressing the probability of remaining in intermediate states may help reducing the occurrence of this problem, in our case Fig. 8.6 b.

An essentially analogous problem is investigated in the context of DNA folding, where one of the objectives is to increase the yield rates of the self-assembly process [29]. Similarly, a lot of research effort is being devoted to the development of high-yield procedures for integration and mass manufacturing of heterogeneous systems via self-assembly of mesoscopic and macroscopic components [34–36].

8.3.2.5 Mathematical Analysis

We studied the behavior of our self-assembly system using kinetic rate equations derived from analogies with chemical kinetics [16]. For the analysis, the quantity of every intermediate product is represented by a state variable.

a) Sequential self-assembly (E=7V).
The first modules form two-units clusters (X2) which eventually assemble into a six-units cluster (X6).
Once this final state was reached, the whole circular structure rotates counter-clockwise at constant speed.

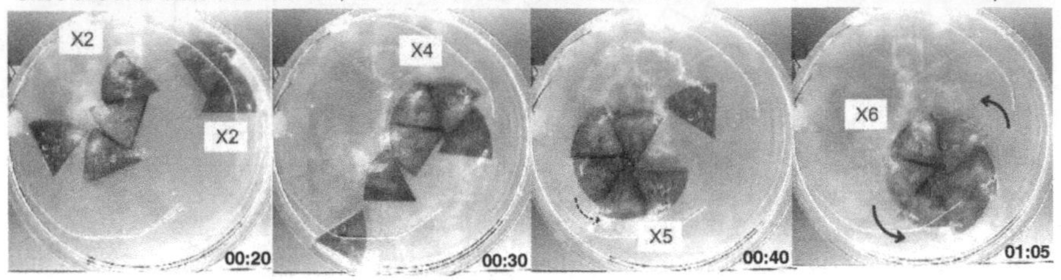

b) One-shot self-assembly (E=8V).
Most clusters are unstable. The six-units cluster (X6), however, is a stable symmetric structure.

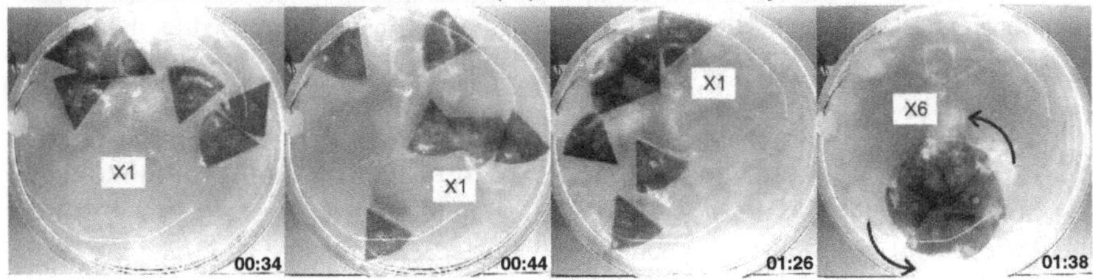

c) Random movements (E=9V).
All kinds of clusters are unstable and quickly disaggregate.

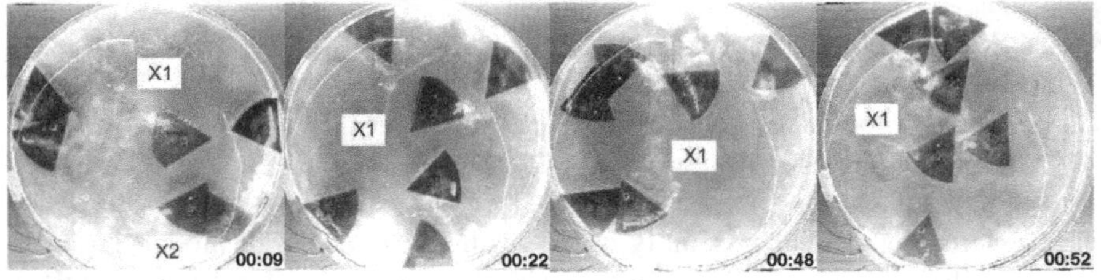

Fig. 8.6 Experimental results. Self-assembly process as a function of the applied electric potential E

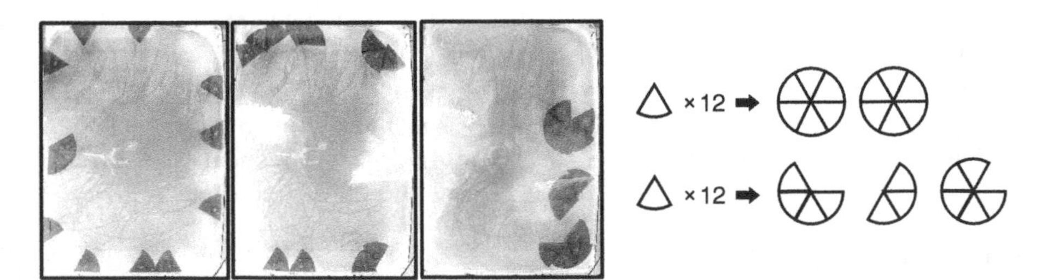

Fig. 8.7 Yield problem and stable clusters. Twelve circular sectors X_1 can aggregate into two X_6 clusters. In most cases, however, the modules organize themselves in more than three clusters. The yield in the upper part of the figure is 100 %, the one in the lower part is 0 %

Figure 8.8 displays the change over time of the number of clusters obtained by solving the system of difference equations with initial condition $X_1 = 100, X_i = 0 (i \in \{2...6\})$ (more detail, see [37]). As can be seen, $X_5(t) > X_6(t)$, which exemplifies the yield problem. The yield rate of each cluster is listed on the top-right of the figure as a circle graph.

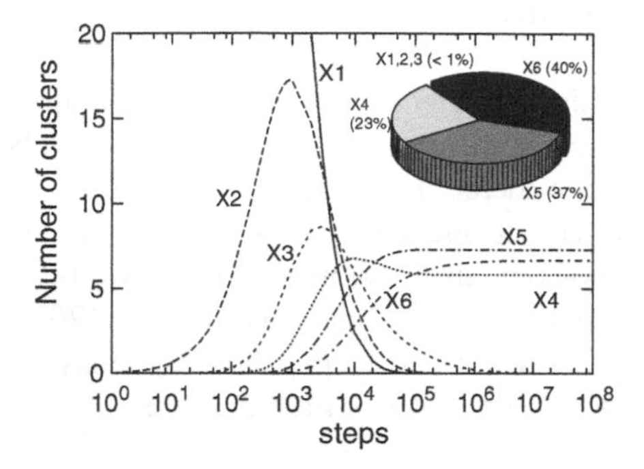

Fig. 8.8 Number of clusters as a function of time. The yield problem is evident from the fact that at steady-state $X_5(t) > X_6(t)$

Figure 8.9a, b shows how shape changes affect the yield of the self-assembly process. The yield rates are normalized by multiplying them by the number of units required to construct a full circle (i.e., in the case of $\alpha = 60°$, the factor is 6; in the case of $\alpha = 180°$, the factor is 2), and plotted as a function of the angle α. As seen in the figure, the yield rate falls off linearly with the angle. This result can be explained by considering that the number of clusters required to form the desired structure is inversely proportional to the angle. An additional point is that clusters formed by an even number of units show better yield rates. In Fig. 8.9b, we plotted the yield rates on a logarithmic scale. As expected, the narrower the angle, the worse is the performance of the system. Interestingly, the relationship between yield rate

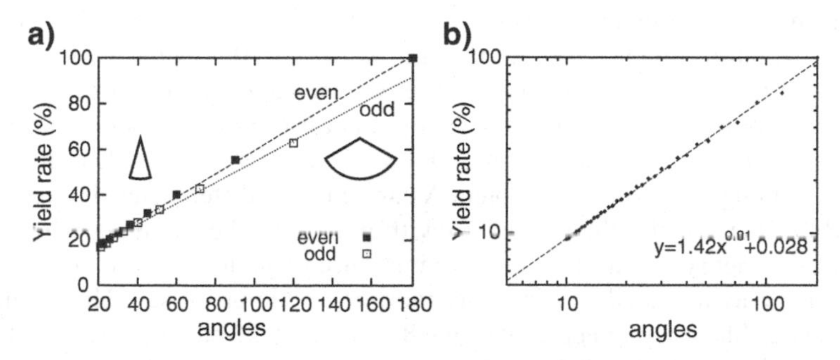

Fig. 8.9 Yield rates as a function of the angle spanned by the pie-shaped wedge; (**a**) linear scale and (**b**) logarithmic scale. Interestingly, the relationship between yield rate and angle follows a power-law with a scaling exponent of 0.81

and angle follows a power-law with a scaling exponent of 0.81. Although this will require additional investigation, we hypothesize that this relationship is the result of the shape of the module (in this case the circular sector with an orthogonally attached magnet).

8.3.2.6 Homogeneous *vs* Heterogeneous

Biological systems consist of many different types of components. We speculate that in any minimal setting, there must exist a suitable level of heterogeneity. In our circular cut model, we considered the angle α to be an adequate parameter for the heterogeneity of the system, i.e. the 60° module and 120° module should be treated as different (heterogeneous) modules. However, once two 60° modules connect, a 120° module forms, which is obviously equivalent to a 120° module. The lesson learnt from this example is that the concepts "homogeneous" and "heterogeneous" cannot be separated from the context in which they are used.

8.3.2.7 Passive or Active?

An additional issue addressed by our model is the distinction between "passive" and "active." For instance, it may be possible to achieve the same behaviors by applying an external rotational magnetic field instead of attaching a vibration motor to each tile. Strictly speaking, active tiles should, therefore, be treated like passive tiles implying that the distinction tends to be trapped by frame of reference problem [38]. The essential difference between the active and the passive system is the type of asymmetries that the two systems can produce.

8.3.2.8 Hierarchical Levels of Functionality

Let us return to the question formulated in the introduction: can we build nonbiological (modular) systems that self-assemble like biological ones? By carefully mapping the interaction networks of the approximately 70 proteins involved in the assembly of the T4 phage, one can observe that the aggregation processes are extremely well organized [1]. For instance, a protein A can only dock on a protein *B* by first coupling with a protein C: $\mathrm{A}+\mathrm{B}+\mathrm{C} \rightarrow \mathrm{AC}+\mathrm{B} \rightarrow \mathrm{ABC}$. It follows that through coupling with protein C, protein A can acquire a different level of functionality, which then enables the interaction with protein B. We assume that this kind of interaction networks can lead to the emergence of multiple levels of functionality, which play a crucial role in many morphogenetic processes. The formation of a propeller-like rotating aggregate (Fig. 8.6 a, b), a bike pattern (Fig. 8.4 c), or a rotation configuration (Fig. 8.4 g) are instances of what one might call "emergent functionality."

8.3.2.9 The Meaning of Morphology

Although the shapes tend to attract attention, an essential role is also played by the distribution of the body from the "origin of the force" (here: magnets). For instance, by shifting the position of the magnets closer to the rounded edge and after perturbing the water, the modules tend to form dimers (two-clusters), orienting the rounded edges towards each other instead of forming a circle (Fig. 8.10 a). This is due to the fact that the configuration is potential, energetically more stable than a full circle (neglecting the dynamic part of the energy). This example seems to indicate that it may not be adequate to perceive this change as a shift of the magnet, but as changing the entire morphology (from the "force source," Fig. 8.10 b). It follows that the reason of the difficulty in achieving lightweight actuators is equivalent to changing the entire morphology.

8.3.3 Connectable Model: with Peltier Connector

To make a step towards highly autonomous lightweight systems, we propose a model incorporating a novel connection mechanism: the water close to the docking interface of one module is frozen to ice, building a local bridge to another module.

8.3.3.1 Peltier Freeze–Thaw Connector

The core of our connector is the Peltier heat pump – a double-faced cooling–heating device that can transfer thermal energy from one side of the device to the other, with consumption of electrical energy. We used the Peltier device to freeze (and thaw) the water between two modules and thus, realize binding (and unbinding) between modules. The polarity of the current applied to the device defines which side is cooled down or heated up. The device consists of different types of semiconducting materials that are connected in series to take advantage of the so-called thermoelectric or Peltier effect. This effect is the direct conversion of an electric voltage into a temperature difference and vice versa, and allows the element to work as a heat pump. Peltier devices are available in various sizes. For our purpose, we used an 8×8 mm element that weighs 0.8 g. Theoretically, the Peltier heat pump can induce a temperature difference of up to 72°C while consuming approximatively 2.60 W. One particular advantage of this type of intermodule connection mechanism is that the absence of mechanical parts makes it scalable. The fact that the connector is devoid of moving parts makes it also intrinsically less prone to failures. One disadvantage is that in order to sustain the connections, energy has to be supplied permanently to the heat pump. For the detachment process, however, there is no need to supply energy because the ice melts when it is not cooled down; moreover, the heat flow from the hot surface supports the thawing process speeding it up.

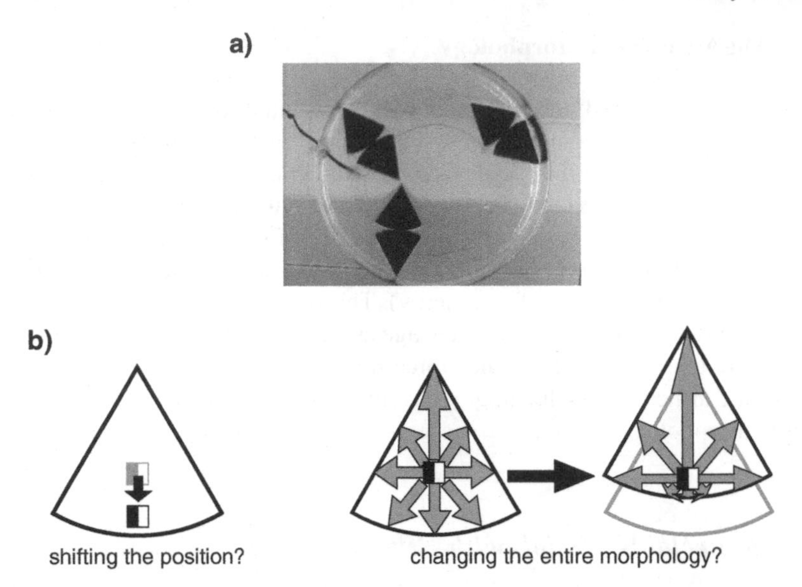

Fig. 8.10 Meaning of morphology. (**a**) Clusters generated by modules with magnets, whose positions were shifted to the rounded edge. (**b**) Shifting the position of a magnet sometimes leads to the result equivalent to the changing the entire morphology

8.3.3.2 Experiment 1: Kite-Shaped Model

The experimental setup was composed of a power supply, a metallic ceiling, a water tank, and six modules immersed partially in water (Fig. 8.11). Each module consisted of a kite-shaped wedge made of flexible plastic (acrylnitrile butadene styrene; ABS) spanning angles of 60° and 30°. The modules (H: 13 mm, L: 30 mm) contained a permanent magnet oriented orthogonally to their main axis to attract or repel other modules (Fig. 8.11 a). A vibration motor was used to endow the modules with a minimal locomotive ability which allowed the modules to move randomly around – vaguely reminiscent of Brownian motion. Rather than using batteries, electricity was supplied to the modules through a pantograph that drew current from a metal ceiling. This solution not only led to lightweight modules ($m = 6.0$ g), but it ensured that all modules received approximately the same amount of energy in a particular experiment.

When an electrical potential was applied to the metallic ceiling plate, current flowed through the pantograph to the vibration motor returning to ground via the platinum electrodes immersed in the water (8 % concentration of electrolyte (salt) was added to the water to make it conductive). To speed up the connection between two modules, the water in which the modules moved was cooled down to approximately −3°C (due to the concentration of salt this was slightly higher than the freezing temperature). Two diodes were used to switch the direction of the current. Current flowed either through the Peltier element or the vibration motor depending on the direction of the voltage applied to the system (Fig. 8.11 c, switch).

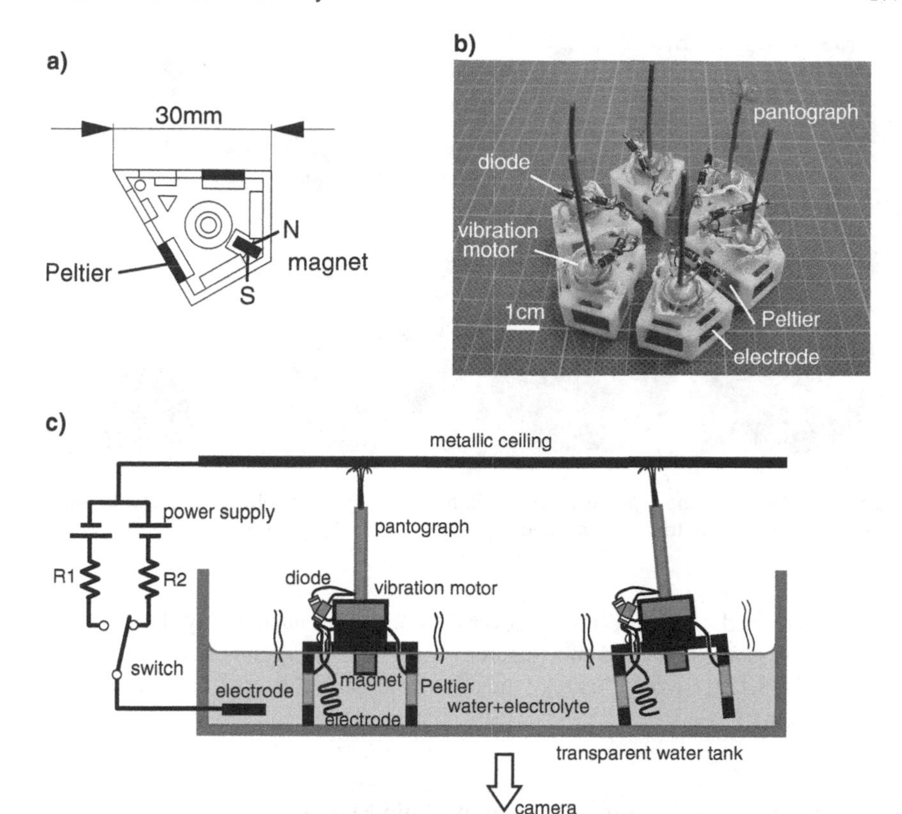

Fig. 8.11 Experimental setup for connectable model. We used the Peltier device to freeze (and thaw) the water between two modules and thus realize binding (and unbinding) between modules. (**a**) Schematic illustration of a module (*bottom view*). (**b**) Picture of six modules. (**c**) Experimental setup with two modules

We first carried out experiments to test the reliability of the connector and to investigate the reconfigurability of our self-assembly system. The result is shown in Fig. 8.12. In the beginning of the experiment, the modules were placed in the arena (Fig. 8.12 a) and arranged by hand to form a hexagonal shape (Fig. 8.12 b). Voltage was applied via the metallic ceiling (Fig. 8.12 c). After 1 min, all six modules were connected to each other forming one unit (Fig. 8.12 d). We then flipped the polarity of the current supplied through the pantograph. The Peltier connectors stopped cooling and the vibration motors started to vibrate causing a disassembly of the hexagonal shape into six separate modules (Fig. 8.12 e). As a result of the vibrations of the motor, the modules moved around in the arena where they eventually got magnetically attracted by another module and started to form triangles (Fig. 8.12 f, g). The experiment was considered completed when the six modules had formed two triangles (Fig. 8.12 h).

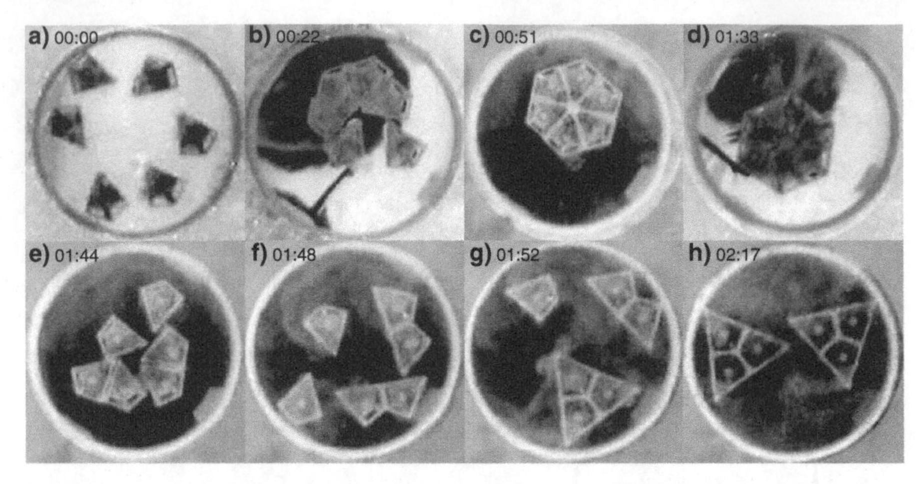

Fig. 8.12 Snapshots of the experiment. A frozen hexagonal module split up into six kite-shaped modules and formed two triangular modules

We conducted the experiment several times. For sufficiently long waiting times T, we always observed two different ways of convergence to the final states: one is in Fig. 8.12 h (two 3-clusters), the other is three 2-clusters (not on the picture, yield problem).

8.3.3.3 Experiment 2: Hinge-Connected Chain Model

The outcome of the experiments described in Sect. 8.3.3.2 led to the question of how to take our modular system to another level of operation. To answer this question, we added permanent magnets to each module. Because with many modules the outcome of this change was not easy to anticipate, we built a new type of module. Each module was physically linked through hinge joints to two modules, forming a chain (Fig. 8.13, $m = 12.8$ g). By taking inspiration from protein folding, we expected a drastic reduction of dimensionality of the search space. The main advantage of this implementation is that it avoids a crucial problem: the increased number of magnets generates undesired configurations. Note that the positions of all the other magnets were replaced and rearranged. Only the center module (colored in red, Fig. 8.13 b) had a large magnet oriented orthogonally to the symmetry axis. The other small cylindrical magnets were oriented vertically – "S"outh poles attracting "N"orth poles and vice versa. As in the modules described in the previous section, diodes were used to direct the current flow. Depending on the polarity of the applied voltage, current only flowed either through the Peltier elements (12 V) or through the vibration motors (10 V).

Snapshots from a representative experiment are shown in Fig. 8.14. At the beginning of the experiment, we arranged by hand two chains of three modules each to form a hexagonal shape (Fig. 8.14 b). Voltage was applied to the system so that

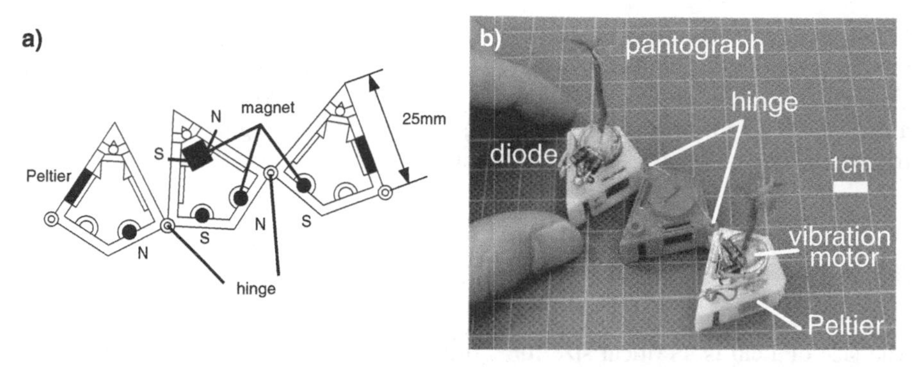

Fig. 8.13 Chain model. (**a**) Schematic illustration of a chain. (**b**) Picture of a chain. Three modules are connected by hinges

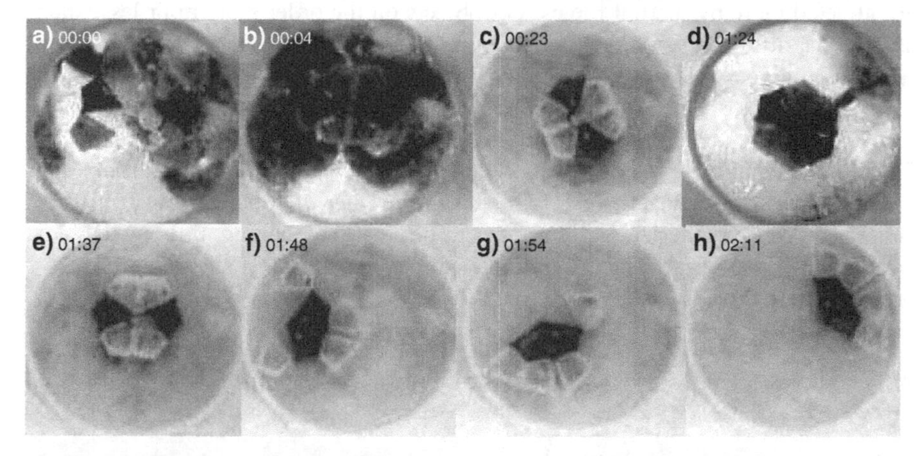

Fig. 8.14 Snapshots of the experiment with chain modules. A frozen hexagonal module split up into two chain modules and two magnetically connected triangle chains were obtained

the Peltier elements were powered (Fig. 8.14 c). After 1 min, an ice layer built up between the modules causing them to attach to each other yielding one single piece (Fig. 8.14 d). We then inverted the polarity of the applied voltage and let the current flow to the vibration motors (Fig. 8.14 e). The ice melted and the modules altered their configuration guided by the magnetic forces (Fig. 8.14 f, g). The transformation was completed in a minute, and two magnetically connected triangle chains were obtained (Fig. 8.14 h).

The success rate of the reconfiguration just described was not as high as expected. We suspect that the low yield rate was mainly due to a design problem: the position of the large magnet was too close to the edge of the module. Therefore, a rather strong movement of the modules was required to induce a disassembly of the initial hexagonal configuration.

The main implication of the two experiments described in this Sect. 8.3.3 is that the restriction of the geometric constraint of modules, in other words, the "dimensionality reduction" of the reconfiguration problems enables the system to transit to a different level of functionality, bringing the modules to magnetically connected triangular clusters, while avoiding undesirable formations (yield problem).

8.3.4 Scale-Free Self-Assembly: Size Matters

The size of 1 cm is a critical size for self-assembly systems. For objects in water at the millimeter scale, viscosity is as important as inertia (the Reynolds number, that is, the ratio of inertial forces and viscous forces, is ≈ 1). It follows that objects smaller than that size are affected more by viscous forces whereas larger objects are affected more by inertial forces. For objects on the order of 1 μm or less, such as bacteria, diffusion is a more effective way of locomotion than active propulsion (e.g., swimming bacteria are slower than diffusing molecules [39]).

In Fig. 8.15 a, 0.5 mm size plastic particles are shown. When mixed with water and shaken, they tend to automatically form three layers which are dominantly affected by (1) static charge on the particles, (2) static charge on the wall, and (3) gravity. As can be seen in Fig. 8.15 b, several number of spherical clusters were generated. We hypothesize that through the mixing and shaking, tiny air bubbles are created in the water. The air around each particle acts as a sticking connector producing attractive forces.

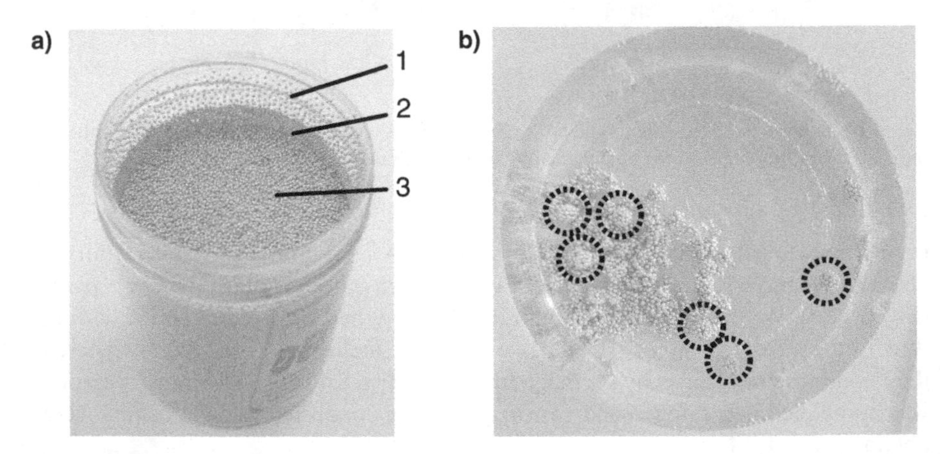

Fig. 8.15 Self-assembly of millimeter scale particles. (**a**) 0.5 mm beads in a cup. (**b**) In water, aggregating into spherical clusters

The implication is that we need to be careful when relating our work to small scales. Although at small scales it might be possible to observe complete bottom-up self-assembly, this might not be the case at large scales.

8.4 Speculations About Life

An important goal of the growing field of self-assembly is the development of a better formal understanding of the specific mechanisms and general principles underlying it. It is clear that the discovery of principles, which hold at all scales will require substantial input from various fields. At the molecular scale, biological systems are one of the examples that achieve robust self-assembly through an intricate web of well ordered reactions. Attention must be paid to the fact that all components are passively interacting even if it looks as if they are actively reacting. To achieve highly autonomous self-assembly modules, the realization of a sufficient number of degrees of freedom that are controllable is necessary for such small scale autonomous-distributed systems. In particular, because of the difficulty in including different types of attractive forces within the same system, realizing a new kind of connection mechanism endows the module with a better means of reacting in the environment. In this sense, the idea of a connector exploiting the thermoelectric effect may open yet another possibility for the state-of-the-art of self-assembly systems.

Figure 8.16 illustrates the level of autonomy realized as a system depending on the fragmentation of information (X-axis) and the total amount of information (Y-axis). We consider life as a kind of self-regulative system. Towards this goal, the system must be "self-observable." More specifically, the system must react differently according to the input depending on its internal state (memory). This is comparable to reactive systems but considered to have a slightly higher autonomy level. We speculate that there must be an adequate level of how to fragment information to the system (neither too heterogeneous nor too homogeneous, X-axis). And the system requires a certain amount of information to realize highly autonomous behavior (Y-axis).

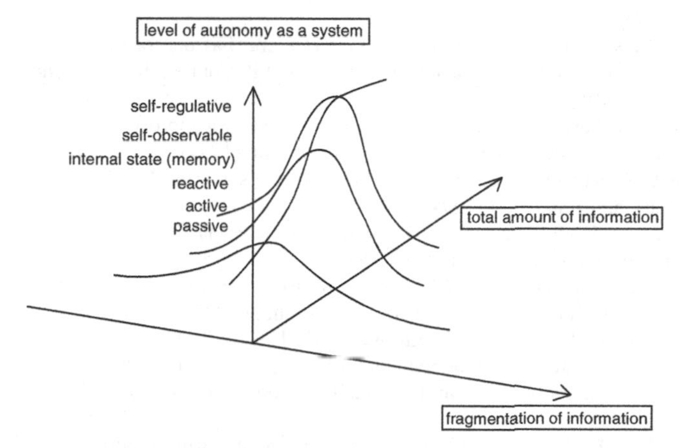

Fig. 8.16 The level of autonomy realized as a system depending on the fragmentation of information (X-axis) and the total amount of information (Y-axis)

Focusing on the mechanisms of living things from a viewpoint of autonomous and distributed systems, it can be noticed that the components which form the morphology are not always highly autonomous [40]. When we discuss autonomy of components in the context of autonomous-distributed systems, the crucial point might not be whether the units are passive or active, but whether the units are passive or "reactive" (Indeed, we could probably obtain the same results by employing modules without vibration motors, but by supplying power through an external, rotating magnetic field, Sect. 8.3.2.3). By drawing from our current experience in designing, constructing, and controlling macroscopic modular systems, we hope that we will be able to derive conclusions about the level of autonomy that is needed to achieve self-assembly. One could say that life is a blind engineer, because the components – e.g., molecules – self-construct into organisms in a completely bottom-up fashion.

Acknowledgments

The authors greatly thank Nathan Labhart, Flurin Casanova, and Marco Kessler for many helpful suggestions. This research is supported by the Swiss National Science Foundation project #200021-105634/1.

References

1. Leiman, P.G., Kanamaru, S., Mesyanzhinov, V.V., Arisaka, F., Rossmann, M.G.: Structure and morphogenesis of bacteriophage t4. Cellular and Molecular Life Sciences **60**, 2356–2370 (2003)
2. Zlotnick, A.: Theoretical aspects of virus capsid assembly. Molecular Recognition **18**, 479–490 (2005)
3. Fukuda, T., Kawauch, Y.: Cellular robotic system (cebot) as one of the realizations of self-organizing intelligent universal manipulator. In: Proc. Int. Conf. on Robotics and Automation, pp. 662–667 (1990)
4. Nakano, K., Uchihashi, S., Umemoto, N., Nakagama, H.: An approach to evolutional system. In: First IEEE Conference on Evolutionary Computation (1994)
5. Chirikjian, G.S.: Kinematics of a metamorphic robotic system. In: Proc. Int. Conf. on Robotics and Automation, pp. 449–455 (1994)
6. Murata, S., Kurokawa, H., Kokaji, S.: Self-assembling machine. In: Proc. Int. Conf. on Robotics and Automation, pp. 441–448 (1994)
7. Murata, S., Kurokawa, H., Yoshida, E., Tomita, K., Kokaji, S.: A 3-D self-reconfigurable structure. In: Proc. Int. Conf. on Robotics and Automation, pp. 432–439 (1998)
8. Murata, S., Tomita, K., Yoshida, E., Kurokawa, H., Kokaji, S.: Self-reconfigurable robot. In: Proc. Int. Conf. on Intelligent Autonomous Systems, pp. 911–917 (1999)
9. Yim, M.: New locomotion gaits. In: Proc. Int. Conf. on Robotics and Automation, vol. 3, pp. 2508–2514 (1994)
10. Kotay, K., Rus, D., Vona, M., McGray, C.: The self-reconfiguring robotic molecule. In: Proc. Int. Conf. on Intelligent Robots and Systems, vol. 1, pp. 424–431 (1998)
11. Rus, D., Vona, M.: Crystalline robots: Self-reconfiguration with compressible unit modules. Autonomous Robots **10**(1), 107–124 (2001)

12. Castano, A., Behar, A., Will, P.M.: The conro modules for reconfigurable robots. IEEE/ASME Transactions on Mechatronics **7**(4), 403–409 (2002)
13. Jorgensen, M.W., Ostergaard, E.H., Lund, H.H.: Modular atron: Modules for a self-reconfigurable robot. In: Proc. Int. Conf. on Intelligent Robots and Systems, vol. 2, pp. 2068–2073 (2004)
14. Zykov, V., Mutilinaios, E., Adams, B., Lipson, H.: Self-reproducing machines. Nature **435**(7039), 163–164 (2005)
15. Penrose, L.S.: Self-reproducing. Scientific American **200–206**, 105–114 (1959)
16. Hosokawa, K., Shimoyama, I., Miura, H.: Dynamics of self-assembling systems: Analogy with chemical kinetics. Artificial Life **1**(4), 413–427 (1994)
17. Hosokawa, K., Shimoyama, I., Miura, H.: 2-d micro-self-assembly using the surface tension of water. Sensors and Actuators A **57**, 117–125 (1996)
18. Bowden, N., Terfort, A., Carbeck, J., Whitesides, G.M.: Self-assembly of mesoscale objects into ordered two-dimensional arrays. Science **276**, 233–235 (1997)
19. Grzybowski, B.A., Mechal Radkowski, Campbell, C.J., Lee, J.N., Whitesides, G.M.: Self-assembling fluidic machines. Applied Physics Letters **84**, 1798–1800 (2004)
20. Grzybowski, B.A., Stone, H.A., Whitesides, G.M.: Dynamic self-assembly of magnetized, millimetre-sized objects rotating at a liquid-air interface. Nature **405**, 1033 (2000)
21. Grzybowski, B.A., Winkleman, A., Wiles, J.A., Brumer, Y., Whitesides, G.M.: Electrostatic self-assembly of macroscopic crystals using contact electrification. Nature **2**, 241–245 (2003)
22. Saitou, K.: Conformational switching in self-assembling mechanical systems. IEEE Transactions on Robotics and Automation **15**, 510–520 (1999)
23. White, P., Kopanski, K., Lipson, H.: Stochastic self-reconfigurable cellular robotics. In: Proc. Int. Conf. on Robotics and Automation, vol. 3, pp. 2888–2893 (2004)
24. White, P., Zykov, V., Bongard, J., Lipson, H.: Three dimensional stochastic reconfiguration of modular robots. In: Proc. Int. Conf. on Robotics Science and Systems, pp. 161–168 (2005)
25. Shimizu, M., Ishiguro, A.: A modular robot that exploits a spontaneous connectivity control mechanism. In: Proc. Int. Conf. on Robotics and Automation, pp. 2658–2663 (2005)
26. Bishop, J., Burden, S., Klavins, E., Kreisberg, R., Malone, W., Napp, N., Nguyen, T.: Programmable parts: A demonstration of the grammatical approach to self-organization. In: Proc. Int. Conf. on Intelligent Robots and Systems, pp. 3684–3691 (2005)
27. Griffith, S., Goldwater, D., Jacobson, J.: Robotics: Self-replication from random parts. Nature **437**, 636 (2005)
28. Mao, C., LaBean, T.H., Reif, J.H., Seeman, N.C.: Logical computation using algorithmic self-assembly. Nature **407**, 493–496 (2000)
29. Rothemund, P.W.K.: Folding DNA to create nanoscale shapes and patterns. Nature **440**(7082), 297–302 (2006)
30. Seeman, N.C.: DNA in a material world. Nature **421**, 427–430 (2003)
31. Shih, W.M., Quispe, J.D., Joyce, G.F.: A 1.7-kilobase single-stranded dna that folds into a nanoscale octahedron. Nature **427**, 618–621 (2004)
32. Winfree, E., Liu, F., Wenzler, L.A., Seeman, N.C.: Design and self-assembly of two-dimensional dna crystals. Nature **394**, 539–544 (1998)
33. Yokoyama, T., Yokoyama, S., Kamikado, T., Okuno, Y., Mashiko, S.: Selective assembly on a surface of supramolecular aggregates with controlled size and shape. Nature **413**, 619–621 (2001)
34. Boncheva, M., Ferrigno, R., Bruzewicz, D.A., Whitesides, G.M.: Plasticity in self-assembly: Templating generates functionally different circuits from a single precursor. Angewandte Chemie International Edition **42**, 3368–3371 (2003)
35. Gracias, D.H., Tien, J., Breen, T.L., Hsu, C., Whitesides, G.M.: Forming electrical networks in three dimensions by self-assembly. Science **289**, 1170–1172 (2000)
36. Wolfe, D.B., Snead, A., Mao, C., Bowden, N.B., Whitesides, G.M.: Mesoscale self-assembly: Capillary interactions when positive and negitive menisci have similar amplitudes. Langmuir **19**, 2206–2214 (2003)

37. Miyashita, S., Kessler, M., Lungarella, M.: How morphology affects self-assembly in a stochastic modular robot. In: IEEE International Conference on Robotics and Automation, pp. 3533–3538 (2008)
38. Pfeifer, R., Scheier, C.: Understanding intelligence. MIT, Cambridge (2001)
39. Hayakawa, J. (ed.): Time of an elephant, time of a mouse. CHUO-KORON-SHINSHA (1992)
40. Alberts, B., Hohnson, A., Lewis, J., Raff, M., Roberts, K., Walter, P.: Molecular biology of the cell. Garland Science, UK (2002)

Chapter 9
Artificial Symbiosis in EcoBots

Ioannis A. Ieropoulos, John Greenman, Chris Melhuish, and Ian Horsfield

9.1 Introduction

Truly autonomous robotic systems will be required to abstract energy from the environment in order to function. Energetic autonomy refers to the ability of an agent, to maintain itself in a viable state for long periods of time. Its behaviour must be stable in order not to yield to an irrecoverable debt in any vital resource, i.e. it must not cross any of its lethal limits [1, 2]. With this in mind, our long-term goal is the creation of a robot, which can collect energy for itself. This energy must come from the robot's environment and must be sufficient to carry out tasks, which require more energy than that available at the start of the mission. In this respect our definition of an autonomous robot is more akin to Stuart Kauffman's definition of an autonomous agent, "a self-reproducing system able to perform at least one thermodynamic work cycle" [3] – but without the burden of self-reproduction!

Building automata is certainly not something new. The first recorded example of an automaton dates back to the first century A.D. when Heron of Alexandria constructed a self-moving cart driven by a counter weight attached to the wheel base [4]. In more recent times, there are of course, some real robots, which already comply with this definition. For example, robots such as NASA's 'Spirit' [5] employ solar panels to power their explorations of Mars and have demonstrated their impressive ability to be self-sustaining. However, there will be numerous domains in which solar energy will not be available such as in underwater environments, sewers or when constrained to operate only in the dark. We are, therefore, interested in a class of robot system, which demonstrates energetic autonomy by converting natural raw electron-rich organic substrate (such as plant or insect material) into power for essential elements of robotic behaviour including motion, sensing and computation. This requires an artificial digestion system and concomitant artificial metabolism or, as in the case of EcoBots-I and -II, a rapprochement between an engineered artefact and a biological system – the robot symbiot.

A. Adamatzky and M. Komosinski (eds.), *Artificial Life Models in Hardware*,

For robots, we have a choice of generating the energy either off-board or onboard; both have advantages and disadvantages. For example, the idea of collecting biomass, in the form of mollusc pests (slugs), by a group of robots and depositing them into a collective external digester unit to generate electricity has been suggested [6–8]. The digester facilitates methanogenesis, and the gas is passed through a methane fuel cell generating electricity, which is stored and in turn 'tapped' by the robots. This idea is somewhat akin to the system employed by leafcutter ants which bring leaf cuttings back to the nest to use as the substrate for the growth of fungus, which is then harvested to provide the colony with energy. Another 'halfway house' mechanism for the robots is to have methane fuel cells onboard and draw off methane from the off-board digester. Such a system has the limitation of reducing the robots' radius of foraging and operation in their mission even though it has been suggested that the 'mobility' radius of this type of robot can be extended using a form of energy sharing based on the principle of trophallaxis observed in some species of social insects [9, 10].

Energy from the environment can come from any organic substrate such as plants (fruits and vegetables), food waste and dead biomass. These substrates can be utilised to produce useful energy in a number of ways. For example, herbivorous animals metabolise raw plant material into biochemical energy that is used for their maintenance. The energy used to maintain animals is derived from electron-rich carbohydrates, which are ultimately oxidised by a chain of metabolic steps. The collective process is called respiration since oxygen O_2 is the end-terminal electron acceptor. The end result of respiration is chemical energy in the form of NADH and ATP, carbon dioxide (CO_2) and water [11]. There are, of course, several organisms that are capable of breaking down carbohydrates in the absence of oxygen by donating electrons to organic electron acceptors (fermentation) or inorganic electron acceptors (anaerobic respiration) [11]. The major inorganic alternative end-terminal electron acceptors are sulphate, nitrate, CO_2 and also metal oxides. For mammalian cells, the types of substrates that can be utilised are limited to sugars, amino acids and organic acids (butyrate) [11]. However, animals can utilise a wider range of substrates by means of digestive transformation, such as hydrolysis of some polymeric substrates into utilisable monomers. Further, some animal species, although lacking the enzymes necessary for the breakdown of many polymeric substrates, are nevertheless able to utilise these through symbiotic associations with microbial cells. Ruminant animals are able to feed on cellulose from hay due to the symbiosis that exists between the anaerobic bacteria in the rumen of the animal and the animal itself. In nature, the symbiosis between animals and microorganisms is an existing successful partnership and, therefore, a good example to exploit in practical engineering.

There are a number of ways that the bacterial metabolism can be utilised to produce useful energy. For example, hydrogen can be produced by microorganisms that can reduce free protons into bimolecular hydrogen H_2. Methane can also be produced by methanogenic microorganisms of the Archaea kingdom for example *Methanobacterium thermautotrophicum* [12–14]. This is mainly the result of the fermentation of simple organic carbon compounds (acetate) or the oxidation of hydrogen. Both methane and hydrogen can then be used as fuels in conventional chemical fuel cells to produce electricity, with conversion efficiencies of the order of 65% and 80% respectively. This is of course a form of indirect energy transduction.

It is also possible to convert the microbial biochemical energy directly into electricity. The technology for performing this is based on the microbial fuel cell (MFC). In this system (as described in the sections to follow), electrons are extracted directly from the microbial metabolic cycle and diverted towards an electrical load via electrodes, thus, establishing a current flow. Direct electricity generation from microbes was seen as an interesting option, worthy of study with some significant advantages over the bio-hydrogen or bio-methane production processes. Moreover, employment of the MFC technology on a robot would result in a kind of 'artificial symbiosis' and represent a novel area for research and development.

On-board biomass converters have the obvious payload disadvantage of carrying the associated extra mass of the converter but, assuming that biomass can be found and ingested, the range is unlimited. Both EcoBots-I and -II, discussed in this chapter, employ this option by incorporating a number of onboard microbial fuel cells. The first robot to generate energy from on-board biomass converters was Wilkinson's Gastrobot [15, 16]. This robot employed MFC to extract energy in the form of electrons from refined biochemical fuel (sucrose).

The implications of adopting such a strategy may have an impact on the manner in which researchers and designers of autonomous systems incorporate this constraint into their mission requirements. There are a number of important issues for robots required to extract food from the environment. Firstly, useful energy will not (for the foreseeable future) be able to be instantly converted from raw substrate and secondly, there will be tasks (particularly those involving effectors or motion), which could not be powered continuously. The net effect is that this class of robot may have to include a 'waiting' behaviour in its repertoire in order to accumulate sufficient energy to carry out a task or sub-task. We refer to this form of behaviour as 'pulsed behaviour'. Thirdly, a robot may need to solve multi-goal action selection problems. In particular, it may be required to exhibit 'opportunistic' behaviour in terms of 'interrupting' its mission to forage or take advantage of energy resources such as a fallen apple. We, therefore, envision truly autonomous robots capable of exhibiting homeostasis, i.e. maintaining a state of internal equilibrium, which is different from its external surroundings by the automatic control over physico-chemical variations, by means of internal feedback and external 'behavioural' mechanisms.

9.1.1 Artificial Symbiosis

Associations between two species of life-form have often been described as symbioses. Although true in the original use of the term, symbiosis is now used to describe an obligate association between two species in which the two life-forms benefit from, but also depend on, each other. This type of interaction is different from proto-cooperation, a term used to describe a non-obligate mutually beneficial association in which neither species requires the other to exist.

There are many other forms of interaction between two species and the whole spectrum of association may be divided into two groups. One, in which neither of the two parties is harmed, and one in which either or both species are harmed. The latter category describes different forms of antagonism with competitive associations in which both species are harmed, amensalism in which one of the species is harmed with no effect on the other and parasitism in which one party benefits by damaging the other. Predation also belongs to this category in which eventually one of the parties is killed by the other. In the former category of associations, apart from symbiosis and proto-cooperation, there is also commensalism, which describes a beneficial interaction for one party with no effect on the other. Neutralism also belongs to this category in which the close spatial association between two species has no effect on each other.

The envisaged system in this work would be one that could potentially mimic symbiosis, which is obligatory and mutually beneficial. Since one part of this association is an artefact, it was decided to use the term *artificial* symbiosis to describe the system. This would come as the result of the integration between the robot and the microbes in the MFCs. The microorganisms, as part of their normal activities would be supplying energy to the robot, which in turn would be performing various tasks, one of which would be to collect more material from which the microbes may extract energy. This seemed to be the most interesting approach in the pursuit for building an autonomous robot.

This chapter, therefore, reports on the development of EcoBots-I and -II, which were proof-of-concept robots powered by the energy generated from MFCs as a result of the internal bacterial metabolism. EcoBot-I was running on MFCs employing *E. coli* monocultures fed with refined sucrose substrates, whereas EcoBot-II employed MFCs with a more complex mixture of microflora, which was fed with unrefined food waste, plant material or even dead insects. Our study aims to address a number of important issues for genuinely energetic autonomous robots. The behavioural repertoire of each of the two robots was a function of the onboard power outputs from the MFCs, therefore, EcoBot-I with a low MFC output, was designed to have a single behavioural task, which was to perform phototaxis, whereas EcoBot-II was designed with a more complicated behavioural repertoire, which involved sensing, processing and communication, on top of the phototactic actuation.

9.1.2 Microbial Fuel Cells

EcoBots-I and -II are powered from MFC, which are bio-electrochemical transducers converting the biochemical energy produced by bacteria metabolising refined or unrefined substrates into electricity. In their simplest form, they consist of two compartments, an anodic chamber (negative, i.e. electron generating) and a cathodic chamber (positive, i.e. electron accepting), both containing electrodes that are separated by a proton exchange membrane. The two electrode terminals are connected to an external electrical load, through which electrons generated at the anode, can flow. The anode compartment accommodates the microbes and the cathode 'closes' the system by accepting the electrons and protons. MFCs have been in existence since the early 1900s, when the first of these devices was reported [17]. Since then, this area has been extensively investigated to improve the understanding of the biochemical, electrochemical, thermodynamic and kinetic mechanisms that cause or restrict various reactions taking place, which generate the electricity.

With regard to the anodic compartment, at least three different types of MFC can be distinguished. These have been termed 'generations' (Gen-I,-II,-III) and the categorisation depends on the electron transfer mechanisms employed between the bacterial cells and the electrode surface [18].

The type of MFC employed in EcoBot-I was of Gen-I, where the biocatalyst in the MFCs' anode was a freshly grown culture of *E. coli* fed with refined sugar and mixed with methylene blue (MB). The latter was used as a redox mediator to extract a portion of the electrons produced from breaking down the sugar in the metabolic pathways of *E. coli*. Electrons are then transferred to the anode electrode and flow through the external electrical circuit to the cathode, thus, producing low levels of electrical current.

The type of MFCs used in EcoBot-II was working on mixed bacterial sludge cultures and are probably a hybrid between Gen-II (sulphate reducing) and -III (anodophiles). In Gen-II MFCs, bacteria metabolise the given substrate(s) and produce biochemical energy used for cell growth and maintenance, as well as waste products. Some bacterial species produce electro-active metabolites (e.g. H_2^-, FeII, pyocyanin, plastocyanine), which act as electron 'shuttles' from the bacterium to the electrode or to other species [19–24]. Thus, natural redox mediators replace the MB used in Gen-I. In addition, anodophilic bacteria [25, 26] attach to the electrode surface directly using this as their end terminal electron acceptor. This mechanism (Gen-III) may account for up to 20% of the activity in a normal sludge MFC, whilst sulphide production may account for the rest [27].

The advantages of using consortia with a large diversity of mixed species are numerous and lie with the wide substrate utilisation capability of such communities. These microcosm systems can be subjected to a cross-feed regime of entirely different substrates and still offer stability and higher levels of power compared to other types. In addition, the fact that these systems can be used for wastewater utilisation, makes them very attractive for scientific research by several workgroups worldwide [28, 29].

Electrons that have been transferred to the electrode surface then flow through the external circuit as current (I) and end up in the cathode. This is due to the electrophilic attraction that exists between the two electrodes. At the electron abstraction stage inside the bacterial cell, hydrogen ions (protons H^+) are released into the anolyte. These species flow through the proton exchange membrane (PEM), into the cathode and hence, the system remains in equilibrium. We have used two cathode systems; the liquid ferricyanide ($K_3Fe^{3-}[CN]_6$) chemical cathode and the oxygen (O_2) gas-diffusion cathode. In the case of the ferricyanide cathode, protons are taken up by the buffer added in the chemical solution up to the saturation point. After this, the continuous incoming of protons into the cathode has a negative effect on the pH of the electrolyte, which eventually degrades completely. As a result, the anode pH starts to decrease, since no more protons can be consumed by the cathodic half-cell and this has a detrimental effect on the overall MFC performance. On the other hand, in the case of the O_2 cathode, protons combine with electrons and O_2 molecules from free air to produce water, thus, never saturating the system (with protons) and as a result oppose any decrease in pH, which would otherwise be lethal.

In this chapter, we explore the use of MFCs as the power generators for robots capable of exemplar tasks of actuation (phototactic locomotion), sensing (temperature) and communication via a radio link. MFCs were bench tested for their ability to consume different types of biomass including plant and insect material. For these preliminary bench experiments, both ferricyanide and oxygen cathodes were tested and compared to the point of system exhaustion. Efficiency of energy conversion was calculated from the maximum available energy deduced using bomb calorimetry, which had to be conducted on the insect and plant material. A second set of experiments were then carried out with the MFCs mounted on and powering EcoBots-I and -II.

Since we are interested in genuine autonomy, in the case of EcoBot-II we eschewed the use of the ferricyanide as the oxidising agent in the cathode. However, one set of short distance experiments using a ferricyanide cathode with flies in the anode was conducted to provide a baseline for comparison. Since the preferred cathodic system was oxygen (free air), a more extensive investigation was undertaken with oxygen cathodes and three types of substrate: sucrose, rotten fruit (peach) and flies. As a final step, we measured the longevity of the oxygen cathode and fly substrate combination by carrying out endurance trials lasting typically up to 12 days.

9.2 Materials and Methods

9.2.1 MFC Setup for Robot Runs

9.2.1.1 EcoBot-I

For EcoBot-I experiments, two stacks of four MFCs, (one of the stacks is shown in Fig. 9.1), were employed onboard (eight in total), connected in series. The total open circuit voltage from the eight MFCs, before connecting them to EcoBot-I was 6.1 V (8×0.8 V).

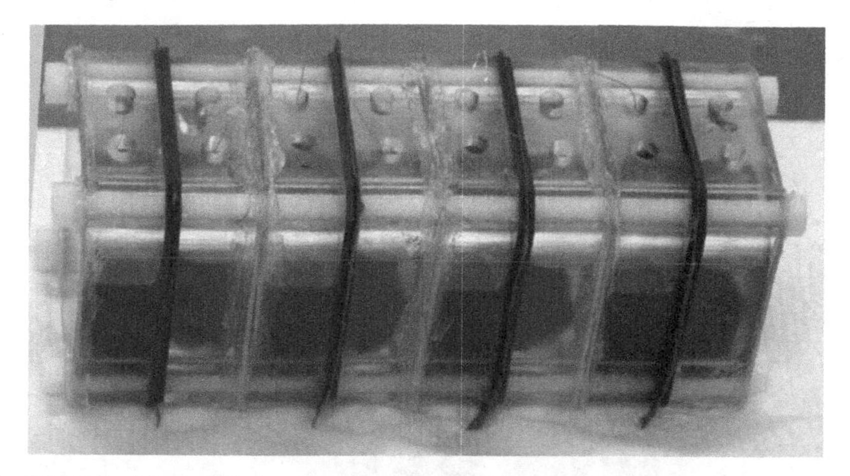

Fig. 9.1 Stack of four MFCs (one of the two stacks employed in the EcoBot-I runs). Dimensions in (mm) are: L (145) × W (50) × H (60). All eight MFCs were connected in series. The *blue chambers* designate the anodes of each MFC with the MB mediator and the *yellow chambers* designate the cathodes of each MFC with the ferricyanide

9.2.1.2 EcoBot-II

For EcoBot-II runs, eight MFCs employing the O_2 cathode were used and as in the case of EcoBot-I, they were connected in series. The total open circuit voltage from the MFCs before connecting them to EcoBot-II was 4.5 V (8×0.6 V). A picture of one of the O_2 MFCs employed onboard EcoBot-II is shown in Fig. 9.2.

9.2.2 Robot Design and Principle of Operation

9.2.2.1 EcoBot-I

EcoBot-I was designed to have a phototactic behaviour by employing the biomimetic approach of cross-wiring the two photo detecting diodes to the two dc geared *escap*® 205 motors, (Portescap, Switzerland), to provide differential drive. The principle of cross-wiring came from the field of human neuroscience, which teaches that each of the brain hemispheres is connected to, and controls, the muscles of the opposite side of the body. This is also true for the vision system, in which case, things in the left field view are interpreted by the right brain hemisphere and vice-versa [30]. The specific motors were chosen due to their high torque characteristics and low input voltage range (max. 5 V after modification), which were ideal for shifting relatively big masses with minimum energy. A circular piece of styrene material having a diameter of 22 and 2.5 cm thickness with two rectangular pockets to accommodate the MFC banks formed the main body of the robot. The styrene material was chosen because of its physical strength and low

Fig. 9.2 MFC employing the O_2 cathode, as used onboard EcoBot-II. Seen in the photo is the open-to-air cathode electrode with microporous filter paper

weight properties. The overall height was 7.5 cm with a total mass of 960 g. This was balanced with the use of two caster wheels. The assembled EcoBot-I (top left), as well as close-up photographs of the photodiodes (bottom left and right), are shown in Fig. 9.3.

System principle of operation

EcoBot-I employed a bank of eight MFCs, the electronic control circuit, two photo detecting diodes and two motors. Energy produced by the MFCs was accumulated in a bank of six capacitors. On reaching the pre-set threshold voltage for discharge (V_{dis}), the energy was released to power the photodiodes and the motors. The energy to the motors was distributed according to the signal from the photodiodes. Once the energy was expended and the pre-set threshold voltage for charge (V_{ch}) was reached, the robot would stop and remain 'idle' until enough energy was accumulated to power the motors and photodiodes again. This gave the prototype robot a form of burst motion phototactic behaviour. In this way, EcoBot-I was managing its energy production, accumulation and consumption in a way that required no human intervention. The circuit involved no batteries since the energy management (accumulation and release) was performed by the bank of six 4,700 μF capacitors (total capacitance = 28.2 mF). Figure 9.4 illustrates the block diagram of the robot's electronic control circuit.

The anolyte was made up of methylene blue MB mediator (0.1 mM), phosphate buffer, *E. coli* in solution and sugar as previously described [31]. The biochemical

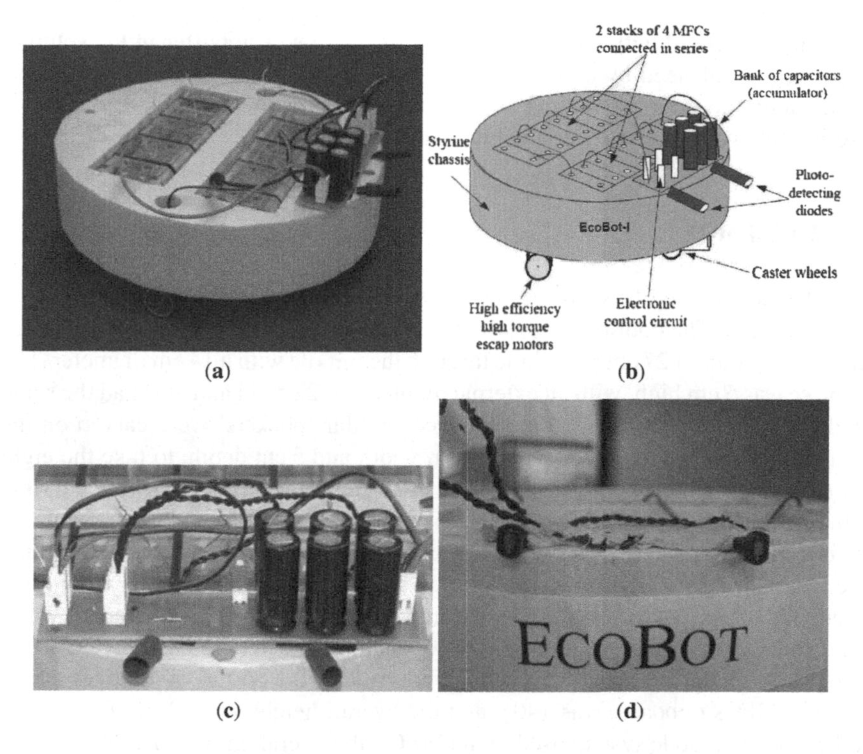

Fig. 9.3 (**a**) Photograph and (**b**) labelled schematic diagram of EcoBot-I fully assembled. The MFCs appear 'plugged in' to the top of the robot chassis. (**c**) Close-up photograph of EcoBot-I electronic circuitry and covered photodiodes and (**d**) close-up photograph of the two photodiodes

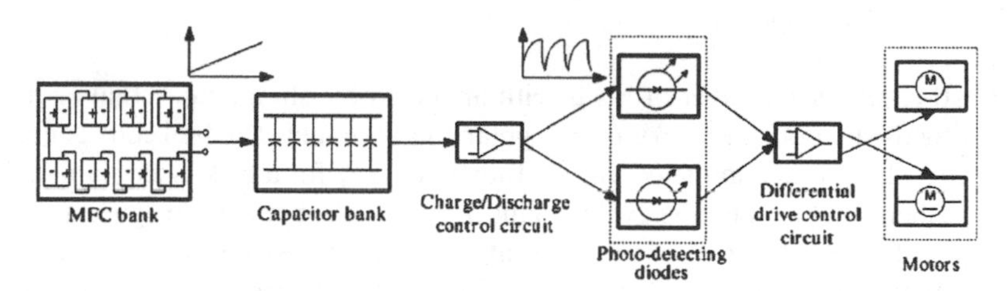

Fig. 9.4 Block diagram representation of the EcoBot-I electronic circuit

reaction at the most negative end of the microbial electron transport chain can be described as follows:

$$NAD^{+} + 2H^{+} + 2e^{-} \Leftrightarrow NADH + H^{+}.$$

In the cathode, incoming electrons from the external circuit reduced the buffered electrolyte in order to complete the system. In the presence of dissolved oxygen, the catholyte is oxidised with some water (H_2O) formation. The buffer in the solution keeps the pH balanced by consuming the hydrogen cations travelling through the PEM, tending to decrease the pH. These processes in both the anode and cathode described above are known as reduction–oxidation (redox).

9.2.2.2 EcoBot-II

EcoBot-II was made of two pieces of lightweight styrene material, placed one on top of the other. The bottom piece, which was the largest, had a height of 2.5 cm, an external diameter of 27 cm and a hole through the middle with a 14 cm diameter. The top piece was 5 cm high, with an external diameter of 22.5 cm and also had the same sized hole cut through its middle. Eight rectangular 'pockets' were carved on the circumference of the top piece with 5.5 cm width and 5 cm depth, to take the eight onboard MFCs, which would be supported by the bottom piece. On the left and right of the underside of the bottom styrene piece, two dc geared *escap*®-205 motors (Portescap, Switzerland) were attached to provide motion. These were the same type motors used for EcoBot-I, which were connected in the same way to provide differential drive. As with EcoBot-I, the motors were modified to operate down to a maximum of 5 V. Balance of the robot was achieved with two caster wheels that were placed on the front and back of the underside of the bottom styrene. The total mass (no MFCs onboard) was 140 g and the overall height was 10 cm. Depending on the type of cathode system used in the MFCs, the overall mass of EcoBot-II with the onboard MFCs was 780 g (O_2 cathode) or 980 g (Fe[CN] cathode). EcoBot-II is shown below in Fig. 9.5.

Principle of system operation

EcoBot-II employs a number of MFC with an anaerobic sludge microbial ecology and raw foods such as insects, fruits or vegetables as the fuel. The fuel cells generate electricity, which is accumulated until sufficient energy for a task is available. The energy is then released and the cycle begins again. The robot was designed to have four types of behaviour: (a) sensing, (b) information processing, (c) actuation and (d) communication. In order to perform phototaxis, two infrared-rejection photodiodes (VTB8440B, Pacer Components, Berkshire, UK) were cross connected with the two motors, to have differential drive and, therefore, steer towards the light. A 1-wire ® low power temperature sensor (DS18S20, Maxim-Dallas) was used for sensing temperature, which was connected to the onboard wireless transmitter (rfPIC12F675, Microchip) for temperature data transmission. The communication system used was the rfPIC kit™ 1 Flash development. The temperature sensor and the communication system were selected amongst others, mainly due to their low input energy requirements and the rfPIC microprocessor for its low 'sleep' mode current consumption. Figure 9.6 illustrates the block diagram of the EcoBot-II circuitry.

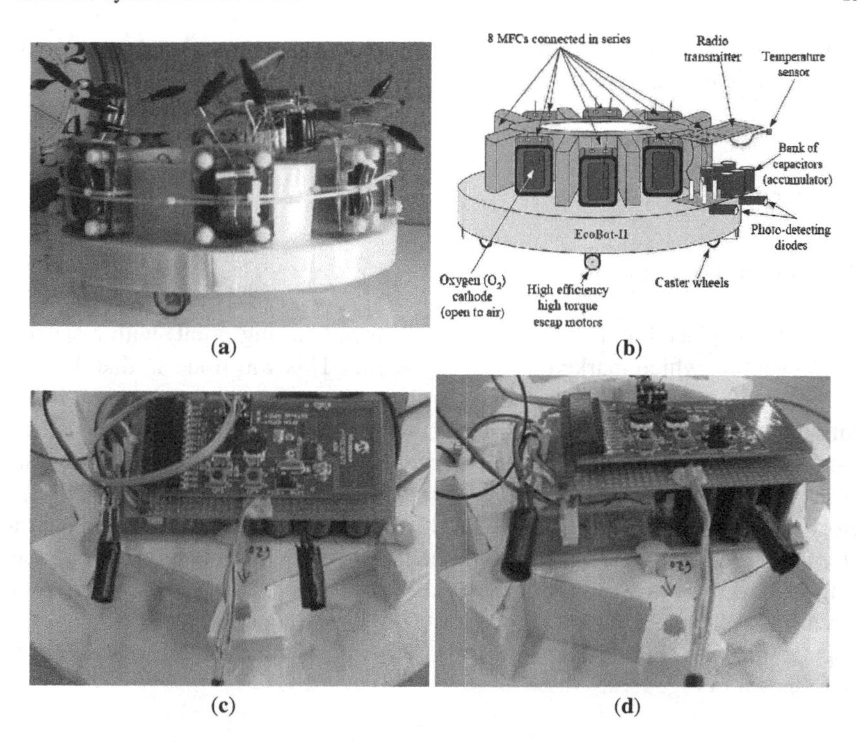

Fig. 9.5 (**a**) Photograph and (**b**) schematic diagram of the EcoBot-II fully assembled. The MFCs appear around the top of the circumference with the oxygen cathode facing outwards and are connected in series. (**c**) and (**d**) Close-up photographs of EcoBot-II sensors

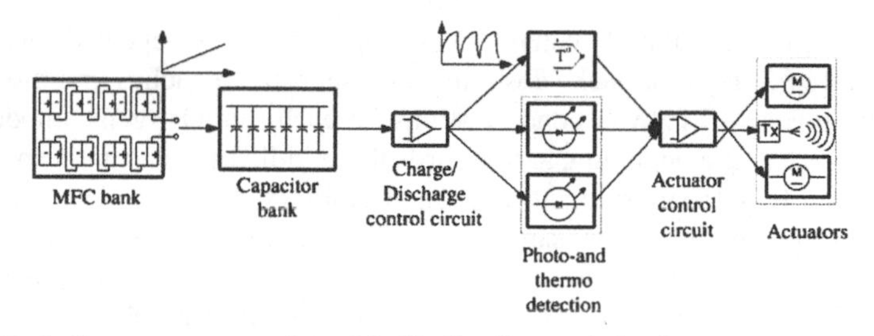

Fig. 9.6 Block diagram representation of the EcoBot II control circuit

As with EcoBot-I, the electrical energy produced by the MFCs was stored into the bank of six capacitors until a preset upper threshold was reached. At this point, the stored energy was released and distributed to the two photodiodes, two motors, temperature sensor and wireless transmitter (actuators). The energy distribution to the motors was governed by the light intensity information received from the photodiodes and the transmitted data was the information sensed by the temperature sensor. This continued until a pre-set lower threshold was reached, in which case the circuit went to sleep-mode until sufficient energy was accumulated in the capacitors. This process was repeated as long as the bacteria were alive.

9.2.3 Experimental Setup

9.2.3.1 EcoBot-I

The robot was placed at a 45° and 50 cm distance (starting point) with respect to the light source, which marked the end of the run. This was done so that the light seeking sequence could be performed, which was a random clockwise turn, until a light source was detected. Once the light source had been detected, EcoBot-I moved in a straight line towards it. At each point the robot stopped to recharge, its position was marked so that the distance between movements could be measured and the trajectory towards the light to be recorded. The total distance, including the radial trajectory covered during the light seeking sequence was approximately 70 cm. Seven EcoBot-I runs were repeated in total. Two of these runs were video recorded and the remaining five were data recorded using the Pico Scope® ADC-100 2-channel computer interface (Pico Technology Ltd). The experimental setup for the EcoBot-I runs is shown in Fig. 9.7a.

9.2.3.2 EcoBot-II

Short distance (50 cm) runs

Experiments, in which the O_2 cathode was employed, were repeated three times over a period of 3 days for each substrate (sucrose, rotten peaches and flies). Substrates were given at 0.1% w/v concentrations for each run. Oxygen cathodes were moistened twice at the beginning and towards the middle of each run. Experiments were started early in the morning of each day (07:30–08:00) when the ambient temperature was approximately the same. Depending on the time taken for the robot to cross the finish line, there was a typical 15–18 h window between each run, in which the MFC were being disconnected from the circuit. The robot was placed at the same start position for all the repeat runs, which was at a 90° and 50 cm away with respect to the light source. Temperature and time data were recorded in real time using a desktop pc. Figure 9.7(b) is a schematic illustration of the experimental setup for these experiments.

In addition to the above experiments, runs using ferricyanide as the catholyte were carried out employed only for the fly substrate (0.1% w/v). These were also repeated three times over a period of 3 days. In these experiments the MFC recovery time window between runs was approximately 22 h.

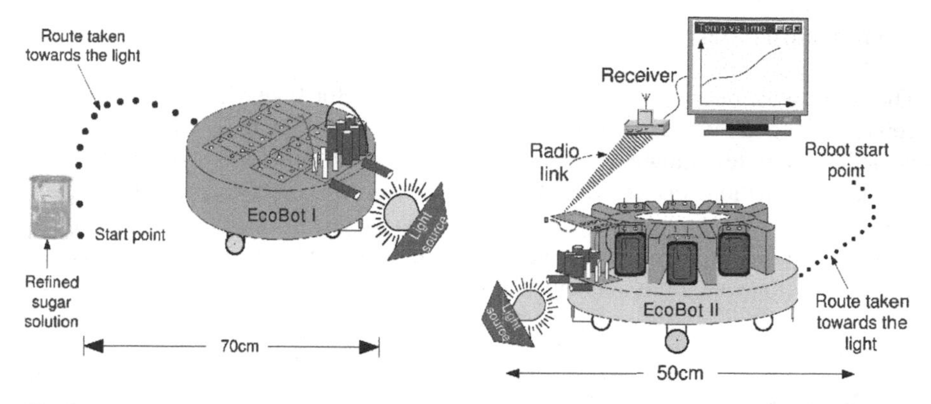

Fig. 9.7 Experimental setup for (*left*) EcoBot-I runs and (*right*) EcoBot-II runs

Long distance (continuous-endurance) runs O_2 plain electrodes

These experiments employed MFCs in which the anodic bacterial cultures were fed with flies and the cathodes were of the O_2 diffusion type. Runs were repeated two times at room temperature and the distance of communication between the EcoBot-II transmitter and desktop pc receiver was restricted to around 20 m. The robot, in both cases, was placed at the same starting position, which was at a 90° and 6 m away with respect to the light source. The bacteria in each MFC were fed with 0.1% w/v flies (equivalent to 1 fly per MFC) at the beginning of the experiment and the O_2 cathodes were moistened once a day in the morning with 3 mL artificial seawater (ASW). Artificial seawater was prepared according to Kester et al. [32]. Rainwater (10 mL per week) was also added to the anode compartment of the MFCs to replenish the water lost through osmotic pressure diffusion that existed between the two half-cells and overall evaporation.

Long distance (continuous) runs with ferricyanide electrodes

This experiment was setup in the same way as the experiments described in the section above but with the open cathodes being moistened with 3 mL of liquid ferricyanide (0.5 M) instead of ASW. As was the case before, the substrate was 0.1% w/v flies and the cathodes were moistened once daily and the EcoBot-II run was carried out only once over a period of 45 h. The objective of this experiment was to investigate the performance of the system, when the cathode was no longer a limiting factor.

Temperature data reception

The temperature data wirelessly transmitted from EcoBot-II were received by the receiver module (rfPIC kit, rfRXD0420, Microchip Technology Inc.), which was connected to a remote computer. The frequency of communication was 433.92 MHz and the data acquisition was performed using Microsoft's® multi-threaded teletyping (MTTTY), which is a virtual RS232 serial port terminal, operating on the TTY communications protocol (baud rate = 9,600). The maximum indoor distance of communication was approximately 30 m.

9.3 Results

9.3.1 EcoBot-I

The power output produced from the eight MFCs was found to be enough to operate the EcoBot-I prototype. In the experimental phase, two different configurations of MFC were compared; eight units in series, against two parallel stacks of four-in-series. The former (1×8) gave an open circuit voltage of 5.4 V and a short circuit current of 15 mA, whilst the latter (2×4) gave an open circuit voltage of 2.5 V and a short circuit current of 22 mA. In practice the eight in series configuration was employed to maintain the voltage above the critical threshold.

In a preliminary test, thin wires were employed to connect an off-board MFC stack to a SolarBot (Solarbotics, Canada, Beam Photo-popper 4.2, 1998), which had the solar panels removed. The rate of burst motion exhibited by the modified

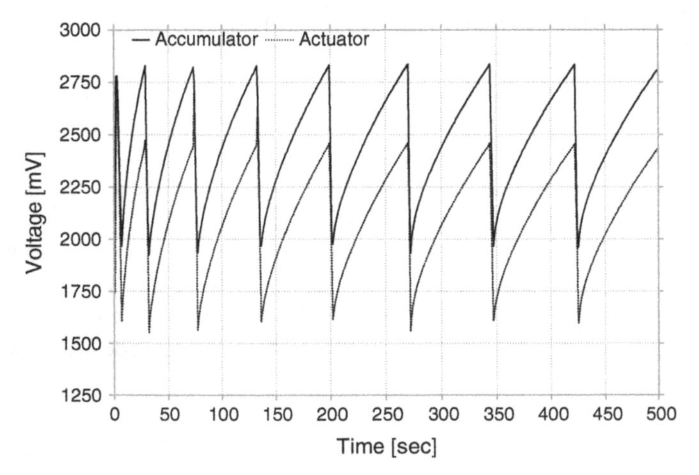

Fig. 9.8 Graphical representation of the charge/discharge cycle for EcoBot-I, across both the accumulator and actuators (motors)

SolarBot when powered by the MFCs was identical to that when powered by the solar panel under normal day-time illumination. Once the SolarBot experiment had proved successful, the next stage was to connect the MFC stack onboard EcoBot-I. Figure 9.8 illustrates the average charge/discharge cycles, in terms of average distance and time for one of the five trials, which is representative for all the EcoBot-I runs.

V_{ch} was set to 1.9 V and V_{dis} was set to 2.83 V. Average charging time t_1 was 30 s with the best times recorded at the beginning of the runs, when the MFCs were rested. Average discharge time was 3 s with the best times recorded once again at the beginning of each run. Knowing the overall capacitance value of the onboard accumulator as well as the average charge/discharge times and the threshold voltages, the average charge/discharge current values can be calculated. Charging current I_{ch} is given by, $I_{ch} = C \times \frac{dV}{dt_1}[A]$, where C is the overall capacitance in Farads [F], dV is the difference between charging and discharging threshold voltages and dt_1 is the difference between charge and discharge times in seconds, therefore;

$$I_{ch} = 28.2 \times 10^{-3} \times \frac{2.82-1.9}{30} = 880\,\mu A$$

Discharging current I_{dis} was calculated using the same equation, hence;

$$I_{dis} = 28.3 \times 10^{-3} \times \frac{2.82-1.9}{2} = 1.3\,mA$$

The average energy (W) produced by the MFCs and stored in the accumulator per charge cycle, with units of joules was calculated using the equation, $W = \frac{1}{2} \times CV^2$ and so $\Rightarrow W = \frac{1}{2} \times 28.2 \times 10^{-3} \times (2.83\text{-}1.9)^2 = 12.2\,mJ$.

The overall mean energy produced during each of the EcoBot-I runs was the result of the energy stored in the accumulator multiplied by the average number of movements. This was calculated to be 12.2 mJ x 27 = 329.4 mJ.

The graph in Fig. 9.9 illustrates the distance covered by EcoBot-I with respect to time. Data points on the curve represent the movement of EcoBot-I towards the light source in a pulsed-mode, as it was running out of energy and recharging to actuate in a repetitive manner. The results shown are the mean with standard deviation calculated from the five trials.

Figure 9.10 is a snapshot from one of the two video recorded EcoBot-I runs. The clock at the top of the picture shows the experimental real time and the black dots behind the robot, illustrate the trajectory that it followed to go towards the light source (bottom right).

On average, the light seeking sequence required four movements on a radial axis. Since only one of the motor was powered during this sequence, the mean distance covered in each movement was 6 cm. The phototactic behaviour required an average of 22 movements on a straight line. During phototaxis, both motors were powered and, therefore, the average distance per movement was 2 cm. The total number of movements in the 70 cm distance was 27 and the total time to reach the goal was 30 min, hence the overall speed of the robot was 1.4 m h^{-1}.

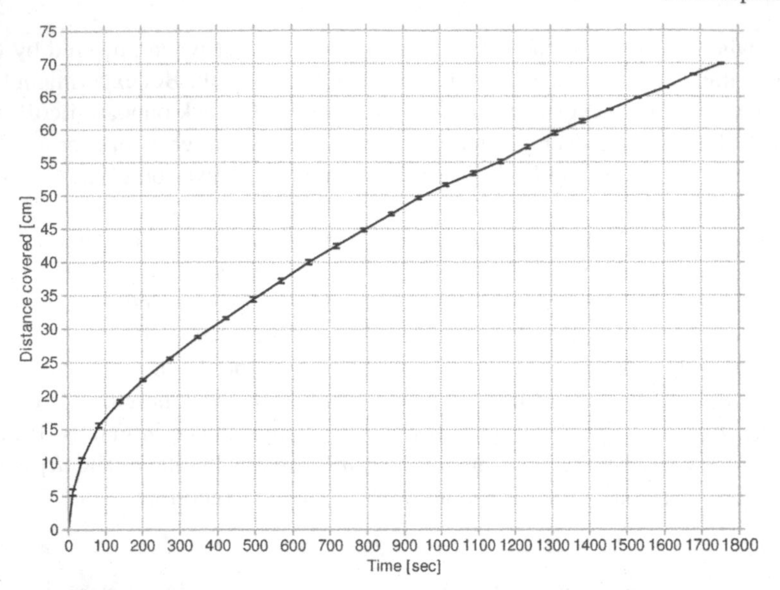

Fig. 9.9 Distance vs. time for EcoBot-I; Data shown are the mean and SD for n=5

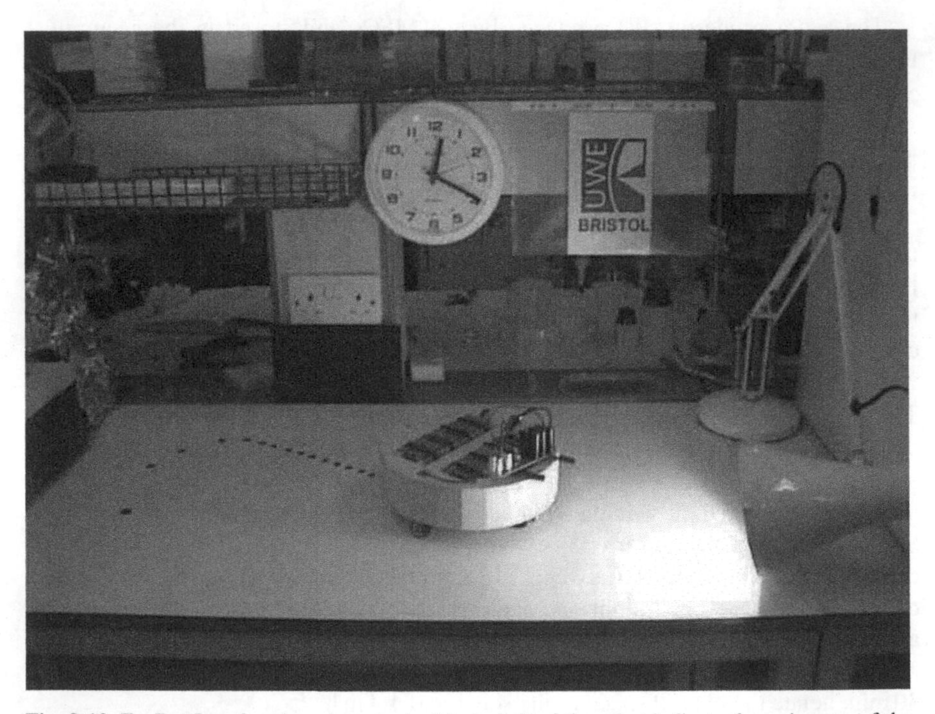

Fig. 9.10 EcoBot-I performing photo taxis. *Marks behind* the robot indicate the trajectory of the phototactic movement. The light source is on the *bottom right* of the picture and the *clock* is showing real time

9.3.2 EcoBot-II

9.3.2.1 Short-Distance Ferricyanide Cathode

The first EcoBot-II runs were carried out with MFCs employing the liquid ferricyanide cathode and fed with 0.1% w/v fly insects. The experiments were taking place once daily, during the same part of the day, when the indoor ambient temperature conditions were similar. Experiments were repeated three times over a distance of 50 cm and the results are shown in Fig. 9.11a, b. As seen, the average total number of movements was 40, and the overall time taken to cover the 50 cm distance was 2 h and 10 min. The sensed temperature increase from approaching the halogen light source was 3°C.

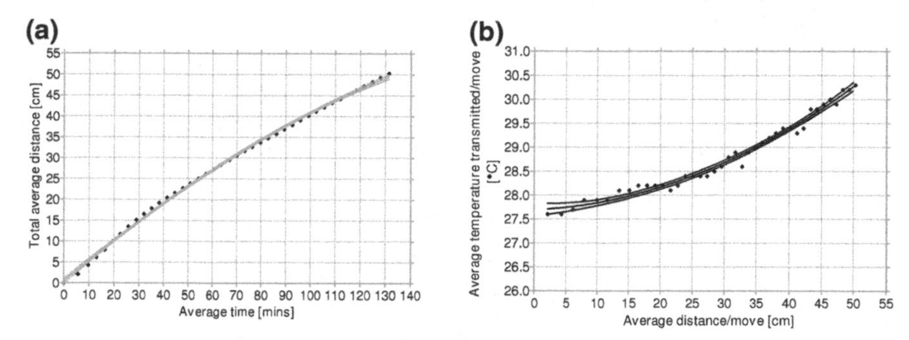

Fig. 9.11 (**a**) Average distance over time and (**b**) temperature gradient over the 50 cm distance from the three EcoBot-II trials running on fly insects and using the ferricyanide cathode. *Closed symbols* are the mean data from the repeat experiments with nonlinear regression fit curve (*solid line*) and 95% confidence interval (CI) (*dotted lines*)

9.3.2.2 Short Distance-Oxygen (O_2)Cathode

The oxygen cathode was the main focus of this part of the work since such a system is more favourable to be integrated with an autonomous agent. Therefore, it was decided to carry out a more extensive investigation by trying both refined and unrefined substrates as fuels. These were sugar, rotten peaches and fly insects and the runs for each substrate were repeated three times over a distance of 50 cm for the same final gram-weight/volume concentrations (0.1% w/v). The results derived from the runs in which the sludge bacteria were fed with sucrose, (0.1% w/v) are shown in Fig. 9.12a, b.

As seen from Fig. 9.12a, the average time taken for EcoBot-II to cover the 50 cm distance and reach the light source was 15 h and 20 min. The temperature gradient (Fig. 9.12b) produced from approaching the halogen light source was of the order of 8°C.

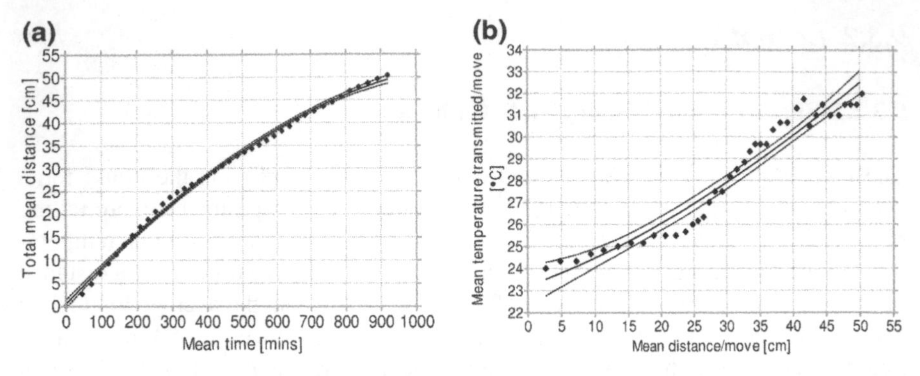

Fig. 9.12 (**a**) Average distance against average time and (**b**) temperature gradient produced for the three EcoBot-II runs. MFCs were fed with 0.1% (w/v) sucrose and employed the O_2 cathode. Data shown are the mean from the three repeats with the non-linear regression curve fit (*solid line*) and 95% CI (*dotted lines*)

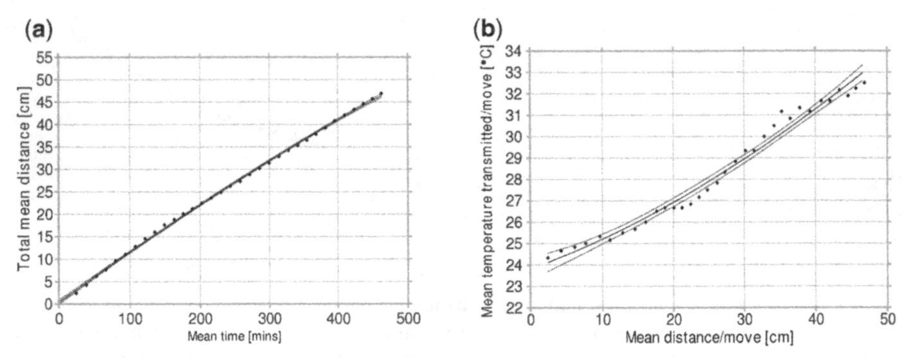

Fig. 9.13 (**a**) Average distance against average time and (**b**) temperature gradient produced for the three EcoBot-II runs. MFCs were fed with 0.1% (w/v) peach portions and also employed the O_2 cathode. Data shown are the mean from the three repeats with the non-linear regression curve fit (*solid line*) and 95% CI (*dotted lines*)

The next substrate that was investigated was of small portions of rotten peach, at the same wet-weight/volume (w/v) concentration (0.1%). The results from the three repeats are shown in Fig. 9.13a, b.

The average total time taken to reach the goal, when the bacteria in the MFCs were fed with portions of rotten peach was 7 h and 43 min. This was approximately 50% of the time taken to cover the same distance when sucrose was used as the fuel. The temperature gradient was of the same order as with sucrose.

Figure 9.14a, b illustrate the results from the three runs in which the sludge bacteria in the MFCs were fed with 0.1% wet-weight/volume flies.

When the fly insects were employed as the fuel, the total average time taken for EcoBot-II to cover the 50 cm distance was 6 h and 12 min. This was 60% and 20% shorter than the average total time taken when sucrose and peach portions

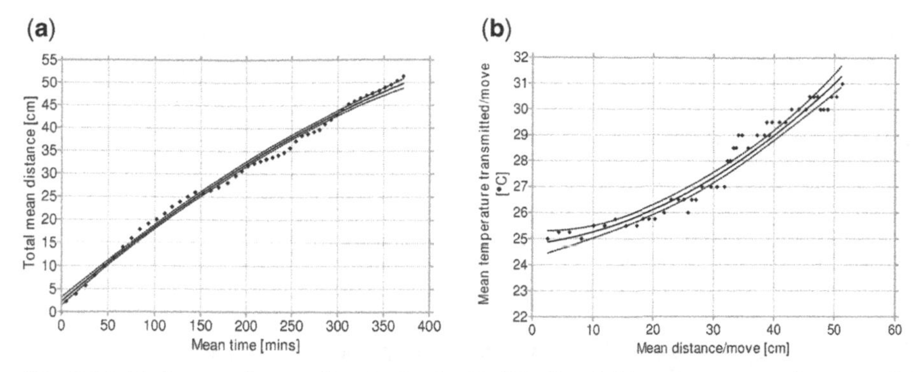

Fig. 9.14 (**a**) Average time vs distance for the EcoBot II and (**b**) temperature gradient over the 50 cm distance for the EcoBot II. MFCs were fed with 0.1% w/v flies and employed the O_2 cathodes. Data shown are the mean from the three repeats with a non linear regression curve fit (*solid line*) and 95% CI (*dotted lines*)

were employed as fuels respectively. Due to variations in the ambient temperature conditions, the gradient produced in these runs was 6°C.

9.3.2.3 Long Distance (Continuous-Endurance) Runs O_2 Plain Electrodes

In these endurance runs, as was the case with previous experiments, the MFCs were exhausted at first and then fed. The preferred substrate for this series of experiments was flies (0.1% w/v), which was in line with the objectives of the work. The robot was left in an open arena, moving towards a light source, sensing and transmitting temperature information to the base station, which was located 20 m away. The light source was placed at a distance of 6 m with respect to the start point and had no effect on the robot's surrounding temperature; EcoBot-II was, therefore, transmitting the ambient temperature. This distance was sufficiently long for investigating the endurance and viability of the onboard MFCs from a single feed of 1 fly per MFC under continuous operation, but short enough to allow phototaxis. The open cathodes were moistened with artificial seawater once in every 24 h for the first 5 days of the experiments. Figure 9.15a, b shows the average distance covered over 11 days and the ambient temperature data transmitted during these runs respectively.

The average endurance period was 11 days with a maximum of 12 and the average distance covered was 2 m with a maximum of 2.15 m. Data points shown on Fig. 9.15 are the data recorded at the end of each day. The average total number of movements towards the light was 150. The long term experiments were stopped when the EcoBot-II recharge time was approaching 24 h.

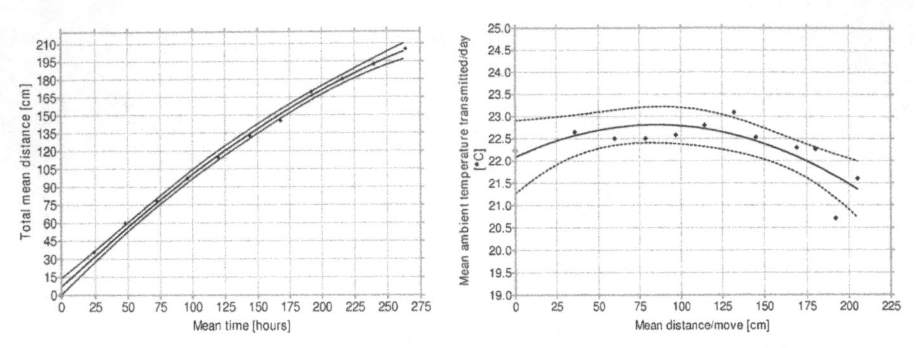

Fig. 9.15 (**a**) Average distance covered over an average period of 11 days and (**b**) temperature data transmitted from the EcoBot II to the base-station receiver for the endurance test. Although the robot was performing phototaxis, the halogen lamps were sufficiently distant not to affect the sensed/transmitted temperature

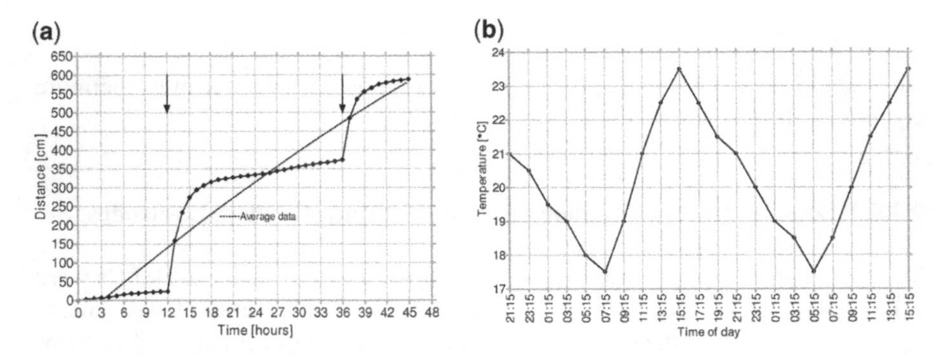

Fig. 9.16 (**a**) Distance covered by EcoBot-II over time; arrows indicate points of open cathode electrode moistening with 0.5 M ferricyanide. (**b**) Ambient temperature information transmitted from EcoBot-II as a function of the time of day. Substrate was 0.1% w/v flies

9.3.2.4 Long Distance (Continuous) Runs Ferricyanide Electrodes

The results from the continuous run in which the open O_2 cathodes were moistened with ferricyanide are shown in Fig. 9.16a, b. The objective of this experiment was to investigate the system performance with a cathode system operating nearer to optimum conditions and thus, being less of a limiting factor.

The total distance covered in the 2-day period was 5.85 m, and the time between movements (charge time) at the point of moistening was reduced to 20 s. Charge time was gradually increased to 7 min and the average distance covered per movement was 1.9 cm. Figure 9.16b, as was the case with the previous continuous runs, illustrates the ambient temperature over the time of day. Towards the end of the experiment, and as EcoBot-II was approaching the light source, it can be seen that the transmitted temperature started to increase.

Table 9.1 Energy efficiency for the different EcoBot II runs. Also shown in the table are the number of times that the robot moved and the time taken for it to reach the goal (short distance runs) or full exhaustion (long distance runs)

Onboard MFC	input [J]	# moves	duration [hrs]	output [J]	η [%]
Short distance (50cm) runs; substrate final concentration 0.1% w/v					
flies-Fe[CN]	701	40	2.19	2.55	0.37
sucrose-O_2	3400	38	15.32	2.43	0.07
peach-O_2	330	38	7.71	2.43	0.75
flies-O_2	701	42	6.19	2.67	0.38
Long distance (continuous) runs; substrate 1 fly/MFC					
flies-Fe[CN] moist	701	296	45	18.51	2.64
flies-ASW moist	701	150	264	9.5	1.35

The EcoBot-II energetics for both the short distance (50 cm) and the long distance continuous runs are summarised in Table 9.1, together with the data recorded when the O_2 cathode was moistened with liquid ferricyanide instead of water. Energy input was calculated from the theoretical maximum calorific value of each substrate and was compared against the sum of energy transfers into the onboard accumulator so as to calculate the EcoBot-II system efficiency.

In the case of the EcoBot-II runs, the output energy was directly related to the sum of energy transfers into the accumulator (hence number of movements). Although the number of movements was almost the same for all short distance experiments, the calculated efficiency was different due to the dissimilarity in the input energy value. The efficiency of the system when fed with sucrose (0.07%) was one order of magnitude lower than that recorded for the unrefined peaches (0.75%) and 20% that recorded for flies (0.39%). The efficiency of the EcoBot-II system when fed with flies was for both the ferricyanide and O_2 cathodes identical. The experimental duration when sucrose was employed as the fuel was comparatively the longest.

During the long distance continuous experiments, the efficiencies recorded were higher. Moistening of the electrodes with ferricyanide resulted in a 2-fold efficiency improvement over artificial seawater (ASW). Consequently, the total number of movements for ferricyanide was twice that for ASW.

On the grounds of behavioural efficiency, EcoBot-II was 3-fold more efficient, in the 50 cm runs, when employing the ferricyanide cathode (since the time required to complete the task was shorter by a factor of 3), than when the O_2 cathode was employed. A similar effect was noted for the continuous long distance runs. In these cases, moistening of the open cathode with ferricyanide enabled EcoBot-II to cover 6 m in 45 h, compared to 2 m in 12 days, when the cathode electrodes were moistened with ASW.

9.4 Discussion

EcoBot-I was shown to partially illustrate self-sustainability, by carrying its power supply, which provided enough energy to perform phototaxis over a period of approximately 30 min and a distance of 70 cm. Although in basic principle this project seemed very similar to Wilkinson's Gastrobot [15, 16], there were several significant differences.

In the first instance, Gastrobot employed two Ni–Cd battery stacks which required initial off-board charging to start the system. These batteries were used to power the various onboard actuators. Neither of the two EcoBots employed any form of conventional power supply, since energy was provided directly by the MFCs. Energy management was performed through a bank of capacitors, which served as short-term onboard accumulators. Hence, there was no need for initial charging to start the system in the case of EcoBots.

Gastrobot employed a 'stomach' at the front-end of the vehicle, which contained the *E. coli* bacterial culture. The bacteria were mixed with hydroxy-1,4-naphthoquinone (HNQ) mediator, and fed with refined sucrose. During the *E. coli* metabolism, the HNQ liquid mediator became reduced (mediator molecules gained electrons) and was driven through microporous filters towards six chemical fuel cells with ferricyanide cathodes. At the anode electrodes, the mediator became oxidised (mediator molecules released electrons) and therefore, electricity was produced at the fuel cell terminals, which was used to recharge the battery stacks. The oxidised mediator was refluent into the fermenter. Neither of the two EcoBots employed a separate bioreactor, since the bacterial cultures were included in the MFC anodes.

On the basis of system design and operation, Gastrobot was bigger and heavier than either EcoBot by a factor of eight, weighing approximately 7 kg (personal communication). The charge/discharge cycle of EcoBot-I (30 s/2 s) was shorter than that of Gastronome (18 h/15 min). In terms of behaviour, EcoBots-I and -II, both weighing less than 1 kg, were performing actuation (phototaxis) and in addition, EcoBot-II was sensing, processing and communicating information about its environment, whereas Gastrobot simply moved forward. Low cost materials were employed in the assembly of both EcoBot versions except for a small amount of platinum wire, whereas Gastronome used reticulated vitreous carbon (RVC) electrodes, titanium end plates and microporous filters. Finally, the performance from both EcoBots was repeatable and consistent.

During the light seeking sequence, the distance per movement covered was longer since the accumulated energy was supplied to only one of the motors. Consequently, the robot was moving on a radial axis, turning towards the light. At the point when the light was detected by both photodiodes, the energy was divided between the two motors, resulting in shorter straight line movements. For EcoBot-I, the task of covering the preset distance (70 cm total – 50 cm straight line) while performing phototaxis was completed in a mean time of 30 min.

In general, the distance per movement covered for EcoBot-I was 3-fold longer than for EcoBot-II even though the control circuit was the same for both robots and therefore, the accumulated energy per charge cycle was of identical magnitude. EcoBot-II moved at a slower rate than EcoBot-I, which means that the charge/discharge cycle for the former was slower and this may be attributed to a number of reasons. Firstly, the distribution of the onboard accumulated energy was dissimilar since, in the case of EcoBot-II, it was divided between sensing, processing, communication and actuation. This suggests that less energy was supplied to the motors (actuators), which was sufficient for a shorter movement than that of EcoBot-I. Furthermore, the open circuit voltage from the ferricyanide cathode MFCs was higher and this may be attributed to the electrochemical properties of the anolytes and catholytes. According to the laws of electrochemistry, the O_2 cathode MFC should have been producing a higher open circuit voltage, since the standard redox potential ($E_0{}'$) for O_2 is +0.82 V, which is twice that of ferricyanide (+0.437 V). In practice this was not the case, since the O_2 cathode was suboptimal and underperforming. This may be due to low O_2 diffusion and electrode reaction rates, which could result in electrode polarisation and therefore further work should be done to improve this system. This was probably when ferricyanide was employed in the initial EcoBot-II short distance experiments, which resulted in the robot being 160 g heavier, the total duration for completing the phototactic task was 1/3 of the most efficient run with the O_2 cathode. The same effect was observed for the long distance EcoBot-II experiments in which the open cathodes were moistened with ferricyanide.

Energy content in raw substrates can be categorised into maximum available energy and maximum metabolisable energy [33] and this is different for different substrates. Maximum metabolisable energy coefficient is given by: MEC = 1−(energy excreted/energy ingested), and even for the same substrates it usually differs for different animals or microorganisms [33]. This suggests that, although a particular substrate may be rich in energy content, the available energy for metabolism greatly depends on the living microorganism and the type of substrate. Potentially, this may prove useful in future developments of autonomous robots with a specially designed diet.

Comparing between the three exemplar substrates employed with the O_2 cathode to power EcoBot-II, flies proved to be better than rotten peach portions and refined sugar. Flies, apart from chitin in their exoskeleton, consist of high protein especially in their eyes and lower body, lipid material as well as numerous vitamins and minerals in the main body and intestines [34]. Bacteria in sludge are quite diverse and therefore, capable of attacking the majority of these multiple substrate classes, thus collectively increasing the number of intake pathways. Rotten fruit utilisation by sludge was not as efficient as flies and this may be attributed to the high acidity with fungal growth conditions, which may be less compatible with anodic digestion and metabolism. Still, it was demonstrated that food waste could be utilised as a fuel to produce useful electricity and although the utilisation efficiency for fruit was not as good as for flies, it was higher than sucrose. Sucrose being refined can be broken down by a single enzyme (sucrose hydrolase SUH), resulting in the need for two sugar transport systems; one for glucose and one for fructose. The limited number of intake pathways used for refined substrates compared with the multitude of enzymes and transport systems found in the sludge consortia that enable the breakdown of complex polymers found in fruits and insects may help explain why sucrose proved less efficient.

The task efficiency of the EcoBot-II, as a system, depended primarily on the number of movements. Hence, the higher the number of movements for the same feed, the higher was the number of energy transfers and thus, the overall energy value. The energy transfers, however, were governed by the operating voltage range (1.9–2.83 V), which required 14 min to accumulate and only 2 s to expend. This means, that the robot would have had to move 40× (for the continuous runs) or 280× (for the short distance runs) more than it did to reach an efficiency of 100%. In terms of time, this would have been several weeks, which would be impossible to maintain for batch-mode MFC systems, without interventions for replenishment or feeding. It is clear that for any design of future EcoBot prototypes, the batch-type MFC should be replaced with a continuous flow MFC, which if given periodic feeding will work for as long as the physicochemical conditions remain favourable; potentially for years.

With regard to the configuration of the fuel cells in a stack, the data suggest that although the control circuitry is 96% efficient in terms of losses, it was inefficient with regards to extracting energy from the MFCs probably due to the MFC series configuration, as it was allowing only a portion of the MFC produced energy to be transferred to the capacitors. The MFCs were primarily connected in series, to step-up the voltage to be within the operating range of 1.9–2.83 V. This was dictated by the minimum voltage rating for the majority of the circuit components and the amount of energy transfer.

The low efficiency of the EcoBot-II system is compensated by the computational ability of this robot. Although slower in charging and only moving 1/3 of the distance/movement than EcoBot-I, EcoBot-II incorporated a level of computational ability that required the employment of a microprocessor unit. This means, that the energy produced by the MFCs was more efficiently expended, since it was distributed to a number of tasks rather than a single task as in the case of EcoBot-I.

During the EcoBot-I and -II experiments, there was lack of vital information available to the operator. Variable parameters such as the status of the charging or discharging voltage were not readily available and could only be determined by manually measuring the terminal voltage across the circuit. Although digital multimeters (DMM) draw very little energy when connected to conventional circuits, in the case of MFCs the current losses incurred due to the DMM connection were more significant. Furthermore, the value of anodic and cathodic pH is also important and was not available. These are variables that should have been monitored and automatically controlled, if there was sufficient energy available for this level of processing power. The EcoBot runs served as a classic example of the underdetermination problem, which in this case seems to be directly related to energy availability.

At these early MFC development stages, each unit provides only tens of microwatts at around 0.7 V. Useful 'working' energy is generated by accumulating the supply from such a relatively small power source. This mode of cyclic 'store and release' activity is referred to as 'pulsed behaviour' and for many autonomous robotic applications it may be acceptable – one can imagine a robotic sensor network where each sensor is active for a relatively short period of time while otherwise accumulating energy for a known period prior to a regular transmission of sensed data. However, a sensor could continue accumulating energy while waiting for some triggering event, which requires reporting immediately. Pulsed behaviour will also have an impact on the temporal dynamics of the way in which robots execute their action selection regime. Doing 'nothing' is an important option; choosing not to carry out actions in the short term gives a robot the opportunity to do things in the longer term, which require large amounts of energy. This could not be achieved unless the robot had 'sacrificed' earlier opportunities for short term gain. At the same time, this will have to be balanced with the energetic requirements of internal 'metabolic' and 'housekeeping' processes such as keeping temperature and pH within lethal boundaries [35] by either internal regulation (e.g. increasing flow rates) or external behaviour (e.g. moving into shadow on a hot day). Thus, as energetically autonomous robots become more sophisticated, there is likely to be an increase in the processing overhead involved with the monitoring of internal and external states as well as planning and action selection – all requiring energy. Moreover, energetically autonomous robots will have to expend energy in obtaining or ingesting their food as well as voiding waste material.

9.5 Conclusions

Batch-mode MFCs have been demonstrated as a possible alternate to conventional power sources for autonomous mobile robots. A stack of eight MFCs was attached to two mobile robots (EcoBot-I and -II) as the sole power source and produced sufficient energy to run the robots through a cycle of tasks repeatedly. Following charging, sufficient energy was accumulated onboard to drive the robot assembly and control circuitry. Typically, 2 s of movement was recorded for both robots following 30 s (EcoBot-I) and 14 min (EcoBot-II) of charging time.

As previously described, the main objective of this part of the work was to build a robot that could be directly and solely powered by the bacterial metabolism taking place in MFCs. This would demonstrate a primitive form of integration between the two technologies towards an artificial symbiosis, which would potentially enable the robot to remain self-sustainable by extracting energy from the environment.

Pragmatically, in dealing with real world issues, Gastrobot and EcoBot-I had three serious problems. Firstly, they both used a refined fuel, not unrefined biomass. Secondly, they both employed artificial mediators (HNQ, methylene blue) in the anodic chamber, which were eventually degraded and could not be easily replaced from the environment. Similarly, they employed ferricyanide as an oxidising agent in the cathodic chamber, which again, was used up and could not be easily replaced from the environment. Although the idea of MFC powered robots utilising unrefined substrates has been mooted, this has never been attempted. EcoBot-II successfully addresses these issues by using unrefined biomass (e.g. dead flies), natural mediators (sulphide system) and air cathodes. However, EcoBot-II does not yet actively 'catch' its food or remove its own waste and these important issues will be addressed in the next phase of research.

EcoBot-II, as a self-sufficient system, was shown to be inefficient (Table 9.1). The efficiency was directly related to the number of movements towards the light, which represented the quantity of energy transfers from the MFCs into the capacitors.

Although low in efficiency, EcoBot-II was shown to exhibit a computational ability of a relatively high level of complexity, since apart from simply steering towards the light, it was sensing, processing and communicating the ambient temperature information.

The integration between a mechatronic system and a microbial ecosystem may be seen as a form of artificial symbiosis based on mutualism, since both systems could be benefiting from this association. In such a system, bacteria could be providing the electrical energy to power the robot and while performing pre-assigned tasks the robot could be collecting food for the onboard microbes and ensuring that the internal incubation conditions are optimal, thus completing the ongoing cycle. EcoBot-II represents an exemplar of microbe–robot symbiosis, which can be described as the system whereby consortia of microbes are well adapted to growing in quasi-steady state within the anodic system that is provided by the robot. The continuous or long-term periodic feeding of nutrients to the microbes in the anodic chamber was not designed into EcoBots-I or -II, therefore, this work only partially demonstrated the notion of artificial symbiosis. A robot that can go through many cycles of feeding, actuation tasks and ejection of waste would be required to demonstrate true autonomy. This will be our task for EcoBot-III.

References

1. Holland, O.: Towards true autonomy. In Proceedings 29th Int. Symp. Robot. (ISR98), Birmingham, UK, 84–87 (1998)
2. McFarland, D., Spier, E.: Basic cycles, utility and opportunism in self-sufficient robots. Robot. Autonom. Sys. 20, 179–190 (1997)
3. Kaufmann, S.: Investigations. Oxford University Press: New York, USA (2000)
4. Webb, B.: The first mobile robot. In Proceedings of TIMR 99, Towards Intelligent Mobile Robots. Bristol, UK (1999)
5. Squyres, W., S., Arvidson, R., E., Bell III, J., F., Bruckner, J., Cabrol, N., A., Calvin, W. et al.: The Spirit Rover's Athena Science Investigation at Gusev Crater, Mars. Science 305, 794–799 (2004)
6. Greenman G., Kelly I., Kendall K., McFarland, D., Melhuish, C.: Towards robot autonomy in the natural world: A robot in predator's clothing. Mechatronics 13, 195–228 (2003)
7. Kelly, I., Holland, O., Melhuish, C.:Slugbot: A Robotic Predator in the Natural World. In Proceedings 5th Int. Symp. Artif. Life Robot. (AROB 5th '00) Human Welfare Artif. Liferobot., Oita, Japan, pp. 470–475 (2000)
8. Kelly I., Melhuish C.: Slugbot: A Robot Predator. In Proceedings Euro. Conf. Artif. Life (ECAL), Prague, Czech Republic, pp. 519–528 (2001)
9. Kubo, M., Melhuish, C.: Robot Trophallaxis: Managing Energy Autonomy in Multiple Robots. In Proceedings Towards Autonom. Robot. Sys. (TAROS 04), Colchester, UK, pp. 77–84 (2004)
10. Melhuish, C., Kubo, M.: Collective Energy Distribution: Maintaining the Energy balance in Distributed Autonomous Robots. In Proceedings 7th Int. Symp. Distrib. Autonom. Robot. Sys., Toulouse, France, pp. 261–270 (2004)
11. Prescott, L. M., Harley, J.P., Klein, D. A.: Microbiology. Brown, London (1995)
12. Boopathy, R.: Methanogenic transformation of methylfurfural compounds to furfural. Appl. Environ. Microbiol. 62, 3483–3485 (1996)

13. Lomans, B. P., Op den Camp, H. J. M., Pol, A., van der Drift, C., Vogels, G. D.: Role of methanogens and other bacteria in degradation of dimethyl sulfide and methanethiol in anoxic freshwater sediments. Appl. Environ. Microbiol. 65, 2116–2121 (1999)
14. Watanabe K, Kodama Y, Hamamura N, Kaku N.: Diversity, abundance, and activity of archaeal populations in oil-contaminated groundwater accumulated at the bottom of an underground crude oil storage cavity. Appl Environ Microbiol. 68, 3899–3907 (2002)
15. Wilkinson, S.: 'Gastronome' – A Pioneering Food Powered Mobile Robot. In Proceedings 8th IASTED, International Conference on Robotics and Applications, Paper No. 318–037, Honolulu, Hawaii, USA (2000)
16. Wilkinson, S.: Hungry for success – future directions in gastrobotics research. Industrial Robot 28(3), 213–219 (2001)
17. Potter, M. C.: Electrical effects accompanying the decomposition of organic compounds. Proc. R. Soc. 84 B, 260–276 (1912)
18. Ieropoulos, I., Greenman, J., Melhuish C., Hart J.: Comparison of three different types of microbial fuel cell. Enz. Microb. Technol. 37, 238–245 (2005)
19. Bond, D. R., Holmes, D. E., Tender L. M., Lovley, D. R.: Electrode-reducing microorganisms that harvest energy from marine sediments. Science 295, 483–485 (2002)
20. Habermann, W., Pommer, E.-H.: Biological fuel cells with sulphide storage capacity. Appl. Microbiol. Biotechnol. 35, 128–133 (1991)
21. Hernandez, M. E., Newman D. K.: Extracellular electron transfer. Cell. Mol. Life. Sci. 58: 1562–1571 (2001)
22. Kim, B. H.: Development of a Mediator-less Microbial Fuel Cell. In Abst. 98th Gener. Meet. Amer. Soc. Microbiol. Washington, D.C., USA, Paper # 0-12, Session 116-0 (1998)
23. Rabaey, K., Boon, N., Siciliano, D., Verhaege, M., Verstraete, W.: Biofuel cells select for microbial consortia that self-mediate electron transfer. App. Environ. Microbiol. 70, 5373–5382 (2004)
24. Sigfridsson, K.: Plastocyanine, an electron-transfer protein. Photosynth. Res. 57, 1–28 (1998)
25. Bond, D. R., Lovley, D. R.: Electricity Production by Geobacter sulfurreducens Attached to Electrodes. Appl. Environ. Microbiol. 69, 1548–1555 (2003)
26. Caccavo, Jr. et al.: Geobacter sulfurreducens sp. nov., a Hydrogen- and Acetate- Oxidising Dissimilatory Metal-Reducing Microorganism. Appl. Environ. Microbiol. 60(10): 3752–3759 (1994)
27. Ieropoulos, I., Melhuish, C., Greenman, J., Hart, J.: Energy accumulation and improved performance in microbial fuel cells. Power Sources 145, 353–356 (2005)
28. Liu, H., Ramnarayanan, R., Logan, B. E.: Production of electricity during wastewater treatment using a single chamber microbial fuel cell. Environ. Sci. Technol. 38, 2281–2285 (2004)
29. Min, B., Logan, B. E.: Continuous Electricity generation from domestic wastewater and organic substrates in a flat plate microbial fuel cell. Environ. Sci. Technol. 38(21), 5809–5814 (2004)
30. Coren, S.: The left-hander syndrome: The causes and consequences of left-handedness. Free, New York, NY, (1992)
31. Bennetto, H.P.: Electricity generation by microorganisms. Biotech. Ed. 1, 163–168 (1990)
32. Kester, D., Duedall, I., Connors, D., Pytkowicz, R.: Preparation of artificial seawater. Limnol. Oceanogr. 12, 176–179 (1967)
33. Weiser, J., I., Porth, A., Mertens, D., Karasov, W. H.: Digestion of chitin by Northern Bobwhites and American Robins. Condor 99, 554–556 (1997)
34. DeFoliart, G. R.:Insects as human food: Gene DeFoliart discusses some nutritional and economic aspects. Crop Prot. 11, 395-399 (1992)
35. Ashby, W.R.: Design for a Brain. Chapman and Hall, London, UK (1952)

Chapter 10
The Phi-Bot: A Robot Controlled by a Slime Mould

Soichiro Tsuda, Stefan Artmann, and Klaus-Peter Zauner

10.1 Introduction

Information processing in natural systems radically differs from current information technology. This difference is particularly apparent in the area of robotics, where both the organisms and artificial devices face a similar challenge: the need to act in real time in a complex environment and to do so with computing resources severely limited by their size and power consumption. Biological systems evolved enviable computing capabilities to cope with noisy and harsh environments and to compete with rivalling life forms. Information processing in biological systems, from single-cell organisms to brains, directly utilises the physical and chemical processes of cellular and intracellular dynamics, whereas that in artificial systems is, in principle, independent of any physical implementation. The formidable gap between artificial and natural systems in terms of information processing capability [1] motivates research into biological modes of information processing. Hybrid artifacts, for example, try to overcome the theoretic and physical limits of information processing in solid-state realisations of digital von Neumann machines by exploiting the self-organisation of naturally evolved systems in engineered environments [2, 3].

This chapter presents a particular unconventional computing system, the Φ-bot, whose control is based on the behaviour of the true slime mould *Physarum polycephalum*. The second section gives a short introduction to the information-processing capabilities of this organism. The third section describes the two generations of the Φ-bot built so far. To discuss information-theoretic aspects of this robot, it is useful to sketch the concept of bounded computability that relates generic traits of information-processing systems with specific physico-chemical constraints on the realisation of such systems in different classes of computational media. This is done in the fourth section. The concluding section gives an outlook on engineering as well as foundational issues that will be important for the future development of the Φ-bot.

A. Adamatzky and M. Komosinski (eds.), *Artificial Life Models in Hardware*,

10.2 *Physarum Polycephalum* as Information Processor

The plasmodium of the true slime mould, *Physarum polycephalum* (Fig. 10.1), is an amoeba-like unicellular organism, whose body size ranges from several 100 μ to a radius of more than 1 m. Despite its large size, the single cell acts as an integrated organism and is known for its distributed information processing.

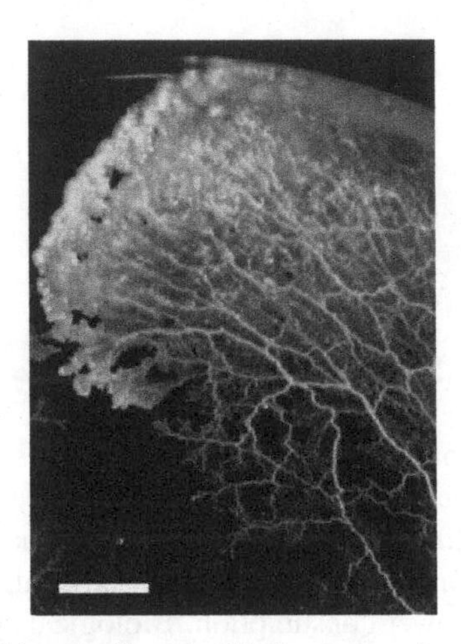

Fig. 10.1 The plasmodium of the true slime mould *Physarum polycephalum* growing on 1.5% agar gel. The *white bar* on the panel indicates 5 mm

A plasmodial cell of *Physarum polycephalum* consists of an ectoplasm tube that encloses an endoplasmic core. The former is a gel membrane layer, while the latter is a more fluid state of the protoplasm [4]. In the ectoplasm tube, cytoplasmic actomyosin periodically aggregates to form sheet-like structures and then unravels into fibrils. These structural changes create a hydrostatic pressure gradient within the cell, and eventually give rise to a flow of ectoplasm shuttling from one location in the cell to other parts of the cell and back. If a cell is not stimulated by any external stimuli, then this contractile rhythm is synchronised throughout the cell. However, when a local part of a cell is exposed to an external stimulus such as food or white light, it leads to de-synchronisation of the rhythm. The frequency of the oscillating rhythm at the stimulated part increases if it is an attractive stimulus, and decreases if it is repulsive. Such local frequency change affects oscillations of other parts through protoplasmic streaming and forms a spatial phase pattern in the cell [5–7]. The emerging global phase pattern eventually determines the direction of migration, i.e. the behaviour of the organism [8].

This mode of information processing affords scalability to the plasmodium. As long as the plasmodium is able to form the phase gradient of the contractile oscillation rhythm within its single-cell body, it reacts to various external stimuli in the same fashion no matter how large it grows. Central to this size-invariant behaviour is the spatial phase pattern of the oscillation rhythm formed within a cell. It emerges from the interaction of the intracellular dynamics of the plasmodium and the environment triggered by a contact with an external stimulus and the plasmodium. Several theoretical models have been proposed to explain the behaviour [9, 10] on the basis of the theory of positional information [11]. Takamatsu, taking a system's perspective, constructed coupled nonlinear oscillator systems from the *Physarum* cell by controlling the cell shape using microfabricated structures [12, 13].

The distributed information processing that arises from the integration of oscillatory patterns in the cell enables the plasmodium to optimise food gathering strategies [14, 15].

It is interesting to note that the information processing in the cell can access information about past states. Nakagaki and his colleagues found if exposed to periodic environmental changes, the plasmodium was able to anticipate the next change by changing its behaviour at the time, a periodic change was next due to occur; the memory persisted over several hours [16].

10.3 Cellular Robot Control

The information-processing abilities of the plasmodium described above, together with the relative ease with which plasmodial cells can be manipulated, facilitate the use of this simple organism in hybrid systems. Using the plasmodial cell may provide adaptive behaviour in engineered systems in the future. Robot control will be a good test bed for exploring this possibility. For this purpose, the contractile oscillation dynamics of the cell were employed to control robots. A *Physarum* plasmodium integrated into a robot is coupled to the robot's environment via bi-directional interfaces. External signals received by the robot are transferred to stimulating signals to the *Physarum* cell as robot brain, and the reaction of the cell is transferred back to the robot and interpreted as source of control signals for the robot. In the following subsections, we will present our cellular robot approach, the Φ-bot project, towards life-like autonomous robots.

10.3.1 The First Generation of the Φ-bot: Tethered Robot Design

The *Physarum*-controlled robot consists of three components: a *Physarum* cell, a hexapod robot and interfaces between them (Fig. 10.2).

10.3.1.1 The *Physarum* Circuit

A *Physarum* plasmodium is confined in a star shape which consists of six 1.5 mm diameter circular wells and six 0.4 mm width channels connecting the circles at

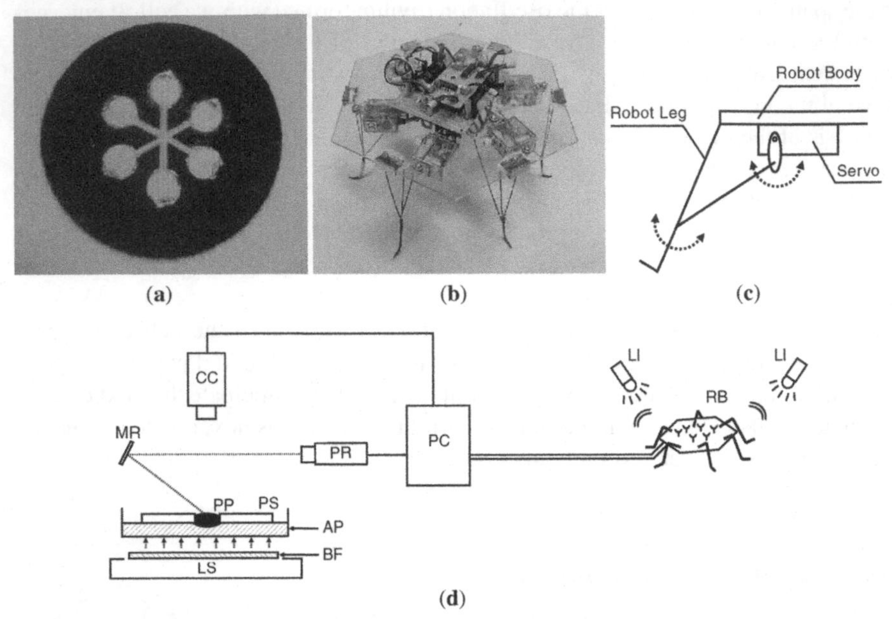

Fig. 10.2 Components for *Physarum* controlled robot. The shape of the *Physarum* cell can be controlled by a dry negative mask on a 1.5% agar gel plate (**a**). Each *Physarum* oscillator controls one leg in a hexapod robot (**b**). Each leg has only one degree of freedom (**c**) and the combination of six leg motions generates various gate patterns (**d**). These two components are interfaced by bi-directional interfaces: Sensors on the robot (RB) detect light inputs (LI) in the environment and transmit signals to the computer (PC). The signals from the sensors are recoded by the PC into a spatial light pattern which is projected as light stimulus with a video projector (PR) via mirror (MR) onto the surface of the plasmodium (PP). Oscillations of the plasmodium (PP), which is patterned on an agar plate (AP) by a plastic sheet (PS), are detected by a CCD camera (CC) as intensity changes in light transmitted from the bandpass filtered (BF) light source (LS)

the centre (Fig. 10.2a). The plasmodium is guided to assume the star-shape on a 1.5 % agar gel by a negative plastic mask with a star-shape cutout. The plasmodium preferentially grows on a moist surface rather than a dry one so that it can absorb water from it. Accordingly, a *Physarum* plasmodium grows on the exposed agar surface and eventually takes a star-shape. We refer to this shape-fixed *Physarum* cell as "*Physarum* circuit". It is known that the part of the plasmodium within a 1.5 mm diameter well shows a synchronised oscillation [12]. The formation and collapse of actomyosin structures within the cell, pump the protoplasm through ectoplasm tubes in the channels of the mask.

The part of the plasmodium inside a well functions as a single oscillator, referred to as "*Physarum* oscillator", and is connected to the other oscillators (wells) through tubes in the channel and therefore, the whole star-shaped cell can be viewed as a coupled nonlinear oscillator system [12].

10.3.1.2 The Hexapod Robot Platform

A hexapod robot is equipped with six leg actuators and six light sensors for omni-directional behaviour and sensing (Fig. 10.2b). Each leg and sensor is designed to correspond to a *Physarum* oscillator so that the *Physarum* circuit can control the robot and interface with the robot's environment.

Each leg is driven by one of six independent servos (Fig. 10.2c), and either pushes or drags the robot body depending on the direction of its swing motion. Although each leg has only a single degree of freedom, in combination they afford numerous gate patterns and provide excellent stability. For example, anti-phase motion between opposite legs (para-position) yields directed motion of the robot.

Each light sensor is directed to a fixed direction with 60° difference between neighbouring sensors and detects light coming only from that direction. A wire bundle is attached on top of the hexagonal body to exchange data between the robot and a PC. The wire does not intercept robot locomotion as it hangs from the wall.

To use the *Physarum* circuit in a robot controller, a bi-directional interface is required for its interaction with the robot. We implemented both directions with optical interfaces mediated by a PC (Fig. 10.2d).

10.3.1.3 Cell-to-Robot Interface

A *Physarum* circuit in a dark chamber is constantly exposed to orange light from the bottom to monitor its activity with a CCD camera. The filtered light contains only the spectral region near 600 nm in which the cell does not respond. Snapshots of the camera image are saved every 2 s. They are used to calculate the relative thickness of each *Physarum* oscillator based on the fact that transmitted light through a *Physarum* circuit is inversely proportional to its thickness. The thickness data from each *Physarum* oscillator is then converted to a motion for the corresponding leg.

10.3.1.4 Robot-to-Cell Interface

If a sensor on the robot detects a light signal in the environment, a strong white spotlight shines on the corresponding *Physarum* oscillator and keeps illuminating it as long as the sensor receives light signals. Unlike the bandpass-filtered light from the bottom, white light does affect the thickness oscillation of the plasmodium [17]. In particular, it triggers changes in oscillation pattern of the *Physarum* circuit and consequently changes the robot locomotion pattern.

10.3.1.5 Results

Without stimulus, the thickness oscillation frequency of the *Physarum* circuit is approximately 0.01 Hz (100 s per period). This value agrees with the literature (e.g. [18]) and thus, can be regarded as the inherent frequency of a *Physarum* circuit. In the presence of light stimuli, we observed several reactions. Figure 10.3 shows two typical reactions of a *Physarum* oscillator: When a white light stimulus is applied to the *Physarum* oscillator, the thickness of the oscillator decreased until the stimulus was removed (Fig. 10.3a), and after some delay the amplitude of oscillation became lower (Fig. 10.3b). At the same time, typically some of the unexposed oscillators increase in thickness and amplitude during the light stimulus.

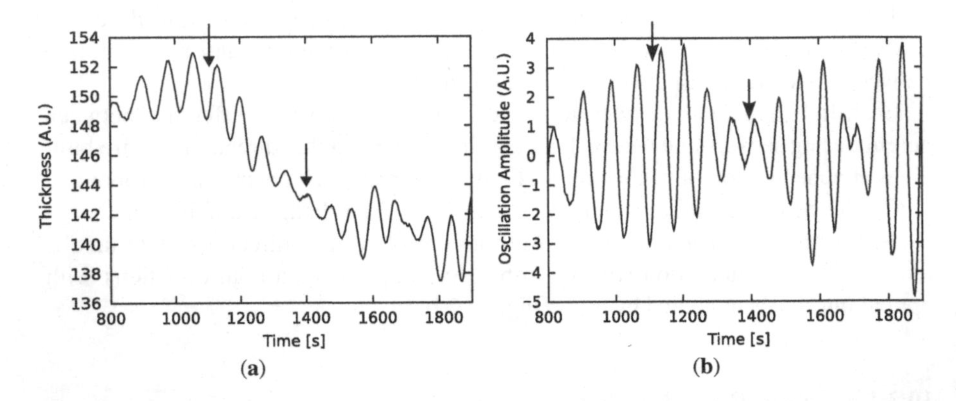

Fig. 10.3 Reactions of a *Physarum* oscillator to light stimulus. Decrease in thickness of the oscillator is observed (**a**) along with reduction of the oscillation amplitude (**b**). *Arrows* in the graphs indicate start (*left*) and end (*right*) of the light exposure

Based on these observations, an update rule for robot legs has been defined as follows:

$$L_i(t) = [L_i(t-1) + \alpha A_i(t) T_i(t)] \pmod{255} \tag{10.1}$$

$$A_i(t) = d_i(t) - d_i(t-\tau) \tag{10.2}$$

$$T_i(t) = \frac{\sum_{n=0}^{99} ((t-n) - \bar{t})(d_i(t-n) - \overline{d_i(t)})}{\sum_{n=0}^{99} ((t-n) - \bar{t})^2}. \tag{10.3}$$

In the above equations, $L_i(t)$ is the position of i-th robot leg, taking discrete values from 0 to 255, and $d_i(t)$ is the relative thickness of i-th *Physarum* oscillator at time t. $A_i(t)$ and $T_i(t)$ are an amplitude and a trend of $d_i(t)$, respectively. $\bar{t}$ and $\overline{d_i(t)}$ are the mean of $(t-n)$ and $d_i(t-n)$ with $n = 0, 1, \cdots, 99$. α is the scaling parameter and τ is the time delay parameter. The above equations representing the motion of each robot leg is determined by thickness changes of the corresponding *Physarum* oscillator in the star-shaped circuit.

When sensors on the robot detect light signals from a single light source, this update rule induces anti-phase motion between the leg on the side of the sensor and opposing legs. This is the case because stimulated oscillators result in smaller $A_i(t)$ and negative $T_i(t)$, whereas other oscillators have larger $A_i(t)$ and positive $T_i(t)$. Eventually this will lead to the robot moving away from the light source.

10.3.2 The Second Generation of the Φ-bot: On-Board Cellular Controller

Although the first generation robot design allows the cell to control a robot, there are several drawbacks. From a practical perspective, the need to keep the plasmodium moist but avoid condensation obstructing the camera view is a challenge. It is also desirable to reduce the size of the interface to a scale, which makes an autonomous robot feasible. Our recent work addresses these issues. To integrate the plasmodium into an autonomous robot it was, therefore, desirable to explore other technologies for monitoring the activity of the plasmodium. A custom circuit board for electrical impedance spectroscopy (EIS) has been developed [19] and mounted on a small wheeled robot platform [20].

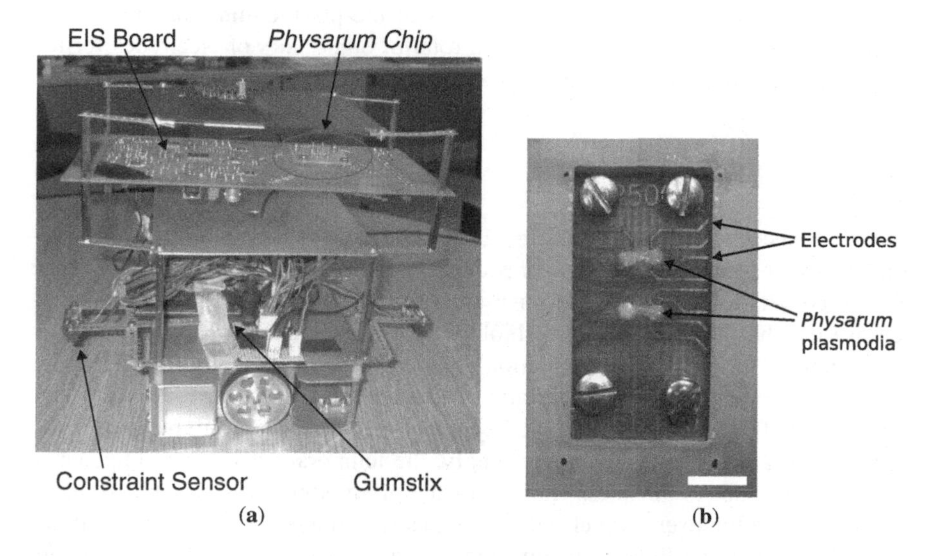

Fig. 10.4 The complete setup of the robotic system (**a**) driven by the *Physarum* plasmodium growing in the *Physarum* chip (**b**). The *white bar* in (**b**) corresponds to 5 mm

Figure 10.4a shows the new setup of the bio-hybrid robotic system. The system consists of four components to be described below: a *Physarum* chip (Fig. 10.4b), an EIS board, a small gumstix computer, and a wheeled robot base. In this configuration, the cell's oscillations are read through impedance measurements and mapped onto the wheel motion of the robot base. The current implementation of the un-tethered robot still lacks the interface between the cell and sensors on the robot, i.e. the robot is driven by the cell's oscillating pattern without any feedback to the cell.

10.3.2.1 The *Physarum* Chip

The *Physarum* chip is a small printed circuit board (PCB) containing two plasmodial cells, one to control the robot and one for reference. To maintain the moisture required by the cell, the PCB is covered on one side with an 1.5% agar gel block, and on the other side with a sheet of the gas-permeable elastomer polydimethylsiloxane (PDMS). The copper side of the PCB with its patterned electrodes faces the agar gel and is insulated from it with laminate. The stack of PDMS–PCB–Agar is clamped with a plexiglass frame. This assembly, referred to as "*Physarum* chip", completely encloses the plasmodial cell and provides the necessary humidity and adequate oxygen supply to keep the cell active for more than 8 h. Each plasmodium is confined in a dumbbell-shaped cut-out of the PCB sheet, as shown in Fig. 10.4b. The dumbbell-shape design, with two 1.6 mm diameter circular holes at a centre-distance of 2.5 mm connected by a 0.4 mm wide channel, follows Takamatsu et al. [21, 22], who studied the oscillation patterns of the plasmodium confined to this shape. The PCB sheet is equipped with a total of eight pairs of electrodes for the two plasmodia samples, with two electrode pairs for each well (Fig. 10.4b).

10.3.2.2 The EIS Board

The impedance measurement circuitry (EIS board) allows to monitor the plasmodium's activity using the electrical impedance spectroscopy technique. Impedance spectroscopy is a high-speed, non-invasive and label-free technique used to characterise the dielectric properties of biological tissues and cells [23]. We employ this technique to monitor the oscillation activity of the plasmodium by measuring the impedance between two local parts of the cell. If a lower frequency AC signal (around 100 kHz) is applied to a *Physarum* cell, a measured impedance value represents the size of the cell. Accordingly, the temporal trace of the impedance values corresponds to thickness changes of the plasmodium cell. The board incorporates two multiplexers, which allows to measure impedance between arbitrary two electrodes on the *Physarum* chip (Fig. 10.4b). In the robot control experiment described below, impedances of the two plasmodia at all eight electrode positions are measured once per second.

10.3.2.3 The Gumstix Computer

A tiny computer on which a customised Linux kernel has been installed (`www.gumstix.com`) mediates between the EIS board and the robot base. The gumstix computer configures the board to retrieve the impedance measurements and saves the data in flash memory. After a signal processing of the measurement data, the computer sends commands to a microcontroller in the wheeled robot base via general purpose input/output ports.

10.3.2.4 The Wheeled Robot Base

A basic design of the robot is adopted from the Braitenberg vehicles [24]. It allows for the *Physarum* chip, the EIS board and the gumstix computer to be mounted, and accommodates the necessary power supplies for the EIS board and the gumstix computer. The base has its own microcontroller that translates simple commands (forward, left and right) from the gumstix computer to the drive level of the stepper motors.

10.3.2.5 Cell-to-Robot Interface

Although the second generation of the Φ-bot still lacks a feedback interface from the robot to the *Physarum* chip like its previous generation, we first ran simple experiments to demonstrate how signals from the cell can be used to drive the robot in the new Φ-bot.

The experiments were run on a 1 m diameter round table. The robot is equipped with two infrared proximity switches and ignores forward commands if one of these switches detects an edge of the table. This effectively constrains the robot to the area of the table that serves as arena for experiments. Position and direction of the robot are tracked by an Ethernet camera mounted above the table using an illuminated target pattern on top of the robot.

A two-step process converts the measured impedance signals into drive commands for the robot base: signal processing to recover the oscillation state of the cell from the impedance measurements, and mapping of the cells oscillation state into actuator commands. First, the moving average over 15 samples ($\approx$15 s) of recorded data is calculated to reduce noise in the impedance measurements. At present the circuitry on the robot has not been optimised to reduce noise, but the signals are strong enough that the simple moving average filter works sufficiently well for the purpose. Next a differential of the averaged signal is computed by subtracting the 15 s delayed signal to remove the long-term trend of the signals [17]. After the signal processing, a command to drive the robot is determined according to the phase relationship between the differential signals from both wells. It is known that the plasmodium, if confined to the dumbbell shape, show in-phase and out-of-phase oscillation patterns between the two wells [22]. Based on this observation, we introduced a simple mapping from oscillations to robot movement: If the two wells are in phase (synchronised mode), the robot takes a random turn. If they are out of phase (phase delayed mode), then it moves straight. The phase relationship is classified by the following simple rule: If the signs of the two differential signals are equal (both oscillations are increasing or both are decreasing), the oscillation state is classified as synchronised mode. If the signs of the differential signals differ, the oscillation state is classified as phase delayed mode. The former is mapped into a random choice of either a "left" or a "right" command, whereas the latter is mapped into a "forward" command. The whole conversion cycle is performed once per second and commands for the robot's actuators are issued accordingly. The update cycle from impedance measurement to change in robot behaviour is once per second.

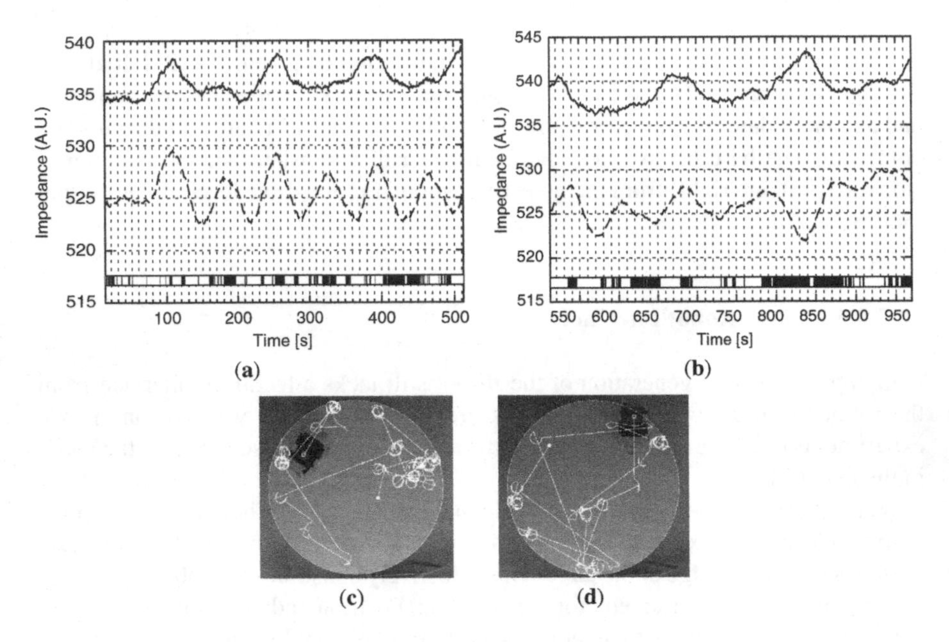

Fig. 10.5 Oscillation of a plasmodium in a *Physarum* chip and the corresponding trajectories of the robot. The moving-averages of the magnitude of the impedance at 100 kHz are plotted for 30–500 s (**a**) and 500–970 s (**b**). The *solid* and *dotted curves* in the plots correspond to oscillations of the plasmodium from right and left wells, respectively. The phase relationship between two wells are shown in the *bottom* of the plots as *black* (in-phase) and *white* (out-of-phase) *vertical lines*. The behaviour of the robot is determined according to the phase relationship as traced in panel (**c**) for 30–500 s and panel (**d**) 500–970 s. A *white solid circle* indicates the start and an *open circle* the end of the trajectory

Figure 10.5a, b shows signals from two consecutive time periods of a single experiment. The two curves show the magnitude of the impedance from left and right circular areas (wells) of the dumbbell shaped plasmodium after a noise filter has been applied.

The trajectories of the robot that results from this mapping are shown in Fig. 10.5. During the time period shown in Fig. 10.5a, the oscillations of the two parts of the plasmodium cell are predominately synchronised and accordingly the robots trajectory shows many random pivot turns (Fig. 10.5c). However, in the period shown in Fig. 10.5b, d, the robot runs straight more often because the oscillation pattern switched to an out-of-phase mode about midway through the period shown in Fig. 10.5b.

Since the current implementation of the Φ-bot has only an uni-directional interface from the plasmodium to the robot, the next step required to complete the second generation of Φ-bot is the implementation of the converse interface from the robot to the cell, i.e. inputs to the plasmodium. This can be achieved by illuminating the cell with white light from suitable LEDs according to signals from sensors on the robot. So far we ran preliminary experiments of white light stimulation to the *Physarum* chip and confirmed the cell shows similar reactions against the stimuli as the *Physarum* circuit of the first generation Φ-bot (Fig. 10.6). Although this part of the interface is still under implementation, our investigations indicate the feasibility of integrating a living cell into the controller of an autonomous robot.

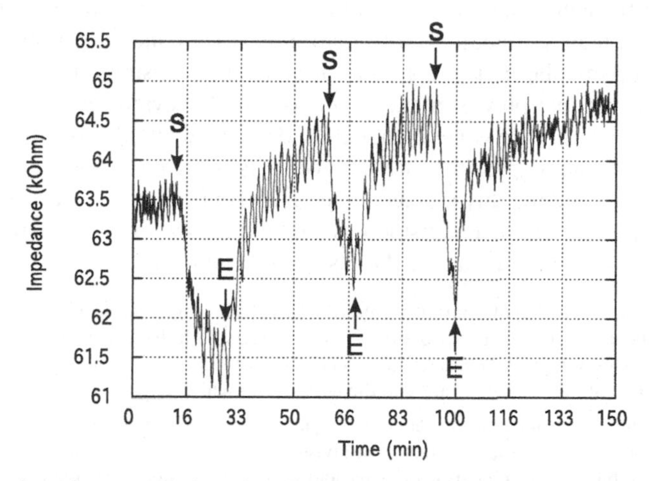

Fig. 10.6 The impedance oscillation of a stimulated well in a *Physarum* chip. When one of two wells in a *Physarum* chip is exposed to white light stimulus, it shows sharp decrease in impedance (corresponding to thickness of the cell within the well). *S* and *E* in the graph indicate start and end of light exposures, respectively

10.4 Computation, Control, and Coordination in the Φ-Bot: Material for a Theory of Bounded Computability

The development of bio-hybrid robots, such as the Φ-bot, is hampered by the fact that available information-theoretic tools in general ignore how nature directly exploits the manifold physical characteristics of its computing substrates. "Information theory" meant here is not just Shannon's statistical theory of communication [25], but the structural science that constructs mathematical models of the form, meaning and use of information, and applies them to empirical phenomena.

In the engineering context of artificial life and robotics, information theory must systematically develop a theory of bounded computability in which generic traits of information-processing systems are related, not with general time and space bounds (as in the theory of computational complexity [26]), but with specific physico-chemical constraints on the realisation of such systems in different classes of computational media (analogously to the theory of bounded rationality [27]). This empirical diversification of information theory would allow the engineering of unconventional computers to utilise empirical knowledge of naturally evolved systems more efficiently since the requirements for particular computational tasks could then be stated directly in terms of physical specifications of computational media [28].

In a theory about possible relations between material media and computational functions of physical systems, the concept of information must integrate the distinction between the behavioural structure of a system, its functional structure, and the structure of its material medium. Information is not a concrete entity that can be localised in a particular part of a system; it is an abstract structure that covers the complex systemic interplay of matter, function and behaviour. Basic information-theoretic concepts that are general enough to describe this interplay for a spectrum of systems as broad as possible, are the concepts of syntax, semantics and pragmatics [29]. They are roughly defined by the philosopher Charles Morris as follows (here "semiotic" should be understood as synonymous with "information theory", and signs be considered fundamental units of information):

> *Pragmatics* is that portion of semiotics which deals with the origin, uses, and effects of signs within the behavior in which they occur; *semantics* deals with the signification of signs in all modes of signifying; *syntactics* deals with combinations of signs without regard for their specific significations or their relation to the behavior in which they occur.
>
> When so conceived, pragmatics, semantics, syntactics are all interpretable within a behaviorally oriented semiotics, syntactics studying the ways in which signs are combined, semantics studying the signification of signs, and so the interpretant behavior without which there is no signification, pragmatics studying the origin, uses, and effects of signs within the total behavior of the interpretation of signs [30].

Syntax, semantics, and pragmatics are, thus, not simply juxtaposed subdisciplines that deal with isolated aspects of signs. Syntactic, semantic, and pragmatic properties of signs are related hierarchically to each other, with pragmatics forming the base since it analyses the overall structure of any behaviour in which signs are involved. Semantics and syntax abstract, to different degrees, from the object of pragmatics: semantics focuses on specifically interpretative behaviour, and syntax puts brackets around signification and analyses all possible methods by which signs can be combined with each other.

In a theory of bounded computability, the general definitions of syntax, semantics and pragmatics must abstract from any particular class of systems, so that the information-theoretic perspective can be applied, not only to human beings, but also to other living systems, to artifacts, and to engineered hybrids, such as the Φ-bot. Syntax, semantics, and pragmatics, thus, denote different ways of structurally representing any kind of information-processing system from a unified point of view, which, nevertheless, should allow for the differentiation between types of computational media.

Though pragmatics is the most fundamental representation of information, the analysis of the Φ-bot in terms of bounded computability shall start with syntax, since – from a general engineering point of view on possible information processing systems – semantics can most easily be described as the study of code-related boundary conditions that must be met by how signs can combine with each other in an information processing system, and pragmatics as the study of behavioural boundary conditions that must be met by how signs can signify something in an information processing system. Applied to the Φ-bot, syntax, semantics, and pragmatics shall each be characterised by a topical "keyword": computation, control and coordination, respectively. This suggests using "Co^3 approach" as a shorthand title of the following remarks on bounded computability.

10.4.1 Computation and the Syntactic Efficiency and Reliability of Computational Media

Information can be represented syntactically by the material structure of a physical system (e.g. a *Physarum* cell). The spatio-temporal organisation of the material components of a system is regarded as an actualisation of the syntactic structure of information in a physical medium. The material structure of the medium actually stands for the syntactic structure of information that is constituted by the set of relations of its elements. The dynamics of self-organisation of the physical medium (e.g. the oscillations of the plasmodium) drives the processing of the syntactic representation of information, but does not require a specific information-theoretic explanation.

Important criteria for classifying computational media from a syntactic perspective are how efficient and reliable these media are in processing syntactic representations of information. Yet to describe a system from the perspective of its efficiency and reliability, presupposes the definition of a purpose that the system shall fulfil. Most generally, this is the purpose of computing a function in the mathematical sense, i.e. to compute, given certain input signals, certain output signals that can then be used for a particular task (e.g. as control signals for the actuators of the Φ-bot). Considering a physical system as a computational medium, any of its physical properties acquires a particular objectifiable function in an engineering sense, namely what it contributes to the computation of a function in the mathematical sense. The syntax of a computational medium, seen from a mathematical perspective, can be categorised by the class of mathematical functions it can compute; this might be done in analogy to the classification of abstract automata [31]. Viewed from an engineering point of view, the syntax of a computational medium can be classified by the efficiency and reliability in which it actually computes the mathematical function it shall compute.

10.4.2 Control and the Semantic Generality of Computational Media

By distinguishing between the mathematical sense and the engineering sense of function, the first step towards the semantic perspective on physical systems as information-processing media has already been taken. Information can be represented semantically by the functional structure of a physical system in the engineering sense. The causal order between the material components of a system is then regarded as an implementation of the semantic structure of information by a physical medium. Yet semantics is not co-extensive with functionality at large, it is code-based functionality. A code, such as the Morse code and the genetic code, connects two syntactic structures with each other. It is a mapping that relates, in case of encoding, each of the possible syntactic elements of a message to a possible element of a signal and, in case of decoding, each of the possible syntactic elements of a signal to a possible element of a message [32]. It is of great importance to emphasise that the semantic relation between signal and message is contingent, which means that it can be chosen by the engineer from a set of possible mappings, all of which are compatible with the natural laws that govern the causal order in the computational medium.

The dynamics of self-organisation of the physical medium implements the semantic structure of information by encoding and decoding syntactic structures in physical processes. From the perspective of semantics, it is necessary to interpret the present state of (a part of) a system as encoding the future state (of another part) of it.

Where is information represented semantically in the interaction loop of the Φ-bot and its environment (see Fig. 10.7)? First and foremost, it is in the code-based functional structure of the artificial control. Encoding happens when stimuli from the robotic light sensors are transduced to white light signals for the cell. Decoding occurs when amplitude signals calculated from measured data of the plasmodium oscillators are processed by the update rule to alter the motion of the robot actuators. The update rule acts as a decoding device that semantically relates a syntactic representation of information (plasmodium oscillation signals) to another syntactic

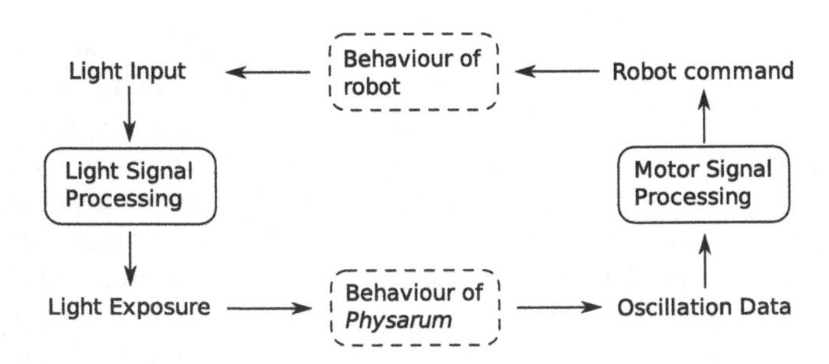

Fig. 10.7 Interaction loop of the Φ-bot. Each part of the diagram corresponds to either syntax (*no box*), semantics (*solid box*), or pragmatics (*dotted box*)

representation of information (robot motor signals). All in all, four syntactic representations of information (external light signals, white light signals for the plasmodium, oscillation signals from the plasmodium, and robot motor signals) are related by two semantic representations of information (namely, the code used for encoding the light sensor data into white light signals and the code used for decoding the oscillation data into motor signals). In-between, the plasmodium connects the two semantic representations by bounded computation.

It is the semantic structure of the robot control device (in short, its control semantics), which gives a particular signification to certain syntactic substructures in the *Physarum* cell. Thereby, the plasmodium is regarded as computing a mathematical function and, by doing so, fulfils an engineered function in the Φ-bot. The control semantics is a necessary condition for the computational use of the syntactic structure of the *Physarum* plasmodium. In code-based mutual constraining, the behavioural changes of the interacting syntactic structures are induced by the rules of the control semantics. Those semantic rules can be described, or implemented, without having a full-fledged theory about the internal workings of the systems involved since the observation of their behaviour suffices to determine the syntactic input into the code.

The set of possible messages that the control semantics, as given by engineered or organic codes [33], can distinguish, is an important semantic criterion for classifying computational media. It measures how general the codes used in a medium to implement semantic representations of information are, i.e. to which degree the codes are able to differentiate between possible messages under given boundary conditions. From the perspective of the computational medium, the control semantics acts as a behaviour amplifier. The microscopic behaviour of, e.g. a *Physarum* cell is amplified to the macroscopic behaviour of a robot by semantic means (analogously to the hierarchical mode of biological information processing [1]). This suggests to think about the generality of codes, i.e. about the degree to which they are able to differentiate between possible messages in relation to given pragmatic boundary conditions, in terms of the graininess of behaviours between which the codes can differentiate. The finer the behavioural differences of the *Physarum* cell that a code can semantically represent are, the more general the code is in respect to this particular control setting.

10.4.3 Coordination and the Pragmatic Versatility of Computational Media

By describing the capability of the control semantics to make distinctions between messages, i.e. the generality of the implemented codes, in terms of behaviour, the first step towards the pragmatic perspective on physical systems as information-processing media has already been taken. The very idea of control is to intervene in a process that, by ensuing interaction, shall produce a particular result [34]. This process is the real-life function that the system in which the controller works, shall fulfil in a certain class of environments. In case of both the Φ-bot and the *Physarum* cell, a particular type of behaviour as regards light sources is to be shown. The Φ-bot does so by an engineered control semantics, the *Physarum* cell by an evolved one.

Information can be represented pragmatically by the behavioural structure of a physical system. The pattern of interaction between the system and its environment is then regarded as an effectuation of the pragmatic structure of information through the agency of a physical medium. The pragmatic structure is constituted by transformations of boundary conditions on coding. When is a message selected for being encoded, when is a signal decoded, and how does the code originate? Generalising the idea that information is constituted pragmatically by the effect of a signal on its receiver [35], the definition of the pragmatic structure of information involves at least two syntactic orders and one semantic mapping.

The dynamics of self-organisation of the physical medium changes internal and external conditions of information processing in the system. From the perspective of pragmatics, it is necessary to interact with the present behaviour of a system in order to let its dynamics lead it to a particular future behaviour.

Where is information represented pragmatically in the interaction loop of the Φ-bot and its environment? First, the behaviour of the robot that results from decoding plasmodium oscillation signals into robot motor signals, changes the boundary conditions on encoding since the effects of the robot's activity on the environment are perceived by the robot's light sensors whose data are then encoded into white light signals for the cell. Second, the plasmodium behaves according to its own dynamics in its direct environment, i.e. in the artificial control. This environment receives the behaviour of the plasmodium in form of oscillation data that is decoded into robot motor signals. The pragmatic interaction of the plasmodium with its engineered environment is, thus, semantically represented in the very same environment and then pragmatically represented by the behaviour of the robot in its real-world environment. The control semantics as given by the two codes described in Sect. 10.4.2 connects the relation of the robot to its real-world environment with the relation of the plasmodium to its artificial environment. The control semantics, thus, maps two different pragmatic representations of information to each other, namely the behaviours of the plasmodium and the robot.

The codes used in the Φ-bot for encoding external stimuli and for decoding internally collected measurement data, are pragmatically connected by the behaviour of the *Physarum* cell, i.e. by how it measurably reacts to the encoded stimuli. To encode information means in pragmatic respect that the control device sets the boundary conditions on how the plasmodium processes syntactic representations of information. To decode information means that the control device semantically represents the pragmatic results of information processing by the *Physarum* cell.

How can this interplay of semantics and pragmatics be described more generally? From a pragmatic perspective, the control semantics can be seen as if it is a signalling convention between two independent information-processing systems that has arisen from a proper equilibrium in the coordination of the behaviours of both systems, here of a *Physarum* plasmodium and a robot, the latter representing the designer's intentions. 'Proper coordination equilibrium' means, a combination of behaviours in which any agent would have a disadvantage, if any one agent had acted otherwise [36]. To put it more realistically, the robot's designer has selected both systems according to their respective capacity to cooperatively realise, in a stable way, the intended real-life function of the Φ-bot by means of the signalling convention that is called 'control semantics'.

The key engineering issue here is how we can exploit the cell's self-organising dynamics to achieve a fully autonomous robot. Arguably, biological cells outperform conventional autonomous robots in many features by exploiting pragmatic versatility, i.e. the high degree to which the cells are able to adapt their behaviour to different environments. An important criterion for classifying computational media from a pragmatic perspective, thus, is how versatile media are in effectuating transformations of the system's behaviour under changing boundary conditions, i.e. to which degree the behaviour of the system is able to adapt itself to different environments. Pragmatic versatility depends on the diversity of environments in which the cooperation of the computational medium and the robot is actualised successfully. For the Φ-bot, the interplay of semantics and pragmatics via its controller is as important as the syntactic efficiency of the *Physarum* cell, since this interplay is the means by which the pragmatic versatility of the plasmodium is also detectable in the behaviour of the Φ-bot.

Coordination between the plasmodium and the robot is relatively easy in case of the Φ-bot since there does not exist a coordination problem with more than one proper coordination equilibrium. If the coordination of a computational medium and a robot has been stabilised by a control semantics in a situation with only one coordination equilibrium, the behaviour of the robot in this situation can, in principle, be anticipated by an external observer. Engineering of control semantics will become more difficult if computational media showing a more unpredictable behaviour than *Physarum* plasmodium, shall be used to control robots whose intended behaviour must be modelled by more complex utility functions than it is necessary for the Φ-bot. Even a *Physarum* cell sometimes shows a behaviour that goes in this direction.

Several researchers have observed that the plasmodium is able to spontaneously change its behaviour pattern against external stimuli to overcome unfavourable conditions [37–39]. For example, Takamatsu found that if the plasmodium is entrained to oscillate at a fixed frequency by periodically-changing the stimulus externally, it spontaneously deviates from the frequency after a certain period even if the stimulus is maintained. She speculated that such spontaneous change might stem from multistability or chaotic behaviour of the plasmodium's dynamics and may contribute to the diversity of behavioural modes of the plasmodium, such as food-searching mode and feeding mode [39].

Although physiological mechanisms underlying such behaviour are yet to be investigated, these observations point to the richness of the cell's internal dynamics. However, they also point to the lack of theory about relations between the dynamics of physico-chemical material structures and their use as computational media. The conventional computing paradigm assumes perfection of each part in a system.

It is, therefore, inadequate when we want to harness the pragmatic versatility of the plasmodium, which results from the richness of its self-organised processing of syntactic representations of information, by a control semantics that allows the robot to adapt its behaviour to a real-world environment.

10.5 Conclusion

In this chapter, we reviewed our cellular robot control approach towards adaptive autonomous robots, followed by a first outline of a theoretical approach to bio-hybrid computing.

We have shown that properties of *Physarum* cells can meet the requirements for the adaptive behaviour generation in engineered systems and how those properties are employed for the control of autonomous robots. Although this work is still ongoing and many problems remain to be solved, we can expect that the use of biological cells will provide several features that re difficult to achieve with conventional technologies, such as self-reconfiguration and self-repair capabilities. A tight integration of cells with existing technologies will require fine control of the micro-environment of the cells. Lab-on-chip technology can be employed to achieve this [40]. In current work, our laboratory investigates, for example, whether the cellular controller can be "switched off" and stored for a long time by keeping it in a dry dormant state and, when needed, easily "powered on" by supplying moisture through microfluidics.

Because unconventional computing devices are radically different from conventional computers, it is inevitable to develop a new theory to capture the informational significance of physical properties of novel computational media. Bounded computability as outlined by the Co^3 approach must be subjected to further discussion and needs to be developed into a full-fledged theory. This presupposes that formal approaches to pragmatic, semantic and syntactic representations of information must be integrated into a general concept of computability under different types of physical constraints. This is, of course, a very challenging task: it amounts to regarding complex adaptive systems as models of information theory (here, 'model' is to be understood in the sense of formal semantics, the mathematical discipline describing the relation between a theory and objects that fulfil this theory). On an even broader horizon, the theory of bounded computability and the engineering of bio-hybrid devices together may make an essential contribution to our general understanding of informational behaviour shown by any kind of adaptive system.

A most important conceptual precondition for reaching such a general understanding of informational behaviour is a consequent relativism in thinking about information. What is to be considered a syntactic, semantic or pragmatic representation of information, depends on the interaction between the described and the describing system. For example, material features of computational media, such as the plasmodium in the Φ-bot, appear, from the perspective of a code-based controller, primarily as constraints on the processing of syntactic representations of information. Information processing by the computational medium has, of course, also internal semantic and pragmatic aspects. There exist, e.g. semantic representations of information in the *Physarum* cell, such as the organic code that structures the expression of genetic information. Yet from the perspective of the controller, those cell-internal aspects are to be considered just as constraints on the efficiency and reliability of processing syntactic representations of information that are actualised in the cell's environment. Therefore, the syntactic effectiveness and reliability of the computational medium show themselves, to the controller, in the pragmatic versatility of the medium, i.e. the degree to which its behaviour is able to adapt itself to changing boundary conditions that bear information syntactically. To describe such complex relations between syntax, semantics and pragmatics, the theory of bounded computability must assume a relativistic attitude towards information.

References

1. Conrad, M.: The importance of molecular hierarchy in information preocessing. In: C.H. Waddington (ed.) Towards a theoreritcal biology, vol. 4, pp. 222–228. Edinburgh University Press, Edinburgh (1972)
2. Adamatzky, A., Costello, B., Asai, T.: Reaction-diffusion computers. Elsevier Science, New York, NY, USA (2005)
3. Zauner, K.P.: Molecular information technology. Critical Reviews in Solid State and Material Sciences **30**(1), 33–69 (2005)
4. Wohlfarth-Bottermann, K.E.: Oscillatory contraction activity in physarum. The Journal of Experimental Biology **81**, 15–32 (1979)
5. Hejnowicz, Z., Wohlfarth-Bottermann, K.E.: Propagated waves induced by gradients of physiological factors within plasmodia of *Physarum polycephalum*. Planta **150**, 144–152 (1980)
6. Matsumoto, K., Ueda, T., Kobatake, Y.: Propagation of phase wave in relation to tactic responses by the plasmodium of *Physarum polycephalum*. Journal of Theoretical Biology **122**, 339–345 (1986)
7. Tanaka, H., Yoshimura, H., Miyake, Y., Imaizumi, J., Nagayama, K., Shimizu, H.: Information processing of Physarum polycephalum studied by micro-thermography. Protoplasma **138**, 98–104 (1987)
8. Matsumoto, K., Ueda, T., Kobatake, Y.: Reversal of thermotaxis with oscillatory stimulation in the plasmodium of *Physarum polycephalum*. Journal of Theoretical Biology **131**, 175–182 (1988)
9. Miura, H., Yano, M.: A model of organization of size invariant positional information in taxis of *Physarum* plasmodium. Progress of Theoretical Physics **100**(2), 235–251 (1998)
10. Miyake, Y., Tabata, S., Murakami, H., Yano, M., Shimizu, H.: Environmental-dependent self-organization of positional information field in chemotaxis of *Physarum* plasmodium. Journal of Theoretical Biology **178**, 341–353 (1996)
11. Gierer, A., Meinhardt, H.: Theory of biological pattern formation. Kybernetik **12**, 30–39 (1972)
12. Takamatsu, A., Fujii, T.: Construction of a living coupled oscillator system of plasmodial slime mold by a microfabricated structure. Sensors Update **10**(1), 33–46 (2002)
13. Takamatsu, A., Tanaka, R., Yamada, H., Nakagaki, T., Fujii, T., Endo, I.: Spatiotemporal symmetry in rings of coupled biological oscillators of physarum plasmodial slime mold. Physical Reviews Letters **87**(7), 078102 (2001)
14. Adamatzky, A.: Physarum machine: implementation of a Kolmogorov-Uspensky machine on a biological substrate. Parallel Processing Letters **17**(4), 455–467 (2007)

15. Nakagaki, T., Kobayashi, R., Nishiura, Y., Ueda, T.: Obtaining multiple separate food sources: behavioural intelligence in the *Physarum* plasmodium. Proceedings of the Royal Society: Biological Sciences **271**(1554), 2305–2310 (2004)
16. Saigusa, T., Tero, A., Nakagaki, T., Kuramoto, Y.: Amoebae anticipate periodic events. Physical Review Letters **100**(1), 018101 (2008)
17. Nakagaki, T., Uemura, S., Kakiuchi, Y., Ueda, T.: Action spectrum for sporulation and photoavoidance in the plasmodium of Physarum polycephalum, as modified differentially by temperature and starvation. Photochemistry and Photobiology **64**(5), 859–862 (1996)
18. Wohlfarth-Botterman, K.E.: Oscillating contractions in protoplasmic strands of *physarum*: Simultaneous tensiometry of logitudinal and radial rhythms, periodicity analysis and temperature dependence. Journal of Experimental Biology **67**, 49–59 (1977)
19. Macey, P.: Impedance spectroscopy based interfacing with a living cell for biosensors and biocoporcessors. Part III Project Report, School of Electronics and Computer Science, University of Southampton (2007)
20. Jones, G.: Robotic platform for molecular controlled robots. Part III Project Report, School of Electronics and Computer Science, University of Southampton (2006)
21. Takamatsu, A., Fujii, T., Endo, I.: Control of interaction strength in a network of the true slime mold by a microfabricated structure. BioSystems **55**, 33–38 (2000)
22. Takamatsu, A., Fujii, T., Yokota, H., Hosokawa, K., Higuchi, T., Endo, I.: Controlling the geometry and the coupling strength of the oscillator system in plasmodium of *Physarum polycephalum* by microfabricated structure. Protoplasma **210**, 164–171 (2000)
23. Coster, H.G.L., Chilcott, T.C., Coster, C.F.: Impedance spectroscopy of interfaces, membranes and ultrastructures. Bioelectrochemistry and Bioenergetics **40**, 79–98 (1996)
24. Braitenberg, V.: Vehicles: Experiments in synthetic psychology. MIT, Cambridge, MA (1984)
25. Shannon, C., Weaver, W.: Mathematical theory of communication. University of Illinois Press, Illinois (1949)
26. Papadimitriou, C.M.: Computational complexity. Addison-Wesley, Reading, MA (1994)
27. Simon, H.: Models of Bounded Rationality, 3 vols. MIT, Cambridge, MA (1982/1997)
28. Tsuda, S., Zauner, K.P., Gunji, Y.P.: Computing substrates and life. In: S. Artmann, P. Dittrich (eds.) Explorations in the Complexity of Possible Life: Abstracting and Synthesizing the Principles of Living Systems, Proceedings of the 7th German Workshop on Artificial Life, pp. 39–49. IOS, Jena, Germany (2006)
29. Artmann, S.: Biological information. In: S. Sarkar, A. Plutynski (eds.) A companion to the philosophy of biology, pp. 22–39. Blackwell, Malden, MA (2008)
30. Morris, C.: Writings on the general theory of signs. Mouton, Den Haag and Paris (1971)
31. Hopcroft, J.E., Motwani, R., Ullman, J.D.: Introduction to automata theory, languages, and computation, 3rd edn. Addison-Wesley, Reading, MA (2007)
32. Cover, T.M., Thomas, J.A.: Elements of information theory, 2nd edn. Wiley, New York (2006)
33. Barbieri, M.: Organic Codes: An introduction to semantic biology. Cambridge University Press, Cambridge (2003)
34. Artmann, S.: Basic semiosis as code-based control. Biosemiotics **2**, 31–38 (2009)
35. MacKay, D.: Information, mechanism and meaning. MIT, Cambridge, MA (1969)
36. Lewis, D.: Convention: a philosophical study, 1st edn. Harvard University Press, Princeton, New Jersey (1968)
37. Aono, M., Hara, M.: Amoeba-based nonequilibrium neurocomputer utilizing fluctuations and instability. In: 6th International Conference, UC 2007, *LNCS*, vol. 4618, pp. 41–54. Springer, Kingston, Canada (2007)
38. Nomura, S.: Symbolization of an object and its freedom in biological systems. Ph.D. thesis, Kobe University (2001)
39. Takamatsu, A., Yamamoto, T., Fujii, T.: Spontaneous switching of frequency-locking by periodic stimulus in oscillators of plasmodium of the true slime mold. BioSystems **76**, 133–140 (2004)
40. Revilla, F., Zauner, K.P., Morgan, H.: Physarum polycephalum on a chip. In: J.L. Viovy, P. Tabeling, S. Descroix, L. Malaquin (eds.) The proceedings of μTAS 2007, vol. 2, pp. 1089–1091 (2007)

Chapter 11
Reaction–Diffusion Controllers for Robots

Andrew Adamatzky, Benjamin De Lacy Costello, and Hiroshi Yokoi

11.1 Introduction

Excitable systems, particularly spatially extended media, exhibit a wide variety of travelling patterns and different modes of interaction. The Belousov–Zhabotinsky [1] (BZ) reaction is the most well known and extensively studied example of non-linear chemical system exhibiting complex behaviour. The BZ reaction is well known to exhibit spontaneous oscillatory behaviour. The mechanistic details are complex but involve a fine balance between an auto-catalytic oxidation process and an inhibitor of the autocatalytic reaction. If the chemical conditions are altered beyond the point where the reaction exhibits spontaneous oscillatory behaviour, then the BZ reaction exhibits a property known as excitability. An excitable system has a steady state and is stable to small perturbations; however, if the perturbations exceed a critical threshold, then the system responds with an excitation event. In the case of a thin layer architecture, this results in a circular wave travelling from the source of the perturbation. Parts of the wave front annihilate when they reach the boundaries or collide with other fronts; however, other parts will propagate across the whole free area of the reactor. Because the BZ system is excitable and due to the properties of wave propagation, it can be considered as a uniform locally connected primitive neural network (a type of neural network similar to that in *Protozoa*). The information processing capabilities of BZ media are relatively well studied in the framework of reaction–diffusion computing [2, 3]. A reaction–diffusion processor is the term used to describe an experimental chemical reactor which computes by sensibly, purposefully and predictably transforming the initial local disturbances in the chemical concentration profiles – the data pattern – into dynamical structures such as travelling excitation waves, or stationary output such as a spatial distribution of precipitate – result patterns – in these processors the computation is implemented via the interaction of either diffusive or phase waves in the chemical system [2, 3]. Reaction–diffusion processors of this general type possess an amorphous structure (absence of any type of rigid hardware-like architecture in the processor's liquid

A. Adamatzky and M. Komosinski (eds.), *Artificial Life Models in Hardware*,

phase) and massive parallelism (a thin layer Petri dish may contain thousands of elementary computing micro-volumes).

The field of reaction–diffusion chemical computing is rapidly expanding and features a number of experimental prototypes capable of solving a wide range of problems: computational geometry, image processing, implementation of logical gates and constructing simple memory devices; see detailed overview in [2, 3]. Some of the unique and desireable features of reaction–diffusion computers [2, 3] include (1) massive-parallelism – each micro-volume of the liquid-phase chemical system acts as an elementary processor and there are potentially many thousands of micro-volumes in a centimetre cubed domain of the medium; (2) local connectivity – the chemical medium's micro-volumes automatically update their states (local concentrations of chemical species) depending on the combined state of their immediate neighbours; (3) potentially massive-parallel input and outputs – data are represented by distributed concentration profiles or gradations of the medium's illumination, while results are configurations of precipitate concentration or dissipative excitation patterns that may be recorded optically; (4) self-healing and self-reconfiguration due to the liquid-phase or gel-based structure of the computing substrate and the fact that elementary processing units are identical; (5) dynamically reprogrammable architectures – computation is implemented by travelling waves or wave-fragments and therefore, this site of the medium acts as a momentary wire. However, annihilation of excitation waves occurs enabling input of additional data streams with time. These characteristics of chemical systems make them ideally tailored for the implementation of novel and emerging architectures of robotic controllers and embedded processors for smart structures and actuators.

In the first part of the chapter, we show that chemical reaction–diffusion processors based on the BZ medium as specified in Sect. 11.2, can be successfully used as robot controllers in various experimental setups and applications.

We start our journey by describing approaches to mounting a chemical computer onboard of a mobile robot. This allows for dynamical interaction between the robot, its environment and the chemical computer. In Sect. 11.3, we define the problem of controlling robot navigation with chemical controllers and outline a possible solution, where we describe a mobile robot with an on-board chemical controller, and discuss how we can use the chemical controller to guide the robot.

In further work, we make the transition from guided/manipulated robots to robots capable of manipulating several objects at once. In Sect. 11.4, we couple a simulated abstract parallel manipulator with an experimental BZ chemical medium, whereby, the excitation dynamics in the chemical system are reflected by changing the OFF–ON mode of elementary actuating units. We convert experimental snapshots of the spatially distributed chemical system to a force vector field and then simulate the motion of manipulated objects in the force field, thus, achieving reaction–diffusion media controlled actuation.

The controllers described in Sects. 11.3 and 11.4 are used predominantly in an open-loop device mode, where the robot or robot manipulator is controlled by the chemical medium but does not provide any useful explicit (only implicit, via human operator) feedback to the chemical media controlling it. We specify this as useful because in robot, taxis movement of the robot provides physical disruption of the chemical processor, but the results are yet difficult to translate to sensible feedback. Until our recent paper [4], there were no experimental results concerning the real-time interaction of robotic devices with chemical reaction–diffusion controllers. In that paper, we designed and implemented a series of experiments concerning the two-way interactions between a spatially distributed excitable chemical controller and a robotic hand. The interactions are two-way because the waves travelling in the chemical media control the movement of the fingers and then the fingers physically (disturb the solution) and chemically (disperse activator from small pipettes placed on the fingers) interact with the reaction–diffusion controller. In Sect. 11.5 we present an overview of our results, see details in [4].

Some people may consider that coupling gel or liquid phase chemical processors with rigid silicon hardware and robotic devices are a bit eccentric and predominantly a curiosity-driven activity. They are of course partly right as conventional hardware is not the right application domain for chemical processors or reaction–diffusion processors in general. In fact reaction–diffusion processors would be better employed in amorphous spatially extended, distributed and reconfigurable architectures with distributed sensor and effector arrays. Such architectures are not yet available in a complete form and therefore, to demonstrate the feasibility and future potential of the reaction diffusion based approach, we employed a living analogue of a reaction–diffusion system encapsulated in a functional elastic membrane – plasmodium of *Physarum polycephalum*. The large size of the plasmodium allows the single cell to be highly amorphous. The plasmodium exhibits synchronous oscillation of cytoplasm throughout its cell body, and oscillatory patterns control the behaviour of the cell. The oscillatory cytoplasm of the plasmodium can be seen as a spatially extended excitable reaction–diffusion based medium.

In the Plasmodium, all the parts of the cell behave cooperatively in exploring the domain space, searching for nutrients via an optimized network of streaming protoplasm. Due to its unique features and ease of experimentation, the plasmodium became a test biological substrate for implementation of various computational tasks. The fact that it is encapsulated in an elastic membrane means that plasmodium are not only capable of computing spatially distributed data-sets but also the physical manipulation of elements of the data-sets. If a sensible, controllable and, ideally, programmable movement of the plasmodium could be achieved, this would open the way for experimental implementations of amorphous robotic devices.

In this chapter, we discuss a set of scoping experiments that aim to establish links between Physarum based computing and Physarum robotics. For the purpose of the experiments, we have chosen the surface of water as a physical substrate for the plasmodiums development to study how the topology of the plasmodium network can be dynamically updated, without the requirement for the plasmodium to be stuck to a non-liquid substrate. With this sytem we have studied the possibility of manipulating small objects floating on the surface with the plasmodium's pseudopodia. In Sect. 11.6, we show that the plasmodium of Physarum is capable of computing basic spanning trees and also of manipulating lightweight objects.

11.2 BZ Medium

A thin layer BZ medium can be prepared using a recipe adapted from [5]: an acidic bromate stock solution incorporating potassium bromate and sulphuric acid ($[BrO_3^-] = 0.5$ M and $[H^+] = 0.59$ M) (Solution A); solution of malonic acid (solution B) ($[CH_2(CO_2H)_2] = 0.5$ M), and sodium bromide (solution C) ($[Br^-] = 0.97$ M). Ferroin (1,10-phenanthroline iron-II sulphate, 0.025 M) is used as a catalyst and a visual indicator of the excitation activity in the BZ medium. To prepare a thin layer of the BZ medium, we mix solutions A (7 ml), B (3.5 ml), C (1.2 ml) and finally when the solution becomes colourless ferroin (1 ml) is added and the mixture is transferred to a Petri dish (layer thickness 1 mm). Excitation waves in the BZ reaction can be initiated using a silver colloid solution. This excitable chemical processor is used both on-board the mobile robot and in experiments to control a robotic hand. The processor used to control the actuator array was based on a silica gel sheet of 0.2 mm thickness which had been soaked in 0.01 M ferroin solution for 30 min. The catalyst loaded gel sheet was then placed in a petri dish containing 10 ml of BZ stock solution devoid of the catalyst (see recipe above).

11.3 Robot Taxis

The BZ reaction can be used to assist the navigation of a mobile robot in a real-life environment [6]. Given an on-board thin-layer chemical reactor containing the liquid-phase BZ medium, the chemical medium can be intermittently stimulated to sensibly guide the robot.

To make a BZ controller capable of robot taxis, we prepared a thin layer BZ reaction using a recipe adapted from [5] and detailed in Sect. 11.2. Excitation waves in the BZ reaction are initiated using a silver colloid solution, which is added to the medium via a micro pipette by a human-operator. The chemical controller is placed onboard a wheeled mobile robot (Fig. 11.1a). The robot is about 23 cm in diameter and able to turn on the spot; wheel motors are controlled by Motorola 68332 on-board processor. The robot features a horizontal platform, where the Petri dish (9 cm in diameter) is fixed, and a stand with a digital camera Logitech QuickCam (in 120×160 pixels resolution mode), to record the excitation dynamics. The robot controller and camera are connected to a PC via serial port RS-232 and USB 1.0, respectively.

Because vibrations adversely affect the excitation waves in the BZ system, we endeavoured to make the movements of the robot as smooth as possible. Thus, in our experiments, the robot moves only with a speed of circa 1 cm s^{-1}. We found that rotational movements were particularly disruptive and caused spreading of the excitation waves from the site of the existing wave fronts faster than would be observed if the process was just diffusion driven. Therefore, to minimize this effect we made the robot rotate with a very low speed of around 1 deg s^{-1}. The vibrations and rotational movements are only considered adverse for the purpose of the original

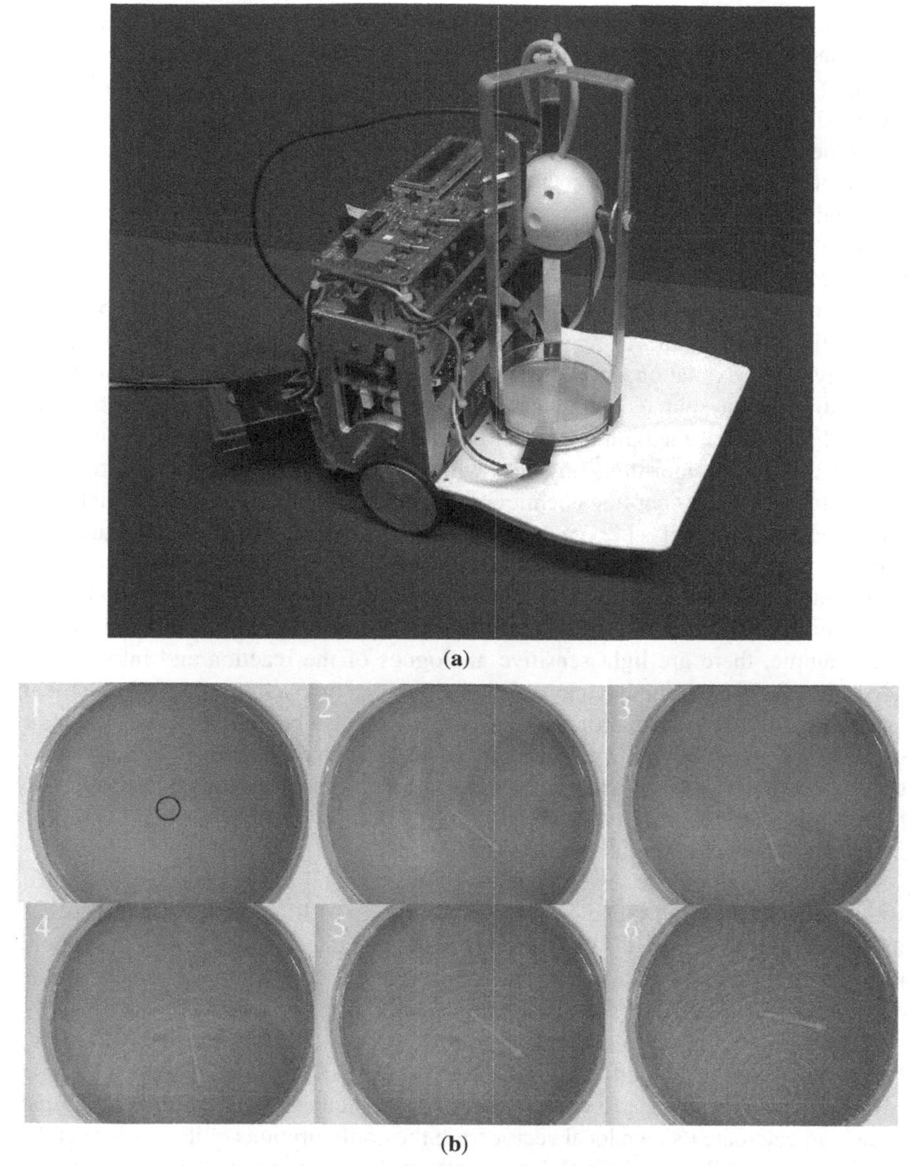

Fig. 11.1 Wheeled robot controlled by BZ medium (**a**) and snapshots of excitation dynamics with the robot's velocity vector extracted (**b**). From [6]

experiments, where we want to establish methods for extracting vectors from the excitation dynamics. In future incarnations, these movements could be used to measure said motion and feedback information to the robot including information about the environment. For example, if the robot collided with an obstacle the physical jolt may be sufficient to reset the excitation dynamics of the chemical controller.

To enhance the images of the excitation dynamics, the Petri dish is illuminated from underneath by a flexible electro-luminescent sheet (0.7 mm thick) cut to the shape of the dish. The sheet, powered through an inverter from the robot's batteries, produces a cool uniform blue light not affecting (at least in conditions of our experimental set-up) physico- chemical processes in the reactor. We found it was preferable to implement experiments in low light to avoid the interference of daylight and lamps.

When the medium was initiated with silver, a local generator of excitation waves is formed (Fig. 11.1b), depending on the amount of activator deployed trains of circular waves (target waves) may evolve. The robot is able to extract the position of the stimulation point, relative to the centre of the chemical reactor by using the topology of the excitation waves. The robot adjusts its velocity vector to match the vector towards the source of stimulation. To guide the robot more precisely, one can excite the chemical medium at several points (Fig. 11.1b). The forced stimulation was simply to test algorithms on board with real chemical controllers and to learn about problems with coupling chemical reactors and robots. Something which was identified as a problem – i.e. physical disruption of chemical waves could actually turn out to be advantageous as it may feed back both direction, velocity and even rotational movement of robot. It should be noted that eventually it was intended for environmental interaction to become the source of stimulation for excitation waves. For example, there are light sensitive analogues of the reaction and information about the robots environment may able to be mapped onto a light sensitive chemical processor and feed information to the robot. It may also be possible to deploy bulk oscillating micro-reactors, which are placed around the robot and are directly electrically coupled to the robots drive motors. The oscillations would be controlled by light levels and if all reactors were coupled, it may be possible to obtain complex information about the robots environment.

The main goal of our research concerning chemical controllers for robots [6] was to develop a stand-alone onboard controller, which could sense and process information from the robot's environment, calculate the current mode of the robots movement and communicate a set of appropriate commands to the robots motor controllers [2, 6].

Therefore, we put the following constraints on a possible solution of the problem. Firstly, an algorithmic solution must be local – if the BZ medium is mapped onto a two-dimensional array, e.g. a cellular automation, then every cell of the array must be able to calculate its own local vector from the configurations of its r-order neighbourhood. Secondly, only spatial data are allowed – a vector towards the position of the stimulation source must be extracted from just one snapshot of the medium's excitation dynamics. In addition, the robot control unit (including sub-systems based in the interfaced PC) is not allowed to remember local or global vectors calculated at previous stages (therefore, we tried to reduce the role of the hardware to interfacing and decoding).

To prove that a vector towards a source of stimulation can be extracted under these three constraints, it is sufficient to demonstrate that the asymmetry of excitation waves is detected from digital snapshots of the medium dynamics.

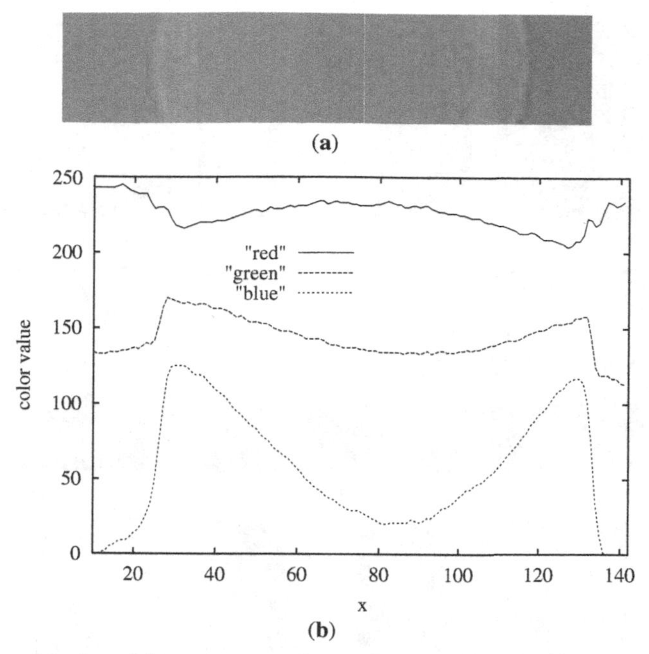

Fig. 11.2 Demonstration of the asymmetry of wave fronts as detected from snapshots of the BZ medium: (**a**) fragment of a snapshot of an expanding target wave; (**b**) values of *red*, *green* and *blue* components of pixels along x-axis cross-section averaged along y-axis. The distribution preserves its form but exact values in each experiment vary depending on the concentrations of reactants. From [6]

As seen in Fig. 11.2, the shape of the excitation wave front is visually represented by a spatial distribution of red, green and blue components of each pixel's value. The distribution of the blue component efficiently characterizes all parts of the wave front, including the zones corresponding to the sharp excitation event and the zone where the medium gradually returns back to the reduced steady state (excitable state) via the refactory state. Therefore, only the blue component of the pixel colours was necessary to reconstruct the stimulation position. All normals to the wave front, calculated from gradients of the blue colour values, are oriented towards a site where the wave originated. Therefore, a simple analysis of the spatial distribution of normals would give the robot an indicative position of the stimulation source.

The algorithm does not work if the source of stimulation is represented by an almost perfect circular wave front, all parts of which are inside the processed zone of the reactor – in this case all local vectors will cancel each other. However, this can be coped with by introducing a breathing receptive field, which changes its size during experiments, or simply by waiting until some parts of the wave front reach the edges of the reactor and disappear. The dynamically changing receptive field could also reduce the influence of phase waves spontaneously generated at the edges of the reactor.

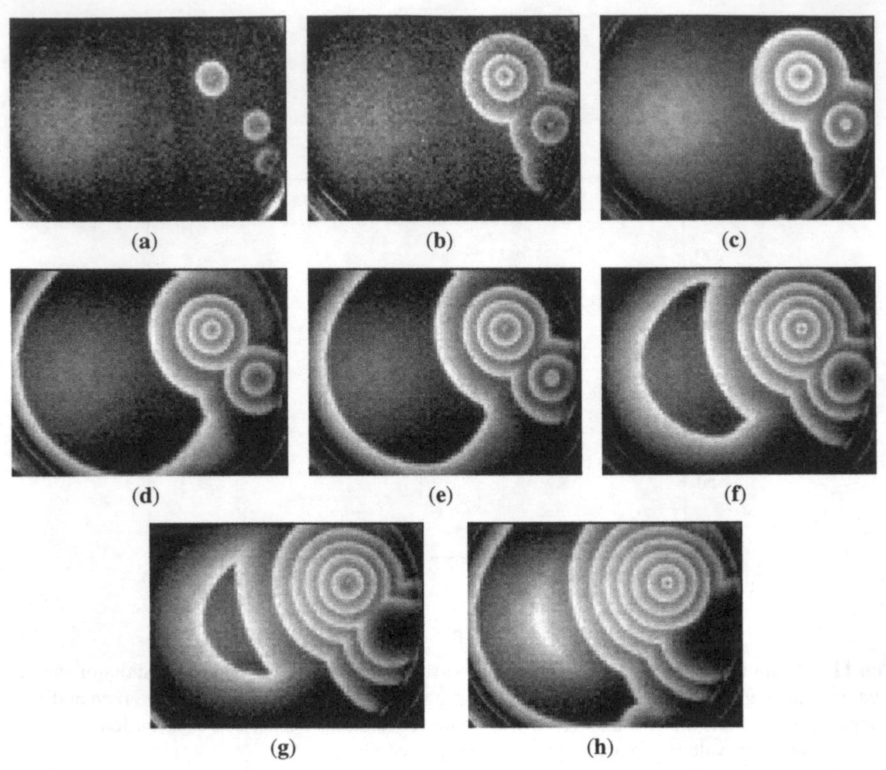

Fig. 11.3 Snapshots of the BZ medium, the intensity of the *blue component* of the pixel colour is displayed. From [6]

Consider the example of global vector extraction from the reactors excitation dynamics in Figs. 11.3 and 11.4. Three sources of target waves were initiated at different time intervals. At first, the robot detects no stimulation because the wavefronts are perfectly closed and circular (Figs. 11.3a and 11.4a). Then the outer wavefront of one of the target waves reaches the edge of the reactor and annihilates by becoming disconnected. The robots control algorithm calculates a vector towards this specific point of stimulation (Figs. 11.3b and 11.4b). At the same time, another wavefront from a secondary source of stimulation also becomes disconnected (in Fig. 11.3a, c–e) causing a deviation in the global vector (Fig. 11.4c–e). The second source of target waves was initiated using a smaller quantity of silver colloid and therefore, only four waves are initiated from the original point of stimulation before the site returns to the steady reduced state. the result is that the global vector rotates back toward the first source of stimulation (Fig. 11.3f–h).

In real life situations, robots are often required to choose between several sources of stimulation. How could this type of decision-making be implemented in an on-board chemical controller? Given two sources of point-wise stimulation, the robot

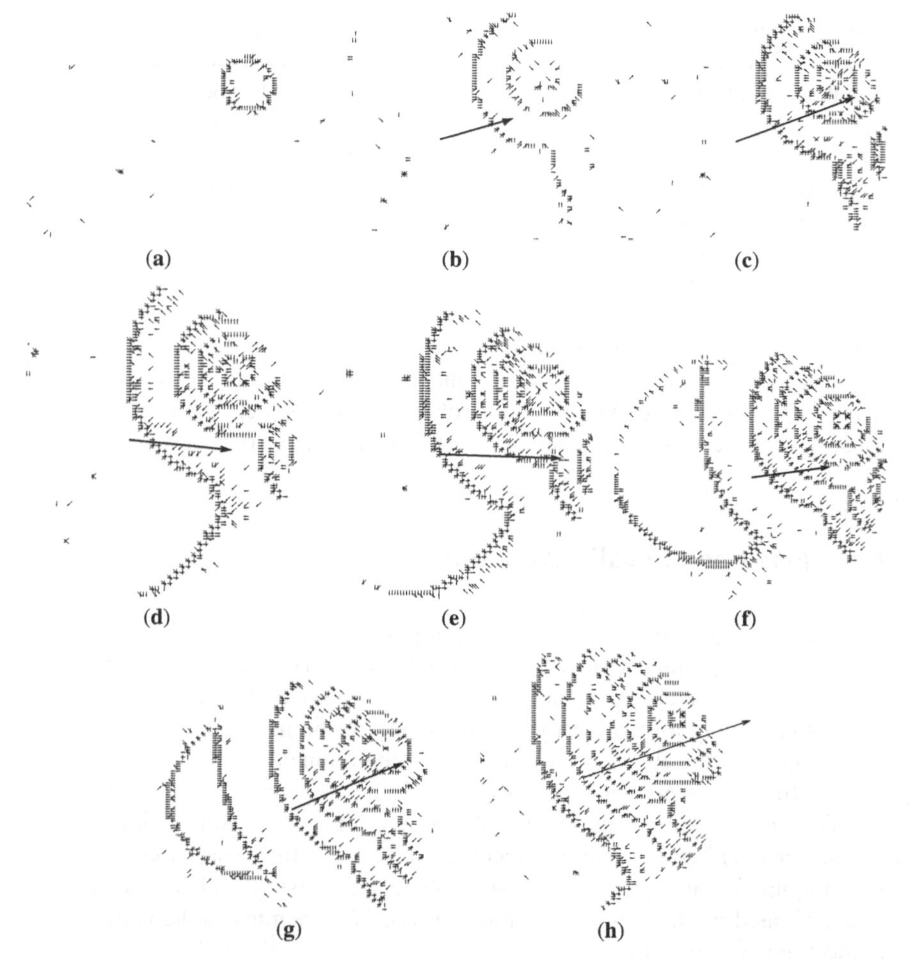

Fig. 11.4 Local vector fields and global vectors (*shown in bold*) extracted from snapshots of BZ medium in Fig. 11.3. From [6]

usually chooses the older source. This is because, as discussed, the oldest stimulation point is represented by predominantly broken wavefronts that have a greater degree of spatial asymmetry and usually a greater number of wave fronts when compared to the more recent sources of stimulation (Fig. 11.5a, b). However, when the older source of excitation disappears or if the newer target wave has a higher frequency of wave initiation, then this will be selected as the current orientation target. Therefore, it can be said that the onboard chemical controller is capable of memorizing the temporal order of the stimulations. Moreover, a fine balance could be achieved between the intensity and time tag of multiple stimulations to achieve a more precise trajectory for the robot's motion.

Once stimulated, the robot moves in circles because it recalculates the position of the stimulation source every time regardless of its previous position and does not remember previously extracted vectors. This circular motion has been described previously in simulation experiments [2] and is a typical behavioural mode for mobile excitable lattices with a wide interval of node sensitivity. Relatively straightforward motion could be achieved by stimulating the chemical reactor twice and positioning these points of stimulation at opposite sites of the reactor, so that a bisector of these two points gives the orientation of the velocity vector. Also the circular motion of a robot can be prevented by increasing the bromide ion concentration and thus, making it impossible for a generator of target waves to persist in the reactor. In this case, just a few wave fronts will travel across the reactor and disappear. Obviously in this case, the robot must analyse snapshots of the medium after longer delays, e.g. once per 10–15 min. To slow down or stop the robot we could 'block' the velocity vector with several sources of wave generation, as shown in Fig. 11.6.

11.4 Open-Loop Parallel Actuators

How can a reaction–diffusion medium manipulate objects? To find out, we coupled a simulated abstract parallel manipulator [7] with an experimental BZ chemical medium. The simulated manipulator is a two-dimensional array of actuating units, each unit can apply a force vector of unit length to the manipulated objects. The velocity vector of the object is derived via the integration of results of all local force vectors acting on the manipulated object.

To demonstrate the potential of the BZ system for manipulating simple objects, we built an interface between the recordings of space-time snapshots of the excitation dynamics and simulated physical objects. At first we calculated the force fields generated by the mobile excitation patterns and then tested the behaviour of an object in these force fields.

This coupling of an excitable chemical medium with a manipulator allows the conversion of experimental data relating to wave movement in the BZ medium into a force vector field and subsequently assess the imparted motion of objects in the changing field, thereby, achieving reaction–diffusion medium controlled actuation [8].

The chemical processor used to perform actuation was prepared following the recipe in Sect. 11.2, see also [5], based on a ferroin catalyzed BZ reaction. A silica gel plate is cut and soaked in a ferroin solution. The gel sheet is placed in a Petri dish and BZ solution added. The dynamics of the chemical system are recorded at 30 s intervals using a digital camera.

The cross-section of the BZ wavefront recorded on a digital snapshot shows a steep rise of red colour values in the pixels at the wavefront's head and a gradual descent in the pixels spanning the wavefront's tail. Assuming that excitation waves push the object, local force vectors generated at each site – a pixel of the digitized image – of the medium should be oriented along the local gradients of the red colour

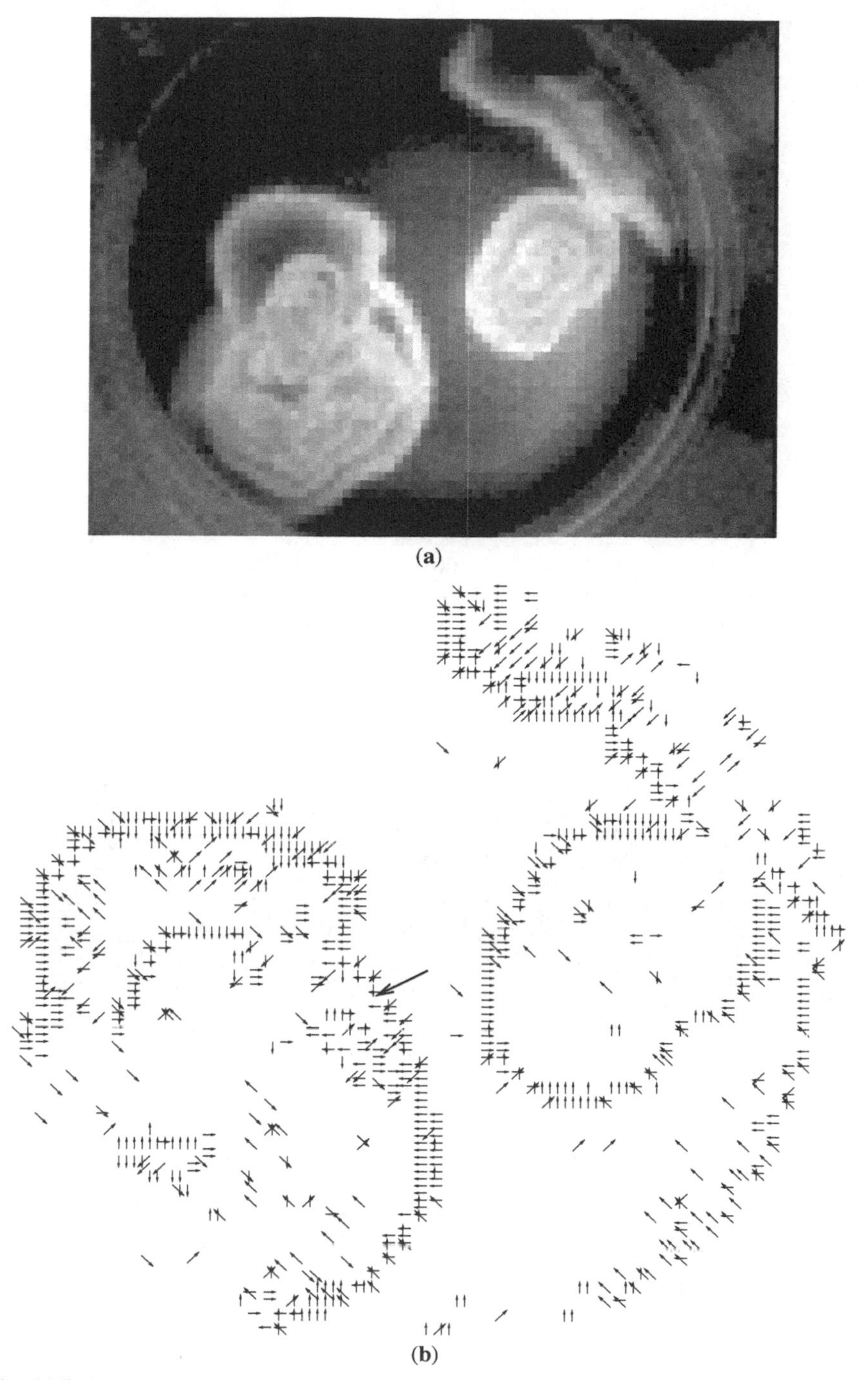

(a)

(b)

Fig. 11.5 An example of the decision making procedure in a chemical processor: selection of the strongest generator of target waves, representing a "strongest" source of stimulation. (**a**) the blue component of an image of the BZ reaction, (**b**) local vector field and global vector (*in bold*). From [6]

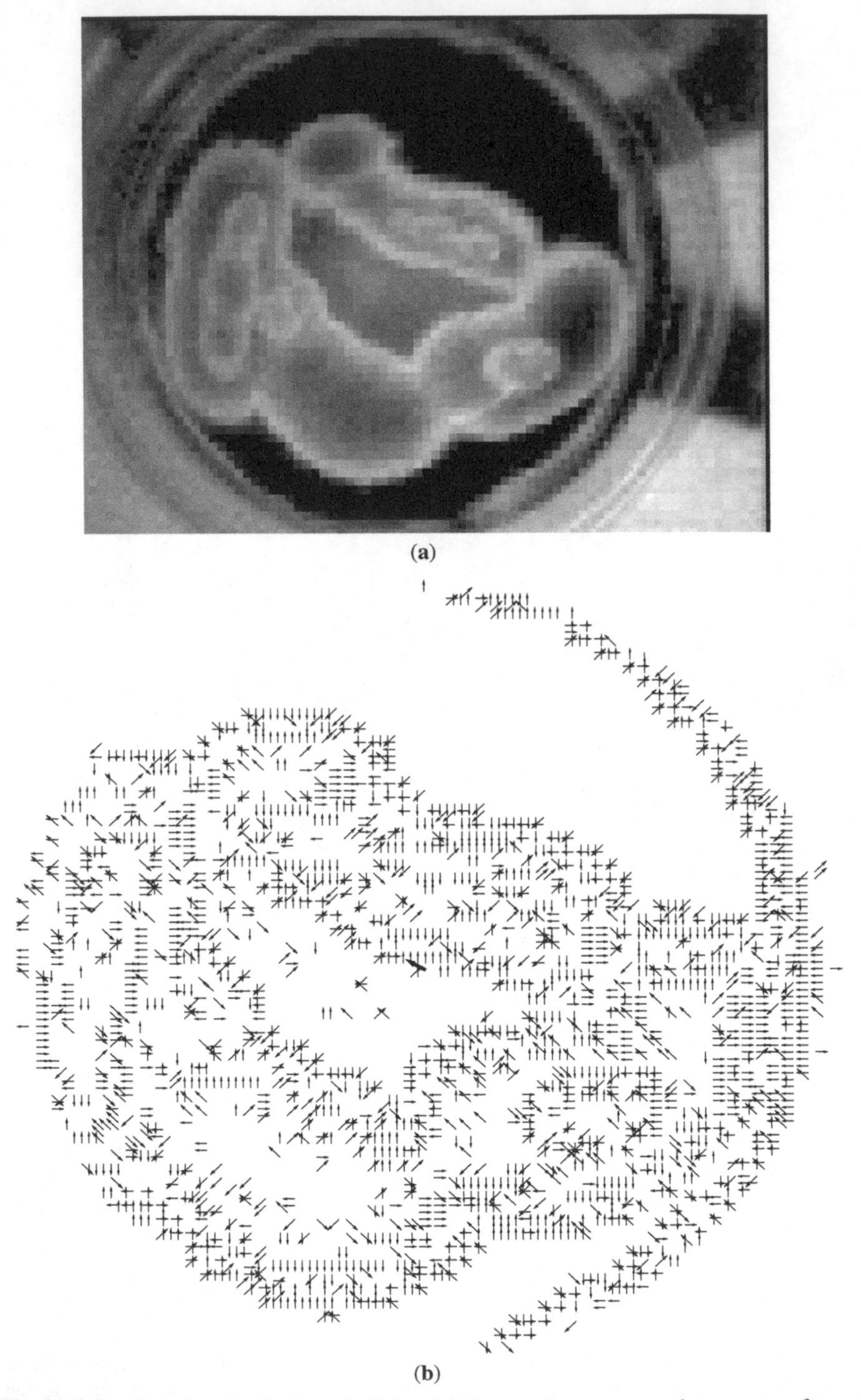

Fig. 11.6 Stopping the robot by "trapping" the global vector between several generators of target waves. (**a**) blue component of BZ image, (**b**) local vector field and global vector (*in bold*). From [6]

values. From the digitized snapshot of the BZ medium, we extract an array based on the red component of the images and then calculate the projection of a virtual vector force at the pixel. An example of force vector field extracted from a fragment of an excitation wavefront is shown in Fig. 11.7.

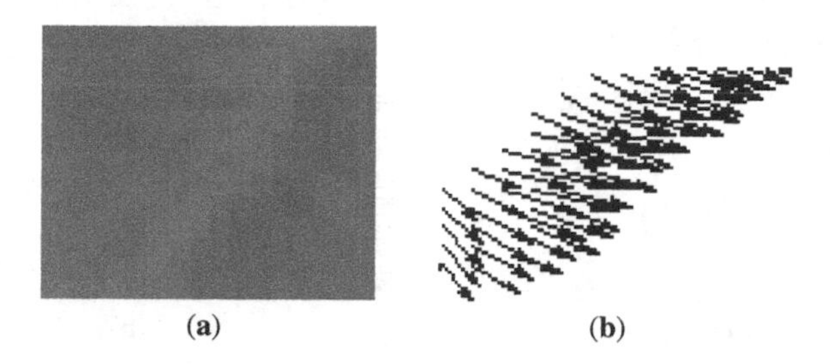

Fig. 11.7 A fragment of an excitation wavefront in the BZ medium (**a**) local force vectors extracted (**b**). From [7]

Force fields generated by the excitation patterns in a BZ system (Fig. 11.8) result in tangential forces being applied to a manipulated object, thus, causing translational and rotational motion of the object [7].

When placed on the simulated manipulating surface, pixel-objects move at random in the resting domains of the medium; however, via random drift each pixel-object does eventually encounter a domain of co-aligned vectors – representing the excitation wave front of the BZ medium – and is therefore, translated along the vectors. An example of several pixel-objects transported on a "frozen" snapshot of the chemical medium is shown in Fig. 11.9.

Rough trajectories of pixel-objects (Fig. 11.9a) show distinctive intermittent modes of random motion separated by modes of directed 'jumps' guided by travelling wave fronts; compare Fig. 11.9a, c. Smoothed trajectories of pixel-objects (Fig. 11.9b) show that despite a strong chaotic component in manipulation, pixel-objects are successfully transported to the sites in the medium where two or more excitation wavefronts meet.

The phenomenon of the manipulation of pixel-objects by dynamically changing patterns of excitation reflects, in principle, that of the stationary configurations (Fig. 11.10). At the initial stage, involving excitation target waves originating from several sources of stimulation, but do not fill the entire manipulation space (see e.g. Fig. 11.10, $t = 0$ min and $t = 1$ min) and therefore, the pixel-objects are transported to the resting parts of the medium, where they either continue drifting at random, or they can follow artefact gradations of colour imposed by the external illumination (see trajectories of two pixel-objects, which started their journeys at the right part of the medium's snapshot. Fig. 11.10, $t = 0$ min). Eventually, they are pushed over the domains corresponding to the bisectors' separating parts of the medium relating to the sources of target waves.

The overall speed of pixel-object transportation and in some cases, even success depended on the frequency of wave generation by the various target waves. As a

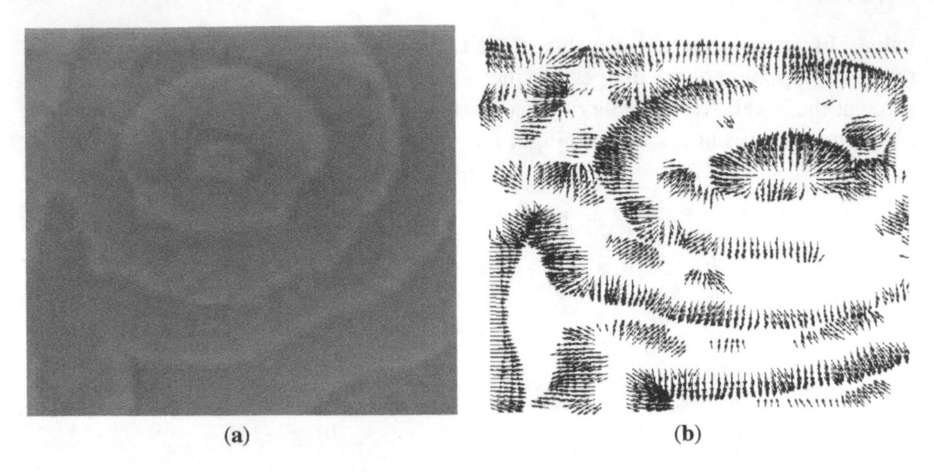

Fig. 11.8 Force vector field (**b**) calculated from BZ medium's image (**a**). From [7]

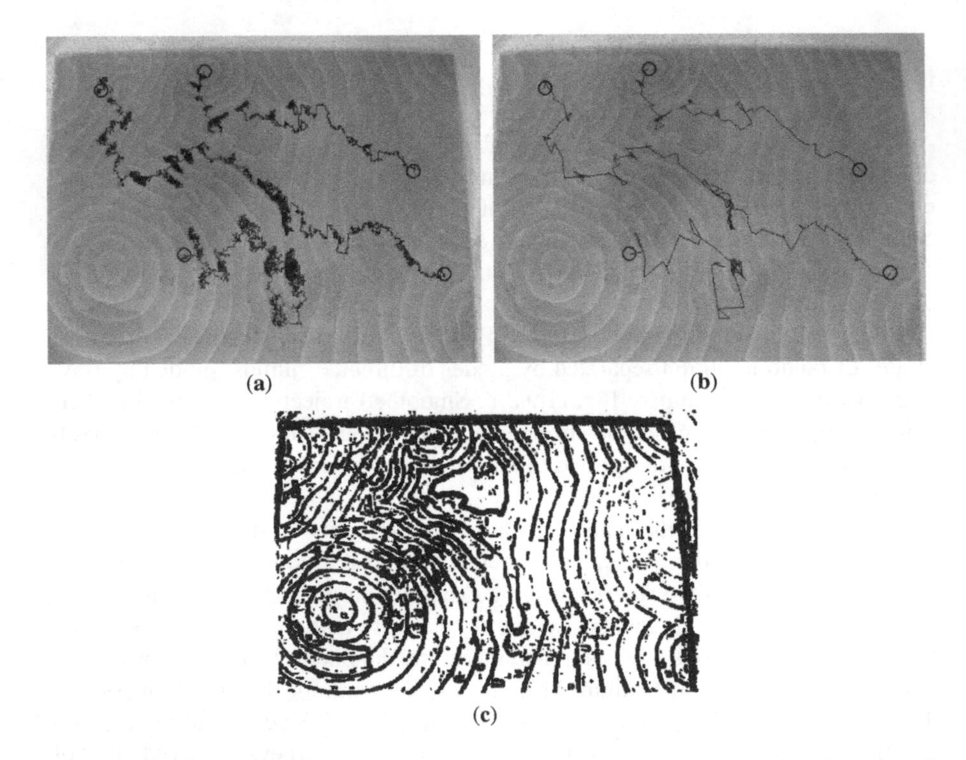

Fig. 11.9 Examples of manipulating five pixel-objects using the BZ medium: (**a**) trajectories of pixel-objects, (**b**) jump-trajectories of pixel-objects recorded every 100th time step, (**c**) force vector field, extracted from the BZ snapshot (**a**), used for manipulating pixel-objects. Initial positions of the pixel-objects in (**a**) and (**b**) are shown by *circles*. Note locally noisy drifting of the pixels. From [7]

rule, the higher the frequency the faster the objects are transported. This is because in parts of the medium spanned by low frequency target waves, there are lengthy domains of the reactor, which is in the resting state, resulting in the absence of force vectors. Therefore, pixel-sized object can wander randomly for a long time before encountering the next wave front (compare the pixel-object transported by low-frequency target waves on the right side of Fig. 11.11a with the pixel-object trajectory on the right side of Fig. 11.11a).

Our method of manipulating pixel-objects also works in the conditions of a 'non-ideal' excitable chemical medium. Thus, in the situation when numerous spiral waves emerge in the system (Fig. 11.11b), the pixel-objects bounce off the spirals; an overall southward drift of the objects in Fig. 11.11b is caused by the tendency of excitation wavefronts to be less fragmented in this direction.

Even when the BZ medium resides in a sub-excitable regime (where sources of excitation cause formation of non-target or spiral waves but mostly soliton-like travelling wave-fragments), the pixel-objects can still be sensibly manipulated (see Fig. 11.11c). The objects are pushed by travelling wave-fragments, however, (as shown in Fig. 11.11c) they can very often slip away.

In general, spatially extended objects repeat the overall pattern of motion observed for the pixel-sized objects; however, due to the integration of many force vectors, the motion of planar objects is smoother and less sensitive to the orientation of any particular force-vector.

In experiments, we found that manipulation results depend on the size of the object, thus for example, see Fig. 11.12, a small square shape (starting its journey at the north–west corner of the chemical reactor, Fig. 11.12a) is accelerated by force-vectors, caused by the excitation waves, and passes the central part of the reactor barely changing its trajectory due to collisions with opposing wavefronts. With increasing size of the object, we take account of larger numbers of local vector-forces, and thus, the objects' trajectory becomes more controllable and in Fig. 11.12b we see that the object is sensibly transported to the 'sink' of the force-field, and stays there "indefinitely".

The response of the object to the excitation wave depends also on the wavelength, the shorter the wavelength the larger the acceleration of the object. Thus in Fig. 11.13, the object starts to travel at the "slope" of the target wave before being "bounced" several times between the oncoming wavefronts, then missing the "sink" of the reactor and travelling through the central part of the reactor between the target waves.

By choosing the positions of the initial stimulation of the excitable chemical controller and the initial position of the manipulated object, we can precisely control the trajectory of the manipulated object. Thus in Fig. 11.14, we demonstrate reflection of the triangular shape by the central part of the target wave (Fig. 11.14a) and also by proximal wavefronts of the generator (Fig. 11.14b).

In [7], we also showed that BZ media exhibiting large populations of spiral wave generators exhibit great potential for the directed transportation of the planar shapes. All objects are usually pushed away, toward the periphery of the spiral wave population; however, they can be easily deviated by carefully positioning sources of target

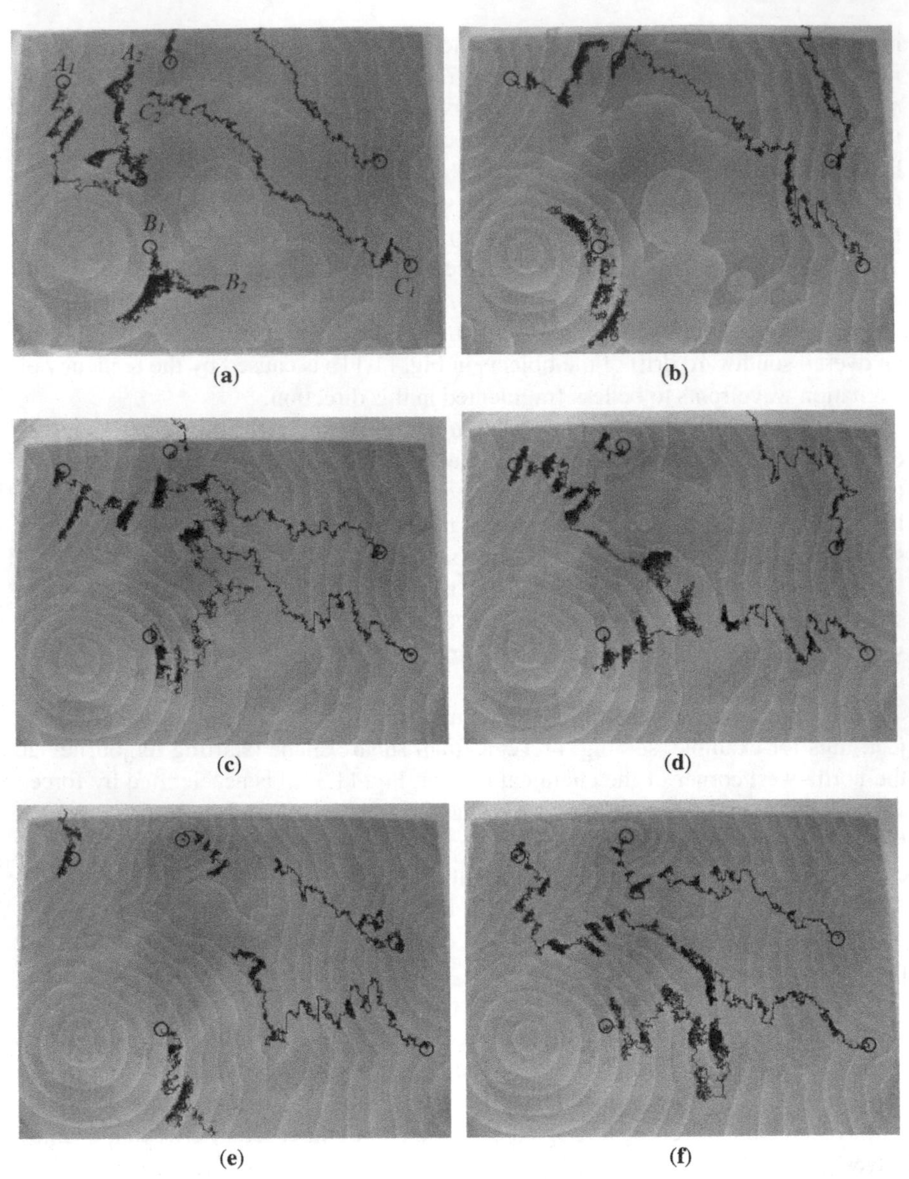

Fig. 11.10 Pixel-objects manipulated by dynamically changing configurations of excitation in the BZ medium. For every snapshot of the medium pixel-objects are transported from the same original points for 10,000 steps of discrete time. Initial positions of pixel-objects are shown by *circles*. Note locally noisy drifting of the pixel-objects. To aid understanding of the pixel-objects motion we labelled, in picture t=0, trajectories of three pixel-objects as $X1$ (start) and $X2$ (end), $X = A, B, C$. From [7]

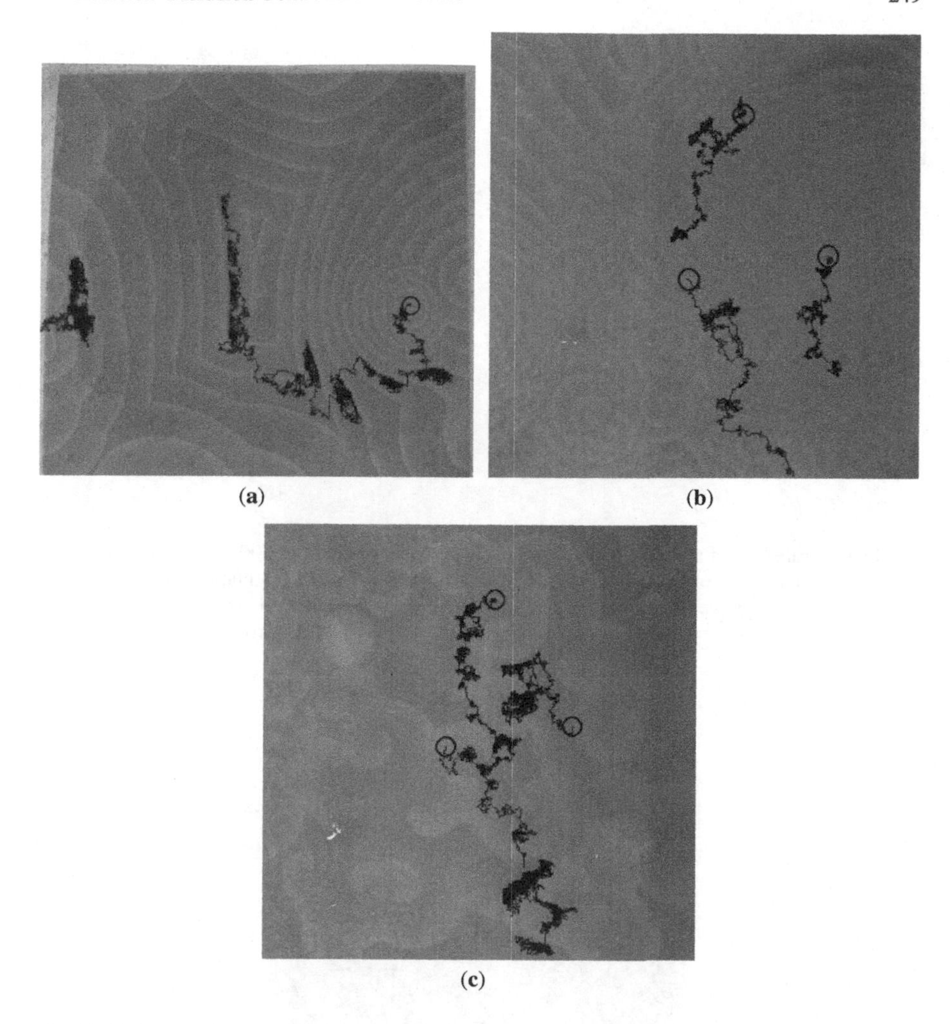

Fig. 11.11 Manipulation of pixel-sized objects in various regimes of excitation in BZ-medium. Initial positions of pixel-objects are shown by circles. Note locally noisy drifting of the pixel-objects in (**a**) and (**b**). (**a**) Example of influencing transportation by frequency of target-wave generation. Both pixel-objects were manipulated for 14,000 times steps. (**b**) Manipulating pixel-objects in a BZ medium with many sources of spiral waves. (**c**) Manipulating pixel-objects by wave-fragments in sub-excitable BZ medium. From [7]

waves. The objects accelerated by populations of spiral waves, can then be confined to a border-line domain between two sources of target waves or even slowed down by a single source of target-waves. A BZ medium in a sub-excitable mode, when localized wave-fragments inhabit the reactor space, may prove to be an invaluable tool for combining manipulation with logical decision making schemes. Each localized wave-fragment acts as a 'force bullet', which can be directed very precisely towards the manipulated object, and the force imparted to the object sends the manipulated shape in the desired direction.

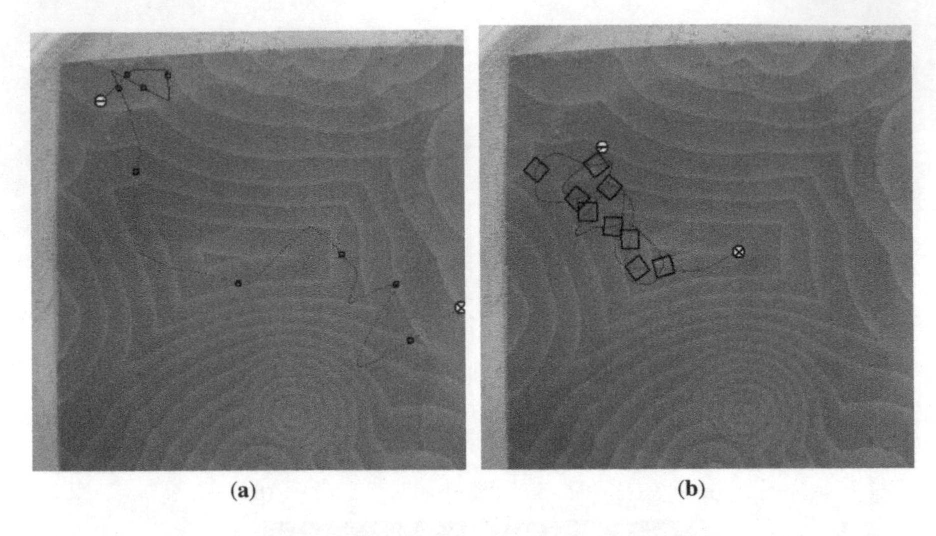

(a) (b)

Fig. 11.12 Snapshots of squares manipulated by the BZ medium during 500 steps. Trajectories of the centre of mass of the squares are shown by the *dotted line*. Exact orientation of theobjects is displayed every 50 steps. (**a**) square 5×5 cell size, (**b**) square 20×20 cell size. Initial position of object is shown by *circle with minus* and final position by *circle with x* . From [7]

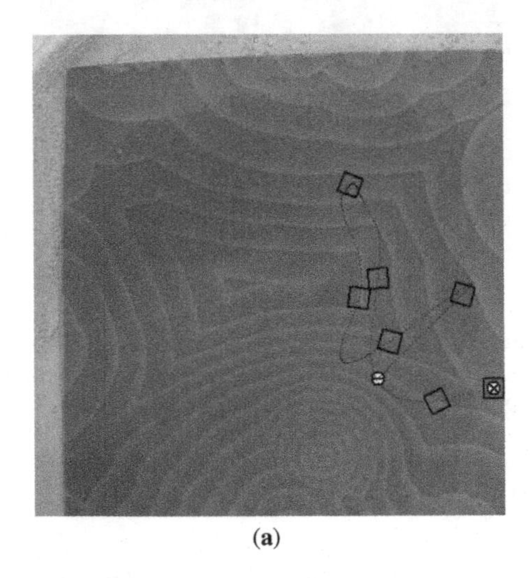

(a)

Fig. 11.13 (**a**) Snapshots of 20×20 cell square manipulated by BZ medium during 500 steps. Trajectory of centre of mass of the squares is shown by *dotted line*. Exact orientation of the object is displayed at every 50 steps. Initial position of the object is shown by *circle with minus* and the final position by *circle with plus*. From [7]

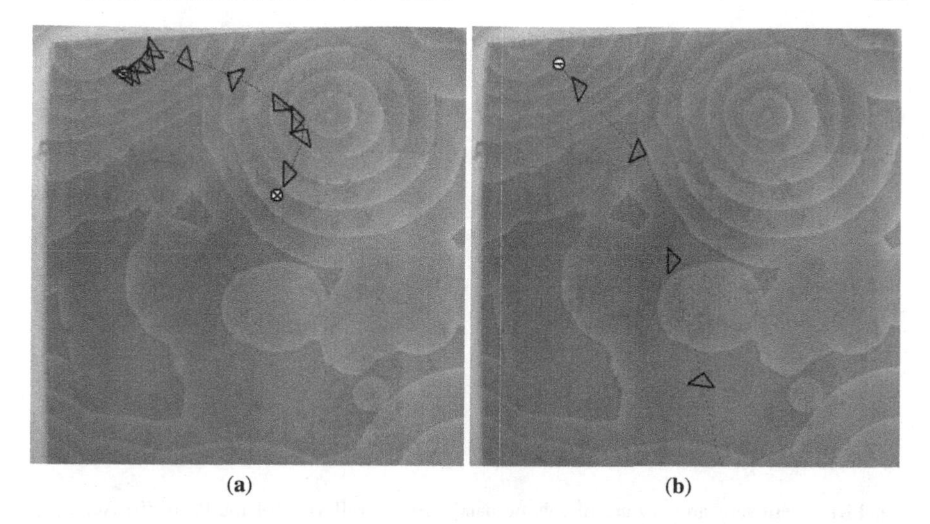

(a) (b)

Fig. 11.14 Snapshots of isosceles, right-angled triangle (size of half square) manipulated by BZ medium during 200 steps. Trajectories of the centre of mass of the triangle is shown by the *dotted line*. The exact orientation of the object is displayed at every 20 steps. Initial position of the object is shown by *circle with minus* and the final position by *circle with plus* . From [7]

11.5 Closed-Loop Control of Robotic Hand

In the previous sections, we described the design of chemical controllers for robots, which can calculate a shortest collision-free path in the robotic arena and guide the robot toward the source of stimulation (taxis). However, the controllers described a lack of feedback from the robot to the controller. In a set of remarkable experiments undertaken by Hiroshi Yokoi and Ben De Lacy Costello [4], it was demonstrated that when a closed-loop interface between the chemical controller and the robot is established, the behaviour of the hybrid systems becomes even more intriguing.

In experiments [4], the robotic hand developed in [9] was used. Each finger of the robotic hand consists of two pivots with springs and wire guides attached (Fig. 11.15a).

We use a spring coil type of stainless steel outer wire as a guide for nylon inner wire to realize the function of the spring. When a finger does not bend the spring coil type of outer wire keeps it straight. If a finger needs torque, the spring coil type of outer wire bends to realize the shortest route of the inner wire; the elastic wires are connected to servomotors (Fig. 11.15b). The motors are controlled by a microprocessor H8Tiny produced by Hitachi. The microprocessor has eight inputs from photo-sensors, and outputs linked to ten servomotors. The microprocessor encodes signals from sensors to run the motor drives. The finger motors are powered by 5 V, and the microprocessors are powered by 9 V. The hand is mounted above the chemical reactor and glass micro-pipettes filled with silver colloid solution were attached to the fingertips to stimulate the chemical medium (Fig. 11.16).

To enable the robotic hand to interact with the chemical processor, we attached glass capillary tubes filled with a solution of colloidal silver to the fingertips, and

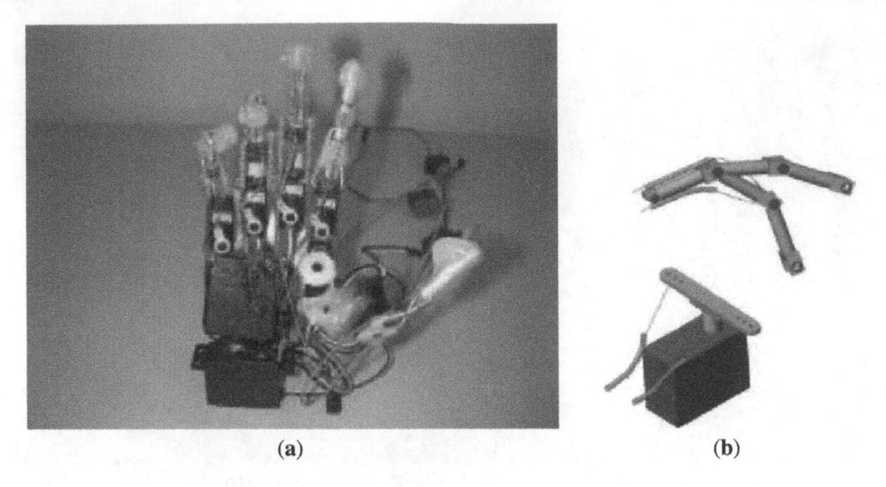

(a) (b)

Fig. 11.15 Hardware architecture of robotic hand. (**a**) Overall view of the hand. (**b**) Actuating fingers via elastic wires (*top*) by servomotor (*bottom*). From [4]

Fig. 11.16 Robotic hand interacts with Belousov–Zhabotinsky medium. From [4]

adjusted the hand above the Petri dish containing the BZ mixture in such a way that when a finger bent by a significant amount the pipettes were immersed just below the surface of the mixture releasing a defined quantity of silver and initiating additional excitation waves in the chemical medium.

To detect excitation wave fronts in the chemical reactor, we used photodiodes (VTB8440B), with a quoted spectral application ranging from 330 to 720 nm (Fig. 11.17). When a wave front travels through the medium and across the sensor, this is detected directly and the sensor's output current increases, this increase is detected by the micro-controller, which starts the servomotor causing the corresponding finger to bend. The fingers are programmed to straighten again a few seconds after bending.

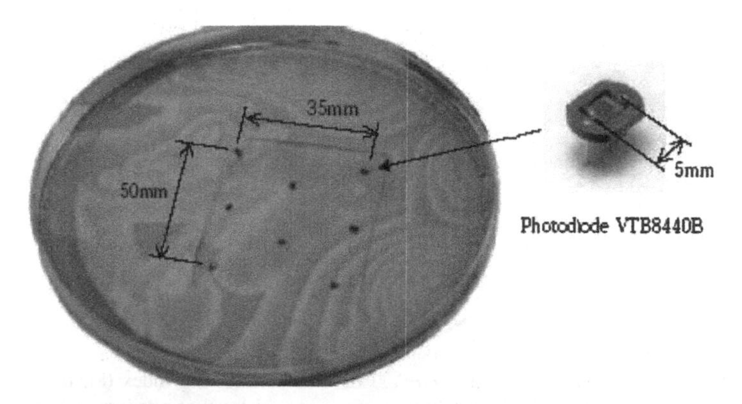

Fig. 11.17 Sensory system of the BZ-Hand. Experimental chemical reactor with photo-sensors underneath (*left*), sensors are visible as black sites on the photograph, and a view of the single sensor (*right*). From [4]

A typical development in the studied system is as follows: when one or more fingers bend for the first time (due to a solitary circular wave initiated at the reactors edge and moving towards the centre of the reactor – thus, crossing all photo-sensors), colloidal silver from the glass micro-pipettes, attached to the fingers, is deposited locally into the medium and initiates a series of excitation waves (Fig. 11.16).

A series of waves is initiated as silver colloid is physically deposited into the system and depending on the amount (proportional to the time the finger is in contact with the surface), it will usually initiate a number of waves at or near the contact point when the system has regained its excitability. As silver diffuses from the sites of contact, multiple excitation sites may be established leading to complex waves or spiral waves (also caused by the disturbance of the reactor surface by the motion of the fingers).

When the excitation waves spread, they stimulate photo-sensors, attached to the underneath of the Petri dish and this causes further bending of the fingers. Let us consider the behaviour of the system when up to three fingers have the ability to move.

The position of the sensors controlling these fingers and the sites where the fingers' touch the medium are shown in Fig. 11.18. When only one finger is operational, it excites (in the first instance) a generalized target wave in the medium. The wavefronts are detected by the sensor and the finger bends delivering more activator (silver) to the system and continuing the excitation process or disrupting existing waves. Thus the system enters a mode of self-excited oscillatory motion (Fig. 11.19a). If two fingers are operational and just one finger bends, then, after some transient period, the robotic hands fingers act in anti-phase because they reciprocally stimulate each other via the wavefronts of the target waves they initiate (Fig. 11.19b).

When both fingers touch the BZ medium at the beginning of the experiment the system starts to exhibit synchronous oscillations (11.19c). This happens because

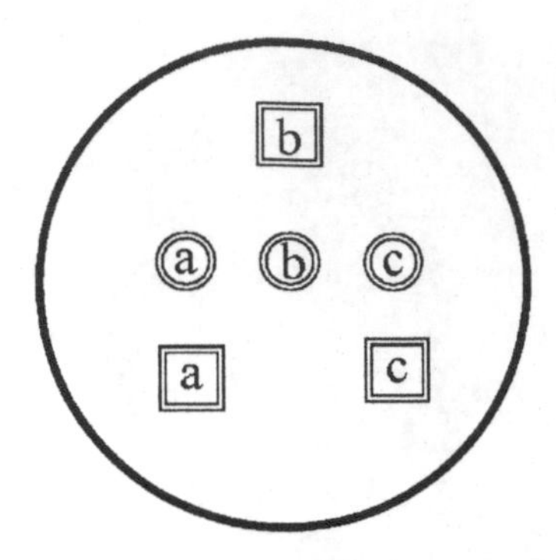

Fig. 11.18 Schematic relative positions (*circles*) of sensors affecting index (a), middle (b) and third (c) digits of the robotic hand, and sites (*rectangles*) where these fingers stimulate the BZ medium

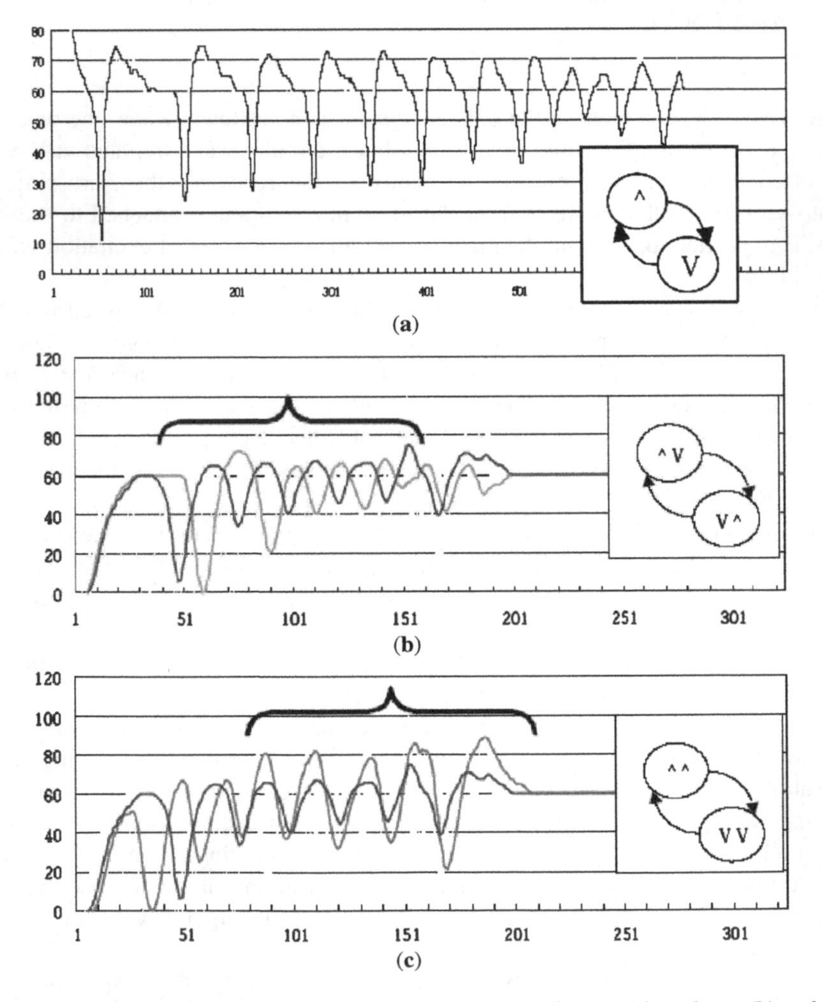

Fig. 11.19 Typical modes of "Fingers – BZ medium" system for one (**a**), and two (**b**) and (**c**), fingers. Voltage on the photo-sensor is shown in vertical axis; time in seconds is shown in horizontal axis. Corresponding movements of the finger are outlined in the insert, *U* means up and *D* down.

the wavefronts of the target waves, initiated by the movement of the fingers, cancel each other out when they collide, and thus, as in the general situation, each finger excites only itself. When three fingers are operated simultaneously, highly complex behaviors are observed (whatever the the starting configuration) – eventually, the system may appear to reach an almost synchronized or repetitive motion, but this is then subject to major transient fluctuations.

See [4] for details of complex dynamics of interactions between BZ medium and the robotic hand.

11.6 Physarum Robots

Plasmodium, the vegetative stage of the slime mould *Physarum polycephalum*, is a single cell, with thousands of diploid nuclei, formed when individual flagellated cells or amoebas swarm together and fuse. The plasmodium is visible with the naked eye. When plasmodium is placed on an appropriate substrate, the plasmodium propagates and searches for sources of nutrients (bacteria). When these sources are located and taken over, the plasmodium forms veins of protoplasm. The veins can branch, and eventually the plasmodium spans the sources of nutrients with a dynamic proximity graph, resembling, but not perfectly matching graphs from the family of *k*-skeletons [10].

The large size of the plasmodium allows the single cell to be highly amorphous. The plasmodium shows synchronous oscillation of cytoplasm throughout its cell body, and oscillatory patterns control the behaviour of the cell. All the parts of the cell behave cooperatively in exploring the space, searching for nutrients and optimizing a network of streaming protoplasm. Because of its unique features and the relative ease of undertaking laboratory experiments, the plasmodium became a test biological substrate for the implementation of various computational tasks.

The problems solved by the plasmodium include maze-solving [11–13], calculation of efficient networks [14] and proximity graphs [10], construction of logical gates [15] and robot control [16].

Can the plasmdoium be considered as a reaction–diffusion amorphous robot? Yes, as shown in more details in [17]. The oscillatory cytoplasm of the plasmodium can be seen as a spatially extended nonlinear excitable media [18–20]. In our previous papers, we hypothesized that the plasmodium of Physarum is a biological analogue of a chemical reaction–diffusion system, encapsulated in an elastic and growing membrane [17]. Such an encapsulation enables the plasmodium to function as a massively-parallel reaction–diffusion computer [3] and also to solve few tasks, which reaction–diffusion computers could not do, e.g. construction of spanning trees [21], and implementation of storage modification machines [22].

Being encapsulated in an elastic membrane, the plasmodium is capable of not only computing spatially distributed data-sets but also physically manipulating elements of the data-sets. If a sensible, controllable and programmable movement of the plasmodium could be achieved, this would open the way for experimental implementation of amorphous robotic devices. There are already seeds of an emerging theory of artificial amoeboid robots [23–25].

We discuss a set of scoping experiments on establishing links between Physarum computing and Physarum robotics. We will show that plasmodium can in principle be used to manipulate lightweight objects in a controllable way according to the positioning of nutrients' sources and can, therefore, be considered an example of a living amoeboid robot. We chose the surface of water as a physical substrate for the development of the plasmodium because in this mode the plasmodium network can be dynamically updated without being stuck to a non-liquid substrate and therefore, small objects floating on the surface can be manipulated by the plasmodium's pseudopodia with ease.

Plasmodium of *Physarum polycephalum* was cultivated on wet paper towels in dark ventilated containers, oat flakes were supplied as a substrate for bacteria on which the plasmodium feeds. We used several test arenas for observing the behaviour of the plasmodium and scoping experiments on the plasmodium-induced manipulation of floating objects. These are Petri dishes with base diameters 20 and 90 mm, and rectangular plastic containers 200 by 150 mm. The dishes and containers were filled to 1/3 depth with distilled water. Data points, to be spanned by the plasmodium, were represented either by 5–10 mm sized pieces of plastic foam, which were either fixed to the bottom of the Petri dishes or left floating on the water surface. Oat flakes were then placed on top of the foam pieces. Foam pieces, where the plasmodium was initially placed, and the pieces with oat flakes were anchored to the bottom of containers. Tiny foam pieces to be manipulated by the plasmodium were left free-floating.

To demonstrate that a substrate is suitable for robotic implementation, one must demonstrate that the substrate is capable of sensing the environment coupled with responsiveness to external stimulus, it must be capable of solving complex computational tasks on spatially distributed data sets and demonstrate locomotion and/or manipulation of objects. We provide basic demonstrations, which may indicate that plasmodium of *Physarum polycephalum* can be successfully used in future experiments on laboratory implementations of amorphous biological robots.

The surface of water has a high surface tension and therefore, it is physically able to support propagating plasmodium, when its contact weight to contact area ratio is small. When placed in an experimental container, the plasmodium forms pseudopodia aimed at searching for sources of nutrients. In most experiments the 'growth part' of the pseudopodia adopts a tree-like structure to achieve fine detection of chemo-gradients in the medium. This also happens to minimize the weight to area ratio of the system. Examples of tree-like propagating pseudopodia are shown in Fig. 11.20ab.

In Fig. 11.20b, we can see that not always pseudopodia grow towards source of nutrients, there is a pseudopodia growing South–west, where no sources of nutrients located. This happens possibly because in large-sized containers volume of air is too large to support a reliable and stationary gradient of chemo-attractants. This may pose a difficulty for the plasmodium to locate and span all sources of nutrients in large-sized containers.

Fig. 11.20 The plasmodium explores the experimental arena by use of propagating tree-like pseudopodia. From [26]

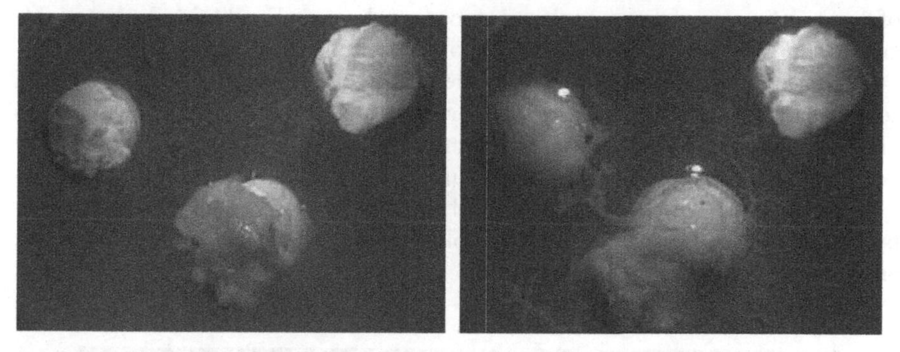

Fig. 11.21 Plasmodium builds links connecting its original domain of residence with two new sites. Initial position of plasmodium is in the photograph on the *left*, propagated plasmodium after 12 h is in the photograph on the *right*. From [26]

In Petri dishes the volume of air is small and, air is fairly static as it is an enclosed system. Therefore, the plasmodium easily locates the various sources of nutrients in its immediate vicinity (Fig. 11.21). Over the time it, therefore, constructs spanning trees where the graph nodes are represented by foam pieces containing oat flakes. In Fig. 11.21, we can see that originally the plasmodium was positioned at the Southern domain. In 12 h, the plasmodium builds a link with Western domain, and then starts to propagate pseudopodia toward the Eastern domain. When the plasmodium spans sources of nutrients, it produces many 'redundant' branches (Fig. 11.21). These branches of pseudopodia are necessary for space exploration but do not represent minimal edges connecting the nodes of the spanning tree. These 'redundant' branches are removed at a later stage of the spanning tree development. A well-established spanning tree of data-points is shown in Fig. 11.22. Initially, the plasmodium was placed in the Western domain, and the plasmodium constructed a spanning tree in 15 h.

We demonstrated that the plasmodium explores the domain space and computes a spanning tree across the surface of water, when placed initially on one of the floating objects. Would the plasmodium be successful if placed directly on the surface of the water? As we can see in Fig. 11.4, the plasmodium works perfectly. We placed a piece of plasmodium on the surface of water (Fig. 11.4, start) and in 3 h the plasmodium forms an almost circular front of propagating pseudopodia, these reach two stationary domains containing nutrients in 8 h (Fig. 11.4).

In usual conditions (on a wet solid or gel substrate), edges of spanning trees presented by protoplasmic tubes, adhere to the surface of the substrate [14, 21, 22]. Therefore, the edges cannot move, and the only way the plasmodium can dynamically update is to make a protoplasmic tube inoperative and to form a new edge instead (membrane shell of the ceased link will remain on the substrate, e.g. see [22]. When plasmodium operates on water surface, cohesion between the water surface and membrane of protoplasmic tubes is small enough for the protoplasmic tubes to move freely. Thus, the plasmodium can make the tubes almost straight and minimize "costs" of the transfer and communication between its distant parts. Two examples of the straightening of the protoplasmic tubes are shown in Fig. 11.24. Such straightening is a result of the tubes becoming shorter due to contraction.

Presence of a contraction may indicate that if two floating objects (both with sources of nutrients) are connected by a protoplasmic tube, then the objects will be pulled together due to shortening of the protoplasmic tube. We did not manage to demonstrate this exact phenomenon of pulling two floating objects together; however, we got experimental evidence of pushing and pulling of single floating objects by the plasmodium's pseudopodia. The plasmodium-induced pushing and pulling are exemplified in Figs. 11.25 and 11.26.

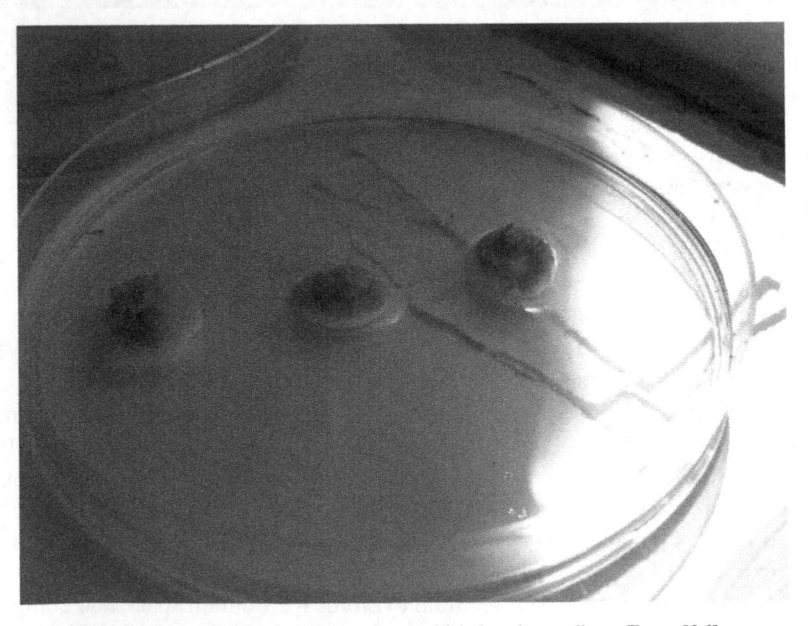

Fig. 11.22 Spanning tree of three points constructed by the plasmodium. From [26]

To demonstrate pushing, we placed a very lightweight piece of plastic foam on the water surface near the plasmodium (Fig. 11.25, 0 h). The plasmodium develops a pseudopodium, which propagates toward the lightweight piece of foam (Fig. 11.25, 5 h). Owing to gravitational force acting on the pseudopodia a ripple is formed on the water surface (Fig. 11.25, 9 h), which causes the piece of foam to be pushed away from the growing pseudopodia's tip (Fig. 11.25, 13 h). Due to the absence of any nutrients on the pushed piece of foam, the plasmodium abandons its attempt to occupy the piece and retracts the pseudopodia (Fig. 11.25, 16 h). The piece remains stationary: but is shifted away from its original position.

In the second example, Fig. 11.26, we observe pulling of the lightweight object. The piece of foam to be pulled is placed between two anchored objects (Fig. 11.26, 0 h). One object hosts the plasmodium and other object has an oat flake on top (i.e. attracts the plasmodium). A pseudopodium grows from the plasmodium's original location toward the site with the source of nutrients. The pseudopodium occupies this piece of foam (Fig. 11.26, 15 h) and then continues its propagation toward the source of nutrients. When the source of nutrients is reached (Fig. 11.26, 22 h) the protoplasmic tubes connecting the two anchored objects contract and straighten, thus, causing the lightweight object to be pulled toward the source of nutrients (Fig. 11.26, 32 h). The pushing and pulling capabilities of the plasmodium can be utilized in construction of water-surface based distributed manipulators [7, 8, 27].

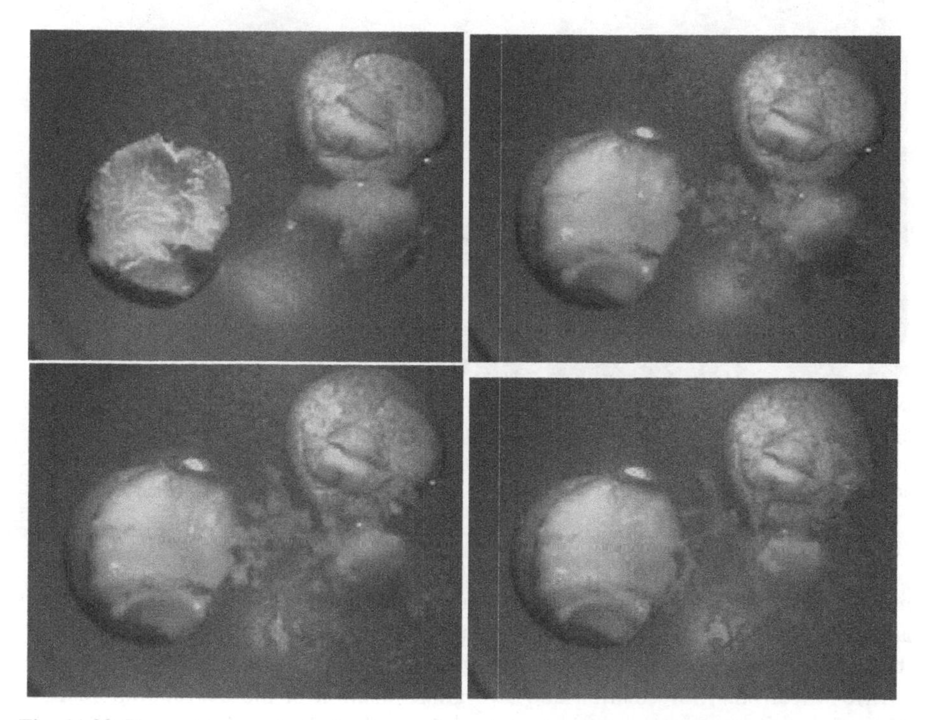

Fig. 11.23 Plasmodium starts its development on the water surface and occupies two sources of nutrients. (**a**) initial position of the plasmodium, (**b**) 3 h, (**c**) 5 h, (**d**) 8 h. From [26]

11.7 Conclusion

In examples of chemical based laboratory prototypes, we demonstrated that chemical reaction–diffusion media are promising, non-silicon distributed controllers for robotics. They are capable of controlling the robot motion in response to external stimulation. They efficiently act as underlying parallel architectures for distributed and intelligent robotic manipulators, which can be integrated into a closed-loop robotic system. Moreover, when encapsulated in an elastic membrane, the reaction–diffusion systems provide an optimal platform for amorphous embedded intelligence.

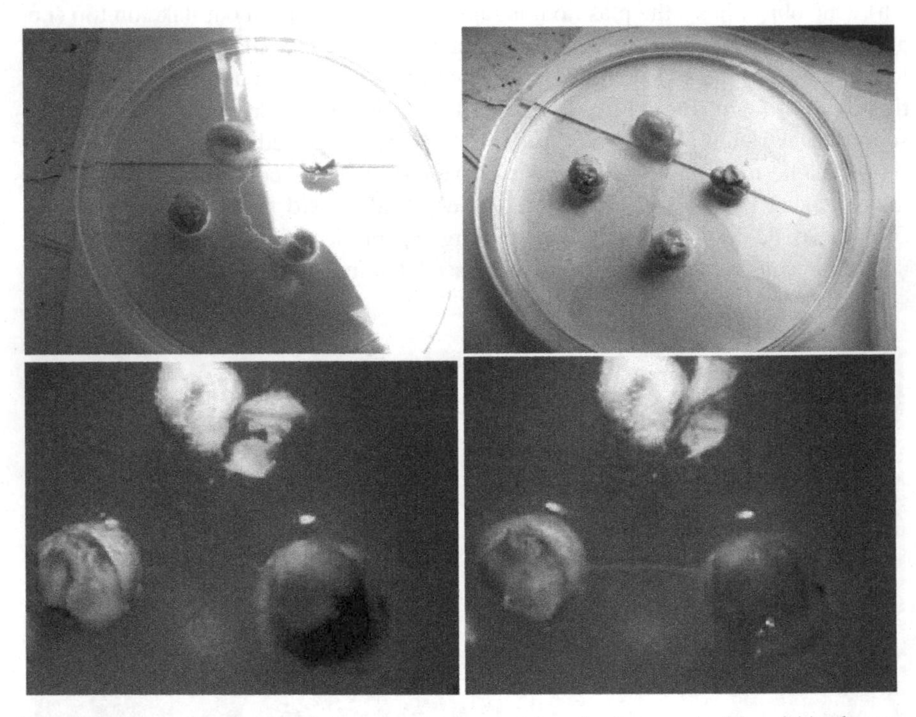

Fig. 11.24 Examples of straightening of protoplasmic tubes. In photographs (**a**) and (**c**) tubes are longer than necessary. In photographs (**b**) and (**d**) the tubes correspond to minimal shortest path between the sites they are connecting. Time interval between the recordings (**a**) and (**b**), and (**c**) and (**d**) is 12 h. From [26]

In experimental conditions, we have shown that it is indeed possible to guide a mobile robot with an onboard excitable chemical processor. To stimulate the robot, and to implement its taxis, we locally perturb concentrations of chemical species in the reactor. Excitation wave fronts propagating in the reactor carry (in their asymmetric wavefronts) information about original position of the stimulation source. Therefore, from images of spatio-temporal dynamics of excitation in the chemical reactor, the robot extracts a direction toward the source of stimulation and implements rotation and forward motion.

We experimentally demonstrated a non-trivial concurrent interaction between an excitable chemical medium and a robotic sensing and actuating device. We proved that it is indeed possible to achieve some sensible control of a robotic hand using travelling and interacting excitation wave fronts in a liquid-phase BZ system. We discovered several patterns of finger movements: self-excited motion, synchronized motion of fingers and reciprocal motion of fingers. To demonstrate viability of the approach, we studied a closed-loop system where the robotic hand excites wave dynamics and these dynamics cause further movement of the hand's fingers.

We presented results of scoping experiments on manipulating simple objects in a simulated massive-parallel array of actuators controlled by experimental excitation dynamics in a BZ based chemical medium. Our basic assumption was that manipulated objects should be pushed by travelling excitation wavefronts. Therefore, we encoded snapshots of gel-based BZ media and converted them into arrays of local force vectors in such a manner that every pixel of the medium's snapshot got its own force vector, the orientation of which was determined by the gradient of the red component (which in turn represents concentrations of chemical species in the medium) in the neighbourhood of the pixel. Thus, e.g. a circular wave in the BZ medium – via the simulated actuator array – transports a manipulated object centrifugally, outward from the the wave's source of origin.

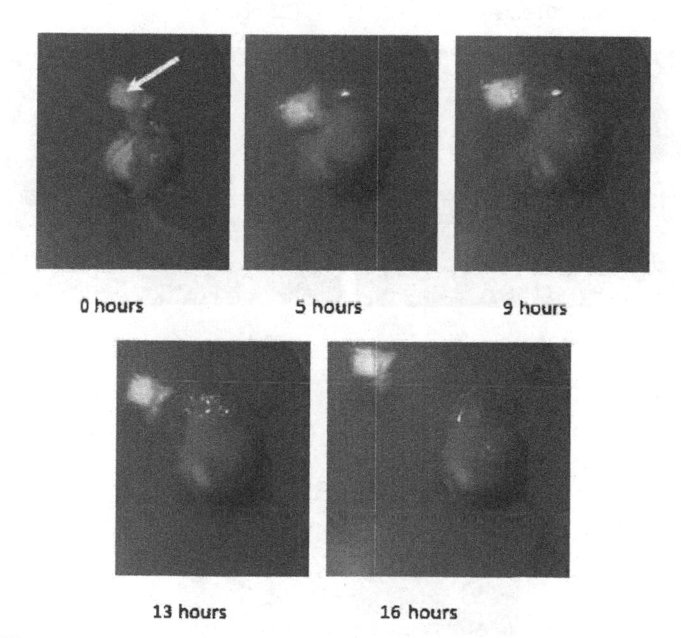

Fig. 11.25 Photographs demonstrate that the plasmodium can push lightweight floating objects. The object to be pushed is indicated by a *white arrow* in the first photograph of the series. From [26]

We also demonstrated that various types of excitation dynamics – target waves, spiral waves, wave-fragments – are capable of the transportation and manipulation of small objects (the size of which is comparable to an elementary micro-volume of the medium) and larger, spatially extended objects, planar shapes. Results on manipulation of spatially extended objects may find practical implementation in developing soft-robots and gel-based actuating surfaces. This may possibly lead to a hybrid amoeba-like robot capable of implementing flexible movement strategies and also grasping and manipulating objects in the environment.

The integration of chemical controllers to hardware components has its drawbacks. Namely, chemical processors are not efficient when integrated in conventional hardware architectures. Any standard wheeled-robot with primitive

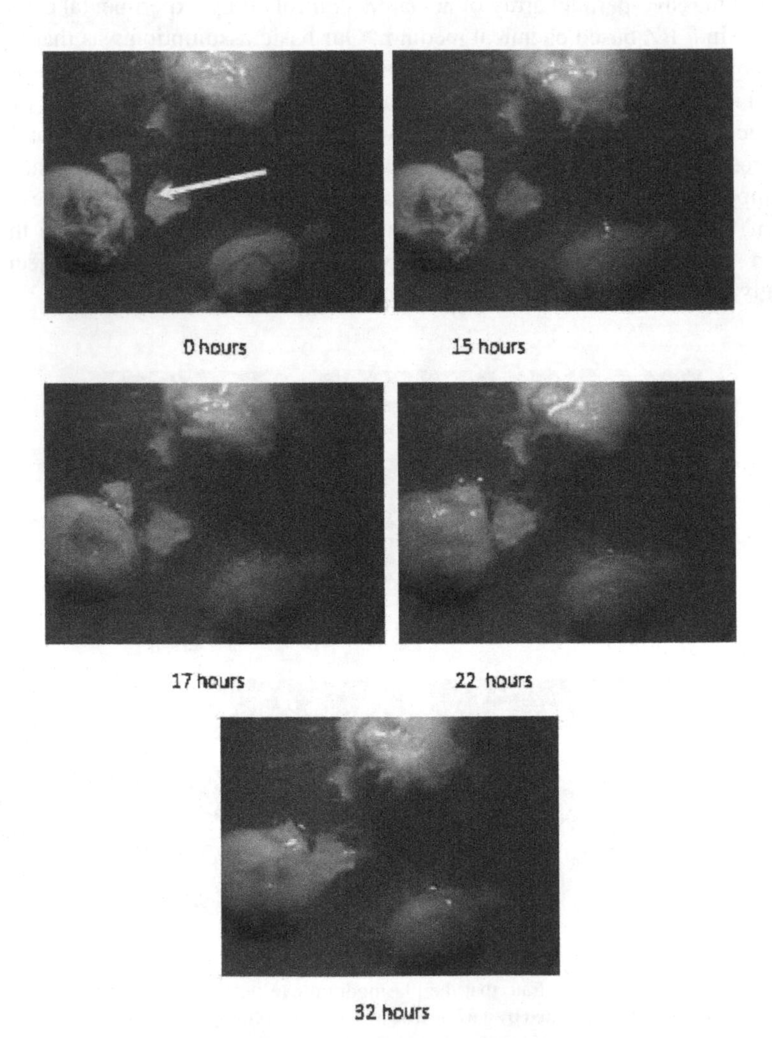

Fig. 11.26 Photographs demonstrate that the plasmodium can pull lightweight objects. The object to be pulled is indicated by white arrow in the first photograph. From [26]

photo-sensors will outperform a chemical-medium-controlled robot in any navigation task. Not only the reaction–diffusion algorithms, when implemented in a chemical medium are slow, but also the analysis of the medium space–time dynamics consumes more computational resources compared to any simple program that can guide a standard mobile robot. However, wave-based control in disordered media will ideally suit amorphous robotic architectures, like those based on electro-activated and oscillating polymers.

The last thesis was proved in our experiments with the plasmodium of *Physarum polycephalum*. Inspired by bio-mechanics of surface walking insects, see e.g. [28–31], our previous studies on implementation of computing tasks in the plasmodium [17, 21, 22, 32] and our ideas on design and fabrication of biological amorphous robots [33], we explored the operational capabilities of the plasmodium of *Physarum polycephalum* on the surface of water. We showed that the plasmodium possesses the essential features of distributed robotics devices: sensing, computing, locomotion and manipulation.

We demonstrated experimentally that the plasmodium (1) senses data-objects represented by sources of nutrients, (2) calculates shortest path between the data-objects, and approximates spanning trees where the data-objects are nodes (in principle, a spanning tree of slowly moving data-objects can be calculated as well), (3) pushes and pulls lightweight objects placed on the water surface. The findings indicated that the plasmodium is a prospective candidate for the role of spatially extended robots implemented on biological substrates.

References

1. Zaikin, A.N., Zhabotinsky, A.M.: Concentration wave propagation in two-dimensional liquid-phase self-oscillating system. Nature 225, 535 (1970)
2. Adamatzky, A.: Computing in Automata Media and Automata Collectives. IoP (2001)
3. Adamatzky, A., De Lacy Costello, B., Asai, T.: Reaction Diffusion Computers. Elsevier (2005)
4. Yokoi, H., Adamatzky, A., De Lacy Costello, B., Melhuish C.: Excitable chemical medium controlled by a robotic hand: closed loop experiments. Int. J. Bifurcat. Chaos 14, 3347–3354 (2004)
5. Field, R. J., Winfree, A. T.: Travelling waves of chemical activity in the Zaikin–Zhabotinsky–Winfree reagent. J. Chem. Educ. 56, 754 (1979)
6. Adamatzky, A., De Lacy Costello, B., Melhuish, C., Ratcliffe, N.: Experimental implementation of mobile robot taxis with onboard Belousov–Zhabotinsky chemical medium. Mater. Sci. Eng. C 24, 541–548 (2004)
7. Adamatzky, A., De Lacy Costello, B., Skachek, S., Melhuish, C.: Manipulating objects with chemical waves: Open loop case of experimental Belousiv-Zhabotinsky medium. Phys. Lett. A, 350(3-4), 201–209, 6 February (2006)
8. Skachek, S., Adamatzky, A., Melhuish, C.: Manipulating objects by discrete excitable media coupled with contact-less actuator array: Open-loop case. Chaos Solitons Fractals 26, 1377–1389 (2005)
9. Ishikawa, Y., Yu, W., Yokoi, H., Kakazu, Y.: Development of robot hands with an adjustable power transmitting mechanism, In: Cihan H. Dagli et al. Eds. Proc. Intell. Engineer. Syst. Through Artificial Neural Networks 10, 631–636 (2000)
10. Adamatzky, A.: Does Physarum follow Toussaint conjecture? Parallel Processing Letters (2008)

11. Nakagaki, T., Yamada, H., Toth, A.: Maze-solving by an amoeboid organism. Nature 407, 470–470 (2000)
12. Nakagaki T.: Smart behavior of true slime mold in a labyrinth. Res. Microbiol. 152, 767–770 (2001)
13. Nakagaki T., Yamada H., Toth A.: Path finding by tube morphogenesis in an amoeboid organism. Biophys. Chem. 92, 47–52 (2001)
14. Nakagaki, T., Yamada, H., Hara, M.: Smart network solutions in an amoeboid organism. Biophys. Chem. 107, 1–5 (2003)
15. Tsuda, S., Aono, M., Gunji, Y.-P.: Robust and emergent Physarum computing. BioSystems 73, 45–55 (2004)
16. Tsuda, S., Zauner, K. P., Gunji, Y. P.: Robot control with biological cells. Biosystems, 87, 215–223 (2007)
17. Adamatzky, A.: Physarum machines: encapsulating reaction-diffusion to compute spanning tree. Naturwisseschaften (2007)
18. Matsumoto, K., Ueda, T., Kobatake, Y.: Reversal of thermotaxis with oscillatory stimulation in the plasmodium of Physarum polycephalum. J. Theor. Biol. 131, 175–182 (1998)
19. Nakagaki, T., Yamada, H., Ito, M.: Reaction-diffusion-advection model for pattern formation of rhythmic contraction in a giant amoeboid cell of the Physarum plasmodium J. Theor. Biol. 197, 497–506 (1999)
20. Yamada H., Nakagaki T., Baker R. E., Maini P.K.: Dispersion relation in oscillatory reaction-diffusion systems with self-consistent flow in true slime mold. J. Math. Biol. 54, 745–760 (2007)
21. Adamatzky A. Growing spanning trees in plasmodium machines. Kybernetes, 37, 258–264 (2008)
22. Adamatzky, A.: Physarum machine: implementation of a Kolmogorov-Uspensky machine on a biological substrate. Parallel Process. Lett. 17 (2008)
23. Shimizu, M., Ishiguro, A.: Amoeboid locomotion having high fluidity by a modular robot. Int. J. Unconventional Comput. (2008), in press
24. Yokoi, H., Kakazu, Y.: Theories and applications of autonomic machines based on the vibrating potential method, In: Proc. Int. Symp. Distributed Autonomous Robotics Systems, 31–38 (1992)
25. Yokoi, H., Nagai, T., Ishida, T., Fujii, M., Iida, T.: Amoeba-like robots in the perspective of control Architecture and morphology/materials, In: Hara, F. and Pfeifer, R. (Eds.) Morpho-Functional Machines: The New Species, Springer, Tokyo, 99–129 (2003)
26. Adamatzky, A.: Towards Physarum robots: computing and manipulating on water surface (2008) `arXiv:0804.2036v1` [cs.RO]
27. Hosokawa, K., Shimoyama, I., Miura, H.: Two-dimensional micro-self-assembly using the surface tension of water. Sens. Actuators A 57, 117–125 (1996)
28. McAlister, W. H.: The diving and surface-walking behaviour of Dolomedes triton sexpunctatus (Araneida: Pisauridae). Animal Behav. 8, 109–111 (1959)
29. Suter, R. B., Wildman, H.: Locomotion on the water surface: hydrodynamic constraints on rowing velocity require a gait change. J. Exp. Biol. 202, 2771–2785 (1999)
30. Suter, R. B.: Cheap transport for fishing spiders: the physics of sailing on the water surface. J. Arachnol. 27, 489–496 (1999)
31. Suter, R. B., Rosenberg, O., Loeb, S., Wildman, H., Long, J. Jr.: Locomotion on the water surface: propulsive mechanisms of the fisher spider, Dolomedes triton. J. Exp. Biol. 200, 2523–2538 (1997)
32. Adamatzky, A., De Lacy Costello, B., Shirakawa T.: Universal computation with limited resources: Belousov-Zhabotinsky and Physarum computers. Int. J. of Bifurcation and Chaos (IJBC), 18(8), 2373–2389 (2008)
33. Kennedy, B., Melhuish, C., Adamatzky, A.: Biologically Inspired Robots in In: Y. Bar-Cohen (Ed) Electroactive polymer (EAP) actuators – Reality, Potential and challenges. SPIE, (2001)

Index